BARRON'S REVIEW COURSE SERIES

Let's Review: Biology

Scott Hunter

*District-level Administrator
and Former Science Teacher
Schodack Central School District
Castleton-On-Hudson, New York*

*Consultant
New York State Education Department
Bureaus of Science Education, Educational Testing,
and Curriculum Development
Albany, New York*

BARRON'S

All inquiries should be addressed to:
Barron's Educational Series, Inc.
250 Wireless Boulevard
Hauppauge, New York 11788

Library of Congress Catalog Card No. 87-33277

International Standard Book No. 0-8120-3877-0

Library of Congress Cataloging-in-Publication Data

Hunter, Scott.
    Let's review: biology / Scott Hunter.
        p.   cm.
    Includes index.
    ISBN 0-8120-3877-0
    1. Biology——Outlines, syllabi, etc.   I. Title.
QH315.5.H86 1988
574——dc 19   87-33277
                                                CIP

PRINTED IN THE UNITED STATES OF AMERICA
456   800   987

PREFACE

FOR WHICH COURSE
CAN THIS BOOK BE USED?

This book is designed to be used as a review text for the New York State course in Regents Biology. The material presented closely mirrors the revised Syllabus in Regents Biology (1982), which is used throughout New York State as the basis of a comprehensive course of study in biology on the secondary level.

Moreover, because of the comprehensive nature of the material in this book, it can be used to supplement virtually any secondary-level college-preparatory course in biology taught anywhere in the United States, using any major textbook.

WHAT SPECIAL FEATURES
DOES THIS BOOK HAVE?

The arrangement of topical information in this book parallels, for the most part, that of the New York State Syllabus in Regents Biology. However, some changes intended to improve the logical flow of this information have been introduced. An effort has been made to present all understandings required for the Regents Examination, while minimizing extraneous material that might tend to distract or confuse the student.

Throughout the book, New York State Syllabus understandings subject to testing on Part I of the Regents Examination are presented in regular, unmarked text. Understandings subject to testing in Part II are presented in special bordered text with the Part II area in question indicated by a two-letter code as follows:

> BC = Biochemistry (Units 1, 2, and 3)
> HP = Human Physiology (Unit 3)
> RD = Reproduction and Development (Unit 4)
> MG = Modern Genetics (Unit 5)
> EC = Ecology (Unit 7)

A special Part III section has been added to help the student to achieve at the maximum level on the Laboratory Skills part of the Regents Examination.

Question sets provide opportunities for student review. Items in these sets are actual Regents and Regents-type questions arranged to correspond to the arrangement of topics in the text.

Each unit ends with a review test that allows the student an opportunity for comprehensive review. Part II questions in each set are marked with their appropriate two-letter codes. Answer keys are provided at the end of the book for odd-numbered questions in each question set and unit test, so that students may easily monitor their own progress.

An extensive glossary following the text presents an alphabetical listing of all major New York State Syllabus terms and their definitions. Terms selected for inclusion are those likely to appear as correct answers or as distractors (i.e., incorrect choices) in multiple-choice Regents Examination questions. For each glossary term, a notation indicates the unit and area (CORE, BC, HP, RD, MG, or EC) in which it first appears. A special glossary of prominent scientists mentioned in the syllabus is also included for easy reference. A comprehensive index makes it easy for students to find complete textual discussion of any topic.

Finally, five full-length Regents Examinations for the past three years (January and June exams) provide students with ample opportunity to test their knowledge of biology and to practice their question-answering skills before taking the Regents Examination.

WHO SHOULD USE THIS BOOK?

A student in the New York State Regents Biology course will find this book to be a valuable supplement to his/her regular textbook. In addition, it will prove useful in preparing for year-round tests and the year-end Regents Examination in Biology. It will help to clarify exactly the behavioral objectives required to maximize achievement on the Regents Examination.

A student in any secondary-level biology course anywhere in the United States can use this book profitably in much the same way. Its clear, concise style helps to clarify concepts, and the absence of extraneous material enables the student to concentrate on required information.

For teachers in New York State public schools, this book will provide an excellent source of review material to prepare their students for the New York State Regents Examination in Biology. All biology teachers will be able to use this book as a companion to their regular textbooks, and will find that their students will gain considerable self-confidence and facility in test-taking through its use. Some school systems may wish to use this book as the primary text for their courses in college-preparatory biology, in view of the comprehensive treatment of the subject matter presented herein.

SCOTT HUNTER

CONTENTS

HOW DO LIVING THINGS SHOW SIMILARITY AND DIVERSITY?

KEY IDEAS	This unit introduces some of the basic concepts in biology that will aid you in your study throughout the Regents biology course. The concepts of life function, biological classification, cell study, and basic biochemistry are the key ideas stressed in this unit.

KEY OBJECTIVES

Upon completion of this unit, you should be able to:

☐ Define life as a function of the life processes carried on by living organisms.

☐ Describe the basis of the five-kingdom system of biological classification, and describe the major characteristics of each of these kingdoms.

☐ Describe the characteristics of the cell that enable it to operate as the basic structural and functional unit of living things.

☐ Recognize the major types of chemical elements and compounds common to living things, and describe some of the chemical reactions in which they operate.

☐ Describe some of the major tools and techniques used by biologists to study cells.

☐ Recognize that within the diversity of living things there is an underlying pattern of unity based on the cell and its functions.

I. THE CHARACTERISTICS OF LIFE

A. Definition of Life

There is no one best definition of life that is accepted by all scientists. Most biologists agree, however, that life is defined by the properties and functions characteristic of living things. These **life functions** include:

- Nutrition
- Transport
- Respiration
- Excretion
- Synthesis
- Regulation
- Growth
- Reproduction

B. Life Functions

Life functions are the various activities carried on by living things. Increasingly, the functions of living things are understood to be the same as the functions of the cells that comprise them. Although these life functions are many and diverse, they can be summarized in the following easily understood list:

1. **Nutrition** includes the activities by which living things obtain food and process it to supply the energy and materials necessary for the growth and repair of their bodies.

2. **Transport** is the process by which substances are absorbed and released by living things. Transport also includes the mechanisms used to circulate materials within organisms.

3. **Respiration** includes the processes by which the energy stored in food is converted to a form that can be directly used by the cells of living things. This energy is used for the maintenance of all life processes.

4. **Excretion** is the process by which the waste products of other life functions are removed from living things.

5. **Synthesis** involves the chemical processes by which large molecules are built from smaller molecules.

6. **Regulation** includes the processes by which living things respond to changes within and around them. It results in the coordination and control of all life processes.

7. **Growth** involves increase in cell number and cell size. Cell growth requires the manufacture of chemical components. Growth results in increase of organism size.

8. **Reproduction** refers to the means by which new cells arise from preexisting cells. When cells in a multicelled organism divide to produce new cells, the result may be the growth of the organism or the repair of damaged tissues. Cell division may also result in the production of new organisms of a species.

C. Metabolism

When all life functions are considered together, the chemical and biological activities are known collectively as **metabolism**. Each of the life functions defined above has an underlying chemical (metabolic) process upon which it depends. Metabolic processes are essentially chemical in nature. As such, they use chemical reactants, result in chemical products and wastes, and are affected by various physical and chemical conditions that exist in the cell.

D. Homeostasis

Homeostasis refers to the condition of balance and dynamic stability that exists within organisms under normal conditions. This stability depends on the coordination of thousands of chemical reactions occurring simultaneously within cells, and may be easily upset by any alteration of the cell's physical or chemical environment. Examples of homeostasis include the maintenance of temperature and blood sugar levels in human beings.

QUESTION SET 1.1

1. Control of all physiological activities of an organism is necessary to maintain that organism's stability in its environment. This life activity is known as
   1. nutrition
   2. respiration
   3. transport
   4. regulation

2. Which term is used to represent all of the physiological activities carried on by an organism?
   1. regulation
   2. metabolism
   3. homeostasis
   4. synthesis

3. The ability of the human body to maintain a constant body temperature is an example of
   1. transport
   2. metabolism
   3. homeostasis
   4. synthesis

4. A characteristic of all known living organisms is that they
  1. require oxygen for respiration
  2. originate from preexisting life
  3. have complex nervous systems
  4. carry on heterotrophic nutrition

5. In an organism, the coordination of the activities that maintain homeostasis in a constantly changing environment is a process known as
  1. digestion
  2. regulation
  3. synthesis
  4. respiration

6. Which life function provides substances that may be used by an organism for its growth and for the repair of its tissue?
  1. excretion
  2. reproduction
  3. nutrition
  4. regulation

7. In an ameba, materials are taken from its environment and then moved throughout its cytoplasm. These processes are known as
  1. absorption and circulation
  2. food processing and energy release
  3. energy release and synthesis
  4. coordination and regulation

8. For survival, a hummingbird uses a considerable amount of energy. This energy most directly results from the life activity of
  1. transport
  2. excretion
  3. regulation
  4. respiration

9. Which life function is primarily involved in the conversion of the energy stored in organic molecules to a form directly usable by a cell?
  1. absorption
  2. circulation
  3. digestion
  4. respiration

10. Complex molecules are formed from simple molecules by the process known as
  1. digestion
  2. transport
  3. respiration
  4. synthesis

11. The life function by which living things rid their bodies of the wastes generated through metabolic processes is known as
  1. locomotion
  2. nutrition
  3. excretion
  4. synthesis

12. In the life process of reproduction, which of the following occurs?
  1. Electrochemical impulses travel along the membrane of the axon.
  2. Nitrogenous wastes diffuse through the cells of the nephron.
  3. Old, worn-out cells are recycled into amino cells.
  4. New organisms of the parent species are produced.

13. The addition of new cells and corresponding increase in organism size is known as
  1. growth
  2. transport
  3. regulation
  4. respiration

II. DIVERSITY AMONG LIVING THINGS

A. Necessity for Classification

A brief survey of living things reveals that approximately 1.5 million different species of every description exist on earth. **Classification** provides scientists with a means for sorting and grouping these organisms for easier study.

The basis of biological classification is **physical structure**, although other criteria, such as embryonic, genetic, and chemical similarities, are also used. It is known that organisms that are similar in their physical traits are usually similar in other ways. Because of this fact, the characteristics of a large grouping of similar organisms can be learned by studying a few representatives of the group. When a new organism is discovered, it can be readily grouped with other, similar organisms when only a few of its characteristics are known.

The system of classification also serves as a basis for the science of **evolution**. It has been found that organisms sharing many traits in common are likely to share a common ancestry as well. (See Unit 6 for a more detailed treatment of the science of evolution.)

B. A Scheme of Classification

The most widely accepted scheme of classification places every known organism in one of five large groupings known as **kingdoms**. The organisms within a particular kingdom share many broad traits in common, although there can be considerable **diversity** (difference) of form among them. The five kingdoms are as follows:

1. **Monera**
2. **Protista**
3. **Fungi**
4. **Plant**
5. **Animal**

Within each kingdom, organisms having greater similarity to each other than to the organisms in other groups are classified together into **phyla**. Therefore each **phylum** contains groups of organisms showing characteristics distinctly different from those of other phyla.

Phyla are subdivided into still narrower groupings of organisms showing high degrees of similarity and known as **genera**. The members of each **genus** are so similar that they might easily be mistaken for each other by most people.

Each genus is further broken down into **species.** The members of a species are so similar biologically that they can share genetic information and reproduce more individuals like themselves. When these new individuals can, in turn, reproduce, they are known as "fertile offspring."

When a new organism is found, it is classified into one of the five kingdoms discussed above. The criteria used in this classification are as follows:

1. The presence or absence of a **nuclear membrane** (a thin, porous covering over the cell's nucleus) within the cell.
2. The condition of **unicellularity** (single-celled) versus **multicellularity** (many-celled) that exists in the general structure of the organism.
3. The type of **nutrition** employed by the organism.

The following chart summarizes the general criteria used for classification of the five kingdoms, and lists a few of their major phyla and some of their representative organisms. Within this chart, monerans are considered the most primitive; animals are considered the most advanced.

Kingdom	Phylum*	Characteristics	Examples
Monera	Bacteria Blue-green algae	Have primitive cell structure lacking a nuclear membrane.	
Protista	Protozoa Algae	Are predominantly unicellular organisms with plantlike and/or animal-like characteristics.	†Paramecium †Ameba Spirogyra
Fungi		Cells are usually organized into branched, multi-nucleated filaments that absorb digested food from their environment.	Yeast Bread mold Mushroom
Plant		Are multicellular, photo-synthetic organisms.	
	Bryophytes	Lack vascular tissues; have no true roots, stems, or leaves.	Moss
	Tracheophytes	Possess vascular tissue; have true roots, stems and leaves.	Geranium Fern Bean Trees (maples, oaks, pines etc.) Corn

Kingdom	Phylum*	Characteristics	Examples
Animal		Are multicellular, heterotrophic organisms.	
	Coelenterates	Have two cell layers, hollow body cavity.	†Hydra Jellyfish
	Annelids	Have segmented body walls.	†Earthworm Sand worm
	Arthropods	Have jointed appendages, exoskeleton.	†Grasshopper Lobster Spider
	Chordates	Possess dorsal nerve cord, internal skeleton.	Shark Frog ‖Human being

*The phyla presented are those required in the Regents course.
†These organisms are treated in detail in Unit 2 of this book.
‖Human beings are treated in detail in Unit 3 of this book.

C. Nomenclature

The scheme of classification discussed above is the first means that scientists use in organizing living things into easily understood groupings. To keep track of the millions of different species that have been discovered and classified, a system of naming (**nomenclature**) has been developed that uses the scheme of classification as its basis.

This system of naming uses two names, in much the same way that most people have at least two names to help others tell them apart. This naming system, known as **binomial nomenclature** ("two-name naming"), was first devised by **Carolus Linnaeus** in the eighteenth century. The two names used are the **genus** name (always capitalized) and the **species** name (always written in lower case). The language used in the system of binomial nomenclature is **Latin**.

Here is an example of binomial nomenclature:

	Latin	Translation	Common Name
Genus:	*Homo*	Human	
Species:	*sapiens*	wise	= Human being

QUESTION SET 1.2

1. The basis of biological classification assumes that
   1. bryophytes contain more vascular tissues than tracheophytes
   2. chordates are more primitive than coelenterates
   3. all cells contain nuclei enclosed by nuclear membranes
   4. organisms with structural similarity are closely related

2. In attempting to classify a newly discovered organism the following characteristics were noted: multicellular specialized organs and tissues, cell walls, chorophyll-containing plastids. The kingdom into which this organism should be placed is
   1. Animal
   2. Plant
   3. Protista
   4. Monera

3. In which kingdom is an organism classified if it lacks a membrane separating most of its genetic material from its cytoplasm?
   1. Protista
   2. Monera
   3. Plant
   4. Animal

4. Which group of organisms in the animal kingdom is characterized by jointed appendages and exoskeletons?
   1. arthropods
   2. chordates
   3. annelids
   4. coelenterates

5. Fish, frogs, and human beings are examples of
   1. coelenterates
   2. annelids
   3. arthropods
   4. chordates

6. The mosquito *Anopheles quadrimaculatus* is most closely related in structure to
   1. *Aedes sollicitans*
   2. *Culex pepiens*
   3. *Aedes aegypti*
   4. *Anopheles punctulatus*

7. Which is true of organisms that are classified in the same genus?
   1. They must be in the same phylum, but may be of different species.
   2. They must be of the same species, but may be in different phyla.
   3. They must be in the same phylum, but may be in different kingdoms.
   4. They must be in the same kingdom, but may be in different phyla.

8. Which group of terms is in the correct order from most general to most specific?
   1. species, phylum, genus, kingdom
   2. genus, species, kingdom, phylum
   3. kingdom, phylum, genus, species
   4. phylum, kingdom, species, genus

9. In a modern system of classification, two organisms would be most closely related if they were classified in the same
   1. kingdom
   2. phylum
   3. genus
   4. species

10. *Ursus horribilis*, the scientific name for the grizzly bear, designates the bear's
    1. kingdom and phylum
    2. kingdom and species
    3. genus and phylum
    4. genus and species

III. SIMILARITY AMONG LIVING THINGS

A. *Structure of Living Organisms and Methods of Cell Study*

The **cell theory** provides the strongest theoretical support for the concept of the **unity** ("alikeness") of living things. The cell theory makes two very important statements about the unity of life:

- "Cells are the basic units of structure and function of living things." In other words, all known living things are constructed of cells, which act both as units of structure and as units of function (operation).
- "Cells come from preexisting cells." This means that cells do not arise spontaneously, as was once widely believed, but must be produced through the biological process of reproduction.

The cell theory was developed over many years by scientists working in many parts of the world. The major tool of these scientists was the microscope, either in simple form such as that developed by Anton von Leeuwenhoek, or in compound form such as that used by Robert Hooke.

There are some notable exceptions to the cell theory. Three of these are as follows:

1. If all cells must arise from previously existing cells, there is a question as to the origin of the first cell. Where did it come from? Biologists have only part of the answer.
2. Viruses contain simple genetic material and carry out a type of reproduction, but are not composed of cells. There is some question as to the status of viruses as "living" organisms, since they apparently do not carry on life functions other than reproduction and synthesis.
3. Certain cell organelles behave like cells in some respects. Both mitochondria and chloroplasts are known to contain genetic material and to produce new organelles like themselves under certain circumstances.

Methods of cell study have changed over time. Improvements in these methods have greatly increased our understanding of cell structure and function. Among these methods are the following:

1. Instrumentation

1.1. Compound light microscope—still the major tool of cell study employed by biologists all over the world. Compound light microscopes of the type used in most school laboratories produce magnifications of 50× to 500× and are suitable for viewing whole cells and large organelles such as nuclei and chloroplasts. Since this tool is also used commonly in the biology classroom it is important that you understand its parts and workings thoroughly.

- Eyepiece lens (ocular) is the lens closest to the eye during study.
- Objective lens(es) is (are) the lens(es) closest to the object of study.
- Stage is the surface upon which the object is placed for study.
- Coarse-adjustment knob is used to make large changes in the focus of the microscope.
- Fine-adjustment knob is used to make small changes in the focus of the microscope.
- Light source provides light, which must pass through the object in order to make it visible through the microscope.
- Diaphragm is used to make changes in the quality and amount of light passing through the object and entering the objective lens.
- Magnification is the quality of the compound light microscope that makes the image of an object appear larger than the object itself. The magnification of a compound light microscope is determined by multiplying the power of the ocular by the power of the objective lens (e.g., 10× ocular × 40× objective = 400× total magnification).
- Resolution is the quality of the compound light microscope that makes it possible to see as separate objects that are very close together in a microscope field.

The various parts of the compound light microscope are shown in Figure 1.1.

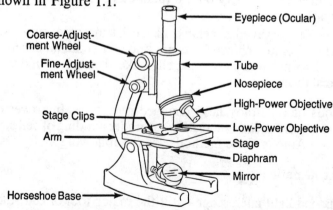

Figure 1.1 Compound Light Microscope

1.2. Electron microscope—an advanced microscope with magnifying powers in excess of 250,000 ×. Electron microscopes permit detailed study of small cell organelles, such as chloroplasts and mitochondria.

1.3. Dissecting microscope—a low-power microscope that gives the viewer a three-dimensional image for purposes of gross dissection. Dissecting microscopes typically produce magnifications of 10 × or 20 ×.

1.4. Ultracentrifuge—a tool used to separate cell parts according to their relative densities.

1.5. Microdissection instruments—tools used to perform minute dissection of individual cells and/or transfer of individual cell organelles.

2.   Measurement (see Figure 1.2)

Cells are too small to measure using standard units of measurement. To accommodate measurement of cells and cell parts, the metric unit known as the **micrometer** (μm) is used. Micrometers are so small that a thousand of them fit into 1 millimeter, and a million are required to make 1 meter. Cells typically are found to have diameters of between 10 and 50 micrometers.

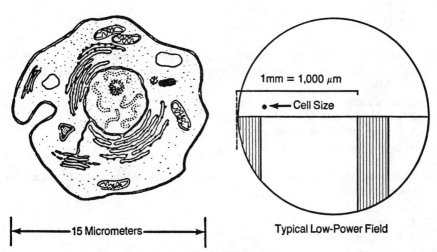

1mm = 1,000 μm

←— Cell Size

├—————————15 Micrometers—————————→| Typical Low-Power Field

Figure 1.2   Cell Measurement

3.   Other Techniques

A variety of other cell-study techniques have taught biologists much about the internal workings of cells and cell parts. One of the most valuable techniques, **staining**, allows normally invisible cell parts to be viewed under the microscope. Stains used in the biology laboratory include **iodine** and **methylene blue**.

QUESTION SET 1.3

1. According to the cell theory, which statement is correct?
   1. Viruses are true cells.
   2. Cells are basically unlike in structure.
   3. Mitochondria are found only in plant cells.
   4. Cells come from preexisting cells.

2. Which statement describes an exception to the cell theory?
   1. Mitochondria and chloroplasts are self-replicating structures.
   2. All cells must come from preexisting cells.
   3. Cells are the basic unit of structure in living things.
   4. Cells are the basic unit of function in living things.

3. A slide of the letters F and R is placed on the stage of a microscope in the position shown in the diagram below. How will the image of the letters appear when the slide is viewed under the low power of a compound light microscope?

   1. RF   3. ꟻЯ
   2. Ⴆꓤ   4. Яꟻ

4. When cells are viewed under the compound light microscope, it may be difficult to see the nuclei in the cells. Which technique would help make the nuclei in the cells more visible?
   1. using a dissecting microscope to view the cells
   2. staining the cells with iodine solution
   3. placing the cells in an ultracentrifuge
   4. placing the cells in a sugar solution

5. To improve the focus when viewing a specimen under the compound light microscope, it would be most reasonable to adjust the
   1. substage illuminator
   2. ocular lens
   3. fine adjustment knob
   4. horseshoe base

6. The internal structures of a mitochondrion and a chloroplast can best be observed by using
   1. an ultracentrifuge
   2. a compound light microscope
   3. microdissection instruments
   4. an electron microscope

7. To transplant a nucleus from one cell to another cell, a scientist would use
   1. an electron microscope
   2. an ultracentrifuge
   3. microdissection instruments
   4. staining techniques

8. An instrument used to collect ribosomes for chemical analysis is
   1. a compound microscope
   2. an ultracentrifuge
   3. a scalpel
   4. an electron microscope

9. According to the chart below, which neuron is the longest?

Neuron	Length of Neuron
A	1.5 micrometers
B	50.0 micrometers
C	0.5 millimeter
D	0.005 millimeter

   1. A
   2. B
   3. C
   4. D

10. A student using a compound microscope estimated the diameter of a cheek cell to be 50 micrometers (μm). What is the diameter of this cheek cell in millimeters?
    1. 0.050 mm
    2. 0.500 mm
    3. 5.00 mm
    4. 50.9 mm

What have we learned about cells through the use of the techniques outlined above? We have learned that the cell is an extremely complex structure composed of smaller functional parts known as **cell organelles** ("little organs"). Each of these organelles has a specialized role to perform in the survival of the cell. Some of the cell's major organelles are as follows:

1. **Plasma membrane**—the outer membrane of the cell, which regulates the absorption and secretion of materials into and out of the cell. The plasma membrane serves as the interface between the cell and its environment, and ultimately controls the nature of the cell's internal environment. (The structure of the plasma membrane is discussed in more detail in Unit 2 of this book.)
2. **Cytoplasm**—a watery medium for the suspension of cell organelles and the diffusion of soluble material throughout the cell. The cytoplasm also serves as the site for the operation of certain of the cell's chemical reactions.

3. **Nucleus**—a spherical organelle, located in the center of the cell, that contains the cell's genetic information in the form of **chromosomes**; allows the free transfer of that genetic information during the processes of synthesis and reproduction.

4. **Nucleolus**—a small organelle, located within the nucleus, that functions in the cell's synthesis mechanism by forming ribosomes involved in the manufacture of proteins.

5. **Endoplasmic reticulum**—a series of intracellular membranes that functions in the cell's synthesis mechanism by housing ribosomes, accepting manufactured proteins, and transporting these proteins to the cell's plasma membrane for incorporation into the membrane or for secretion to the cell exterior.

6. **Ribosome**—a small, dense organelle that serves as a site for the manufacture of protein molecules within the cell; may be attached to the endoplasmic reticulum or may be floating free in the cytoplasm.

7. **Mitochondrion**—a small organelle that contains the enzymes necessary to allow the cell to perform certain aspects of chemical respiration.

8. **Golgi complex**—a series of membrane-bound organelles that functions in the cell's synthesis mechanism by accepting manufactured proteins from the endoplasmic reticulum and transporting these proteins to the cell's plasma membrane for secretion to the cell exterior.

9. **Vacuole**—a membrane-bound organelle containing water and other substances; may serve to store food molecules, nonremovable wastes, or secretion products. Some vacuoles take on specialized tasks in the cell, such as food storage (**food vacuole**) or maintenance of water balance (**contractile vacuole**).

10. **Lysosome**—a specialized vacuole that aids the process of nutrition by carrying digestive enzymes and by merging with food-containing vacuoles; may also help to recycle aging or defective cells.

11. **Centriole**—a cylindrical structure, found primarily in animal cells, that apparently functions in the process of cell division.

12. **Chloroplast**—a pigment-containing structure, found primarily in plant and algae cells, in which the chemical reactions of **photosynthesis** (plant nutrition) occur; the primary pigment is **chlorophyll**.

13. **Cell wall**—a structure, found primarily in plants, that provides mechanical support and protection for the cell. The cell wall is composed of a complex carbohydrate known as cellulose.

A typical animal cell and a typical plant cell are diagramed in Figure 1.3.

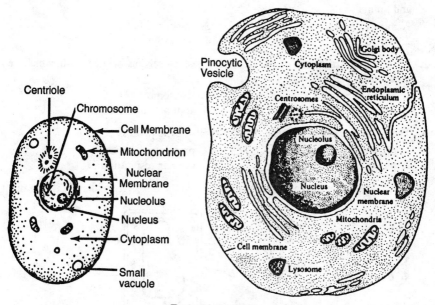

Typical Animal Cell

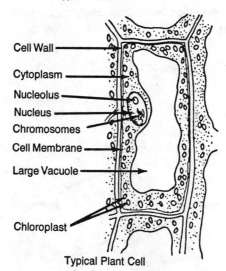

Typical Plant Cell

Figure 1.3   Plant Versus Animal Cells

QUESTION SET 1.4

1. Which cell organelles are considered the sites of aerobic respiration in both plant and animal cells?
   1. mitochondria
   2. centrosomes
   3. chloroplasts
   4. nuclei

2. Intracellular transport of materials is most closely associated with which cell organelle?
   1. cell membrane
   2. cell wall
   3. ribosome
   4. endoplasmic reticulum

3. The ribosome is an organelle that functions in the process of
   1. phagocytosis
   2. pinocytosis
   3. protein synthesis
   4. cellular respiration

4. Which structures in the diagram below enable the observer to identify it as a plant cell?

   1. *A* and *B*
   2. *B* and *C*
   3. *A* and *C*
   4. *B* and *D*

   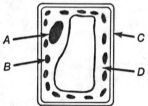

5. Most cellular respiration in plants takes place in organelles known as
   1. chloroplasts
   2. stomates
   3. ribosomes
   4. mitochondria

6. Which organelle is present in the cells of a mouse but *not* present in the cells of a bean plant?
   1. cell wall
   2. chloroplast
   3. cell membrane
   4. centriole

Base your answers to questions 7 and 8 on the diagram of a cell below.

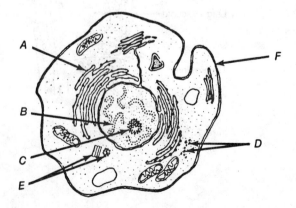

7. Which structures function mainly in reproduction?

1. *A* and *B*   3. *C* and *D*
2. *B* and *E*   4. *A* and *D*

8. Which structures function mainly in transport?

1. *A* and *F*   3. *C* and *F*
2. *B* and *D*   4. *C* and *D*

For questions 9 through 11 choose from the list below the organelle that best matches the description in the question:

ORGANELLE

(1) Food vacuole
(2) Golgi complex
(3) Nucleolus
(4) Lysosome

9. Serves as a "packaging" mechanism for secretions synthesized in the cell.

10. Contains hydrolytic enzymes used in intracellular digestion.

11. Is thought to house RNA used in the process of protein synthesis.

B. Chemistry of Living Organisms

B-1. Elements

The same **elements** that comprise the earth's crust, its water, and its atmosphere also make up the bodies of living things. However, the proportion of elements in living matter is different from the proportion of elements in nonliving matter. Living things are made up of relatively large percentages of the elements carbon (C), hydrogen (H), oxygen (O), and nitrogen (N). Elements found in lower percentages include sulfur (S), phosphorus (P), magnesium (Mg), iodine (I), iron (Fe), calcium (Ca), sodium (Na), chlorine (Cl), and potassium (K).

B–2. Compounds

Compounds are substances composed of two or more different elements bound together chemically. The bonds holding these compounds together may be **ionic** (involving the transfer of electrons between atoms) or **covalent** (involving the sharing of electrons between atoms). See Figure 1.4.

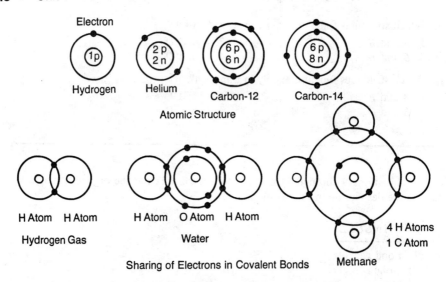

Figure 1.4 Atomic Structure and Covalent Bonding

Two broad categories of compounds are found in nature: **inorganic** and **organic** compounds. Organic compounds contain the elements carbon and hydrogen; inorganic compounds may contain any of the earth's elements, but rarely contain carbon *and* hydrogen together. Living things are composed of both inorganic and organic compounds.

Inorganic compounds commonly found in living things include:

- **Water.** Living things consist of 60 percent to 98 percent water, which acts as the medium for transport and for chemical activities inside the cell.
- **Salts.** Salts are chemicals composed of metallic and nonmetallic ions combined by ionic bonds. They are important for maintaining osmotic balance in the cell, as well as for supplying ions necessary for many of the cell's chemical reactions.
- **Acids** and **bases.** These compounds are important for maintaining the proper balance of hydrogen-ion concentration (acidity/alkalinity) in the cell.

QUESTION SET 1.5

1. Within a cell, which of the following elements is found in the greatest amount?
   1. carbon
   2. calcium
   3. iron
   4. iodine

2. All of the following elements are common to living matter except
   1. hydrogen
   2. oxygen
   3. carbon
   4. uranium

3. All organic compounds found in living things contain
1. nitrogen
2. water
3. sugar
4. carbon

4. The reactions involving most chemical compounds in living systems depend on the presence of
1. sulfur as an enzyme
2. water as a solvent
3. salt as a substrate
4. nitrogen as an energy carrier

5. Which inorganic substance found in living matter aids in the diffusion of gases through a cell membrane?
1. water
2. salt
3. phosphorus
4. iron

6. Which of the following is an inorganic compound that may be found in living matter?
1. glucose
2. hydrochloric acid
3. glycerol
4. hemoglobin

7. The type of chemical bond in which there is a transfer of electrons from one atom to another is known as
1. an inonic bond
2. a hydrogen bond
3. a covalent bond
4. a peptide bond

Organic compounds commonly found in living things include:

- **Carbohydrates**—sugars, starches.
- **Lipids**—fats, waxes.
- **Proteins**—structural proteins, enzymes.
- **Nucleic acids**—DNA, RNA, (Nucleic acids are covered in detail in Unit 5.)

1. Carbohydrates

1.1. Elemental Composition. Carbohydrates are organic compounds, and so contain the elements carbon and hydrogen; in addition, they contain the element oxygen. Carbohydrates generally are found to contain hydrogen and oxygen in the ratio of $2:1$, a ratio different from that in other organic compounds. Like many organic compounds, carbohydrates are actually formed around a "skeleton" of carbon atoms linked together in a chain; some chains are quite short (six or fewer C atoms), while others are very long (thousands of C atoms).

1.2. Molecular Composition. Carbohydrates are found in nature in several different forms:

- The simplest stable carbohydrate is the **monosaccharide** ("simple sugar"). Monosaccharides are formed around a chain of five or six carbon atoms. An example of a monosaccharide is the simple sugar **glucose**, whose formula is $C_6H_{12}O_6$. Glucose is the basic chemical unit of nearly all the more complex carbohydrates. You should be able to recognize the structural formula of a glucose molecule, as shown in Figure 1.5.

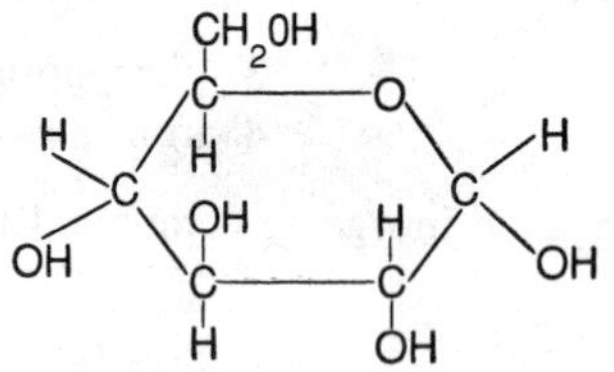

Figure 1.5   Structural Formula: Glucose

- A more complex form of carbohydrate is the **disaccharide** ("double sugar"). Disaccharides are commonly formed around a chain of 12 carbon atoms. An example of disaccharide is the sugar **maltose**, whose formula is $C_{12}H_{22}O_{11}$. Maltose may be formed by chemically combining two molecules of glucose. The chemical process by which this combination occurs is **dehydration synthesis.** Figure 1.6 represents the formation of a molecule of maltose from two units of glucose.

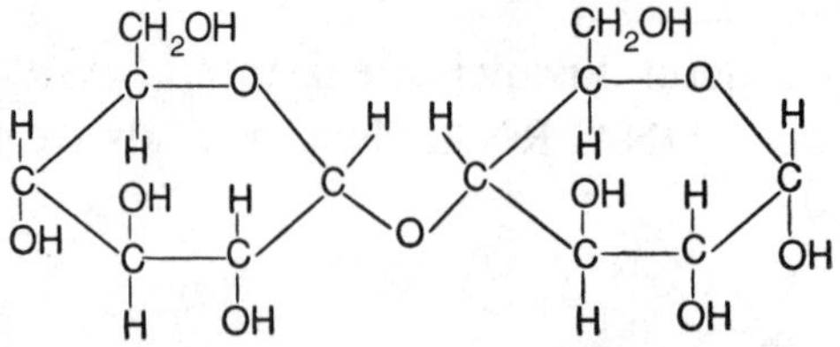

Figure 1.6   Structural Formula: Maltose

- A complex form of carbohydrate is known as a **polysaccharide** ("many sugar"). Polysaccharides may be formed by chemically combining many monosaccharide molecules through the process of dehydration synthesis. The chain of monosaccharide units forming a polysaccharide may be thousands of units long and may be highly branched. **Starch** is a common example of a polysaccharide.

1.3. Examples. Carbohydrates are found in many different forms in nature. Their names commonly end with the suffix "-ose." Here are a few examples of carbohydrates whose names you should recognize:

- Monosaccharides—glucose, fructose, ribose.
- Disaccharides—maltose, sucrose.
- Polysaccharides—starch (amylose), cellulose, glycogen.

1.4. Functions. In all living things, carbohydrates are found as structural components of the cell wall and plasma membrane. They also serve as sources of stored energy in both plants (starch) and animals (glycogen).

2. Lipids

2.1. Elemental Composition. The elements carbon, hydrogen, and oxygen are found in lipids, as they are in carbohydrates. Unlike carbohydrates, lipids have no uniform ratio between the amounts of hydrogen and oxygen in their molecules.

2.2. Molecular composition. Many of the more common lipids are constructed of a unit of **glycerol** (a three-carbon alcohol) combined chemically (by dehydration synthesis) with three molecules of **fatty acid** (a hydrocarbon chain with an attached carboxyl group). Such molecules are known as triglycerides and are constructed as shown in Figure 1.7.

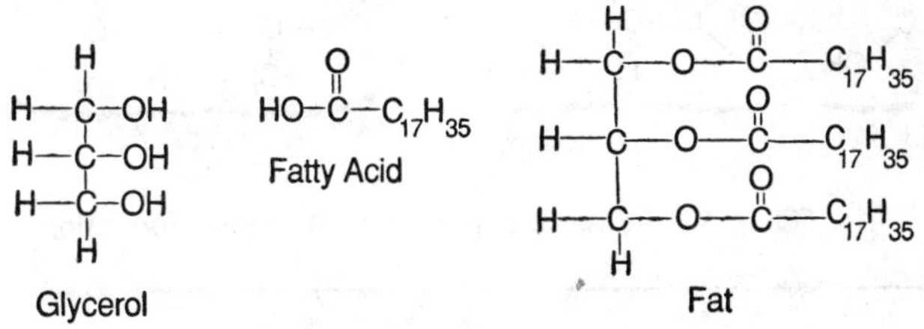

Figure 1.7 Structural Formula: Lipid

2.3. Examples. Typical lipids include animal fats, plant oils, and waxes.

2.4. Functions. Lipids are known to make up a large portion of the plasma membrane and other cellular membranes. They also function as energy-storage compounds for cells.

3. Proteins

3.1. Elemental Composition. The elements carbon, hydrogen, and oxygen are found in proteins, as they are in carbohydrates and lipids. In addition, proteins contain a substantial proportion of nitrogen. Certain proteins may also contain sulfur.

3.2. Molecular Composition. Proteins are compounds made up of repeating units of other compounds known as **amino acids.** Although there are 20 different types of amino acids, they all share a common general structure. All amino acids contain an **amino** group ($-NH_2$) and a **carboxyl** group ($-COOH$) attached to a central carbon atom. A third attached group (**radical**) varies and gives each type of amino acid its unique properties. The generalized structure of an amino acid is shown in Figure 1.8.

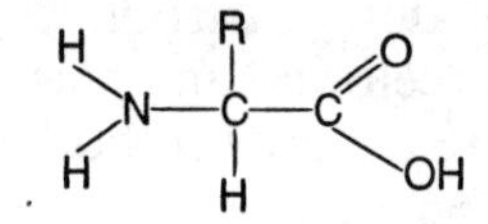

R = variable (radical) group of atoms

Figure 1.8   Structural Formula: Amino Acid

Two amino acid molecules may be joined together chemically by dehydration synthesis to form a **dipeptide.** The term **peptide** refers to the name of the actual chemical bond that joins the two amino acid units together. The formation of a dipeptide is illustrated in Figure 1.9.

Amino Acid *A*   Amino Acid *B*   Dipeptide *AB*

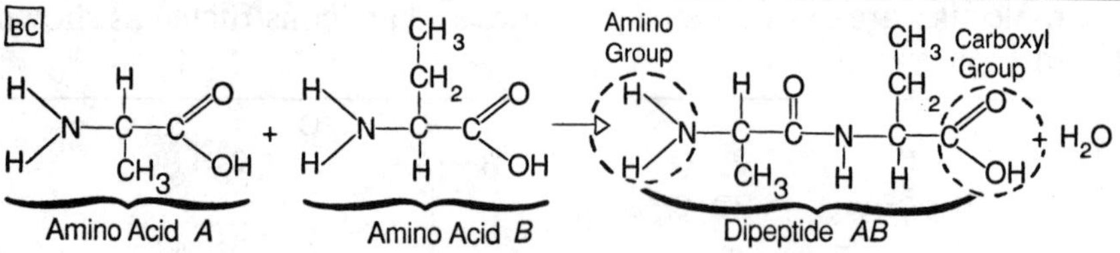

Figure 1.9   Structural Formula: Dipeptide Formation

As more and more amino acids link together by dehydration synthesis, an amino acid chain, known as a **polypeptide**, is formed. Polypeptide chains form the basis of protein molecules. Because of the almost endless variations in which amino acids may be arranged in a polypeptide, proteins are found in thousands of different forms. The extreme variability of proteins is thought to be responsible for the individual variations in living things.

3.3. Examples. A few common examples of proteins are insulin (a hormone), hemoglobin (found in blood cells), and enzymes.

3.4. Functions. The functions of proteins may be classified as either "structural" (forming a part of the cell material) or "functional" (having a role in the chemistry of the cell).

QUESTION SET 1.6

1. The chart below shows the elements present in four different chemical compounds (*A*, *B*, *C*, and *D*). An X indicates the presence of a particular element. Which compound could be a carbohydrate?

Elements	Compound			
	A	*B*	*C*	*D*
Bromine	X			
Carbon		X		X
Fluorine				X
Hydrogen	X	X		X
Lead			X	
Nitrogen			X	
Oxygen		X	X	

1. *A*
2. *B*
3. *C*
4. *D*

2. A common characteristic of carbohydrates, proteins, and nucleic acids is that they
   1. have hydrogen and oxygen atoms present in a 2 : 1 ratio
   2. use dehydration synthesis to combine their basic building blocks
   3. are used as organic catalysts in biochemical reactions
   4. use monosaccharides as their basic building units

3. Which organic compound is correctly matched with the subunit that composes it?
   1. maltose—amino acid
   2. starch—glucose
   3. protein—fatty acid
   4. lipid—sucrose

4. Which represents a carbohydrate molecule?
   1. $C_6H_6O_6$
   2. $C_{12}H_{12}O_6$
   3. $C_6H_{12}O_5$
   4. $C_6H_{12}O_6$

5. Glucose and maltose are classified as organic compounds because both are
   1. carbon-containing substances
   2. composed of simple elements
   3. waste products
   4. artificial sugars

6. Which list of molecules is arranged in order of increasing size?
   1. oxygen, starch, glucose, sucrose
   2. sucrose, oxygen, starch, glucose
   3. oxygen, glucose, sucrose, starch
   4. starch, glucose, sucrose, oxygen

7. Which types of compounds are *not* classified as carbohydrates?
   1. lipids
   2. sugars
   3. starches
   4. polysaccharides

8. Which process produces peptide bonds?
   1. digestion
   2. hydrolysis
   3. dehydration synthesis
   4. enzyme deactivation

9. Which element is present in maltase, but *not* in maltose?
   1. carbon
   2. hydrogen
   3. oxygen
   4. nitrogen

10. Which substances are most commonly used as building blocks in the synthesis of some lipids?
    1. sugars and starches
    2. amino acids and nucleotides
    3. starches and enzymes
    4. glycerol and fatty acids

B-3. Chemical Control

The cell's ability to regulate its own chemical reactions (i.e., to exercise **chemical control**) differentiates it from nonliving things in nature. The cell can be thought of as a complex "chemical factory" in which continual, dynamic chemical activity is occurring. Chemicals metabolized in such cellular reactions are known as **substrates.**

1. The Way in Which Enzymes Function

The controlled nature of cellular chemical reactions is made possible through the actions of the cell's **enzymes.** Enzymes regulate the rate at which the cell's chemical reactions occur. In this role they function as **organic catalysts.** Each chemical reaction in the cell requires its own, specific type of enzyme in order to operate.

1.1. The Structure of Enzymes. Enzymes are protein molecules, and as such are both extremely complex and extremely variable. Certain enzymes also are known to contain nonprotein components known as **coenzymes**; vitamins frequently function as coenzymes. Enzymes are usually named for the particular chemical (substrate) whose reactions they catalyze; enzyme names end in the suffix "-ase."

Substrate	Enzyme Catalyzing
Maltose	MaltASE
Amylose	AmylASE
Lipid	LipASE

1.2. The Active Site. Enzyme molecules are thought to contain specific areas responsible for linking to the substrate molecules. These reacting areas are known as **active sites** and are thought of as pockets or slots on the enzyme molecule into which the substrate molecules fit during the reaction catalyzed by the enzyme (see Figure 1.10 below).

1.3. The Catalytic Action. Although there are many ways to conceptualize the catalytic action of enzymes, the following model, known as the **lock-and-key model**, is widely accepted as being a reasonable summary of this action:

- To begin the process, the enzyme molecule must link with the substrate in a temporary, close physical association known as the **enzyme-substrate complex.**

- This association is thought to occur at the enzyme's active site. The "chemical fit" between this active site and the substrate must be exact in order for the desired reaction to occur, in much the same way that a lock and key must match in order for the lock to be opened (hence the name "lock-and-key" model).

- During the existence of the enzyme-substrate complex, the reaction involving the substrate takes place. The substrate is chemically altered in this reaction, and products are formed. The enzyme molecule, however, is unchanged by the reaction.

- When the reaction is completed, the enzyme and product(s) separate. The enzyme molecule, being unchanged, is free to form additional complexes and to catalyze additional reactions of this substrate.

- A single enzyme molecule may catalyze millions of reactions in this manner during its "lifetime." Eventually, however, the structure of the enzyme molecule begins to deteriorate, and the molecule must be replaced by the cellular process of synthesis.

The catalytic action of enzymes described above is diagramed in Figure 1.10.

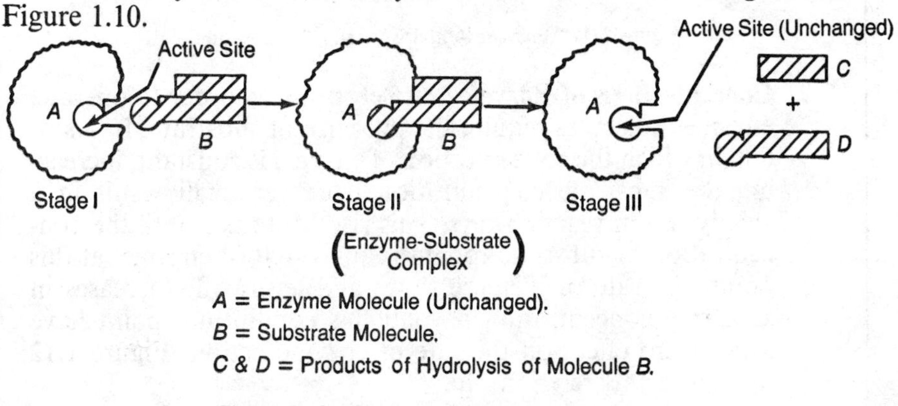

A = Enzyme Molecule (Unchanged).
B = Substrate Molecule.
C & D = Products of Hydrolysis of Molecule B.

Figure 1.10 Catalytic Action Of An Enzyme

2. Environmental Influences on Enzymes

The rate at which enzymes catalyze their reactions changes as conditions inside the cell change. Such conditions as the temperature in which the cell exists, the relative concentrations of enzyme and substrate within the cell, and the cellular acidity/alkalinity (pH) can alter the rapidity with which enzymes work.

BC

2.1. Temperature. The cell and its enzymes are very sensitive to temperature conditions. Extreme cold can slow enzyme action nearly to a halt. As the temperature of the cell rises, its enzymes begin to operate more and more rapidly; an optimum ("best" or "most efficient") temperature allows the most rapid reaction rate. Extreme heat can halt enzyme action by deforming the molecular shape of the enzyme; this distortion is known as **denaturation,** and is a permanent condition that makes the enzyme incapable of further catalytic action. The effect of temperature on a typical enzyme is shown in Figure 1.11.

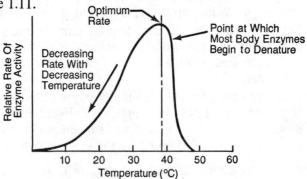

Figure 1.11   Enzyme Activity Versus Temperature

2.2. Concentrations of Enzyme and Substrate. In order to operate, enzyme molecules require the presence of substrate. In a system in which the concentration of enzyme is constant, increasing the substrate concentration from zero will result in a steady rise in reaction rate. This rise continues until the concentration of substrate roughly equals that of enzyme; at this point, the rate of increase slows and levels off. Increases in substrate concentration beyond this equilibrium point have little or no effect on the rate of enzyme action. Figure 1.12 illustrates this relationship.

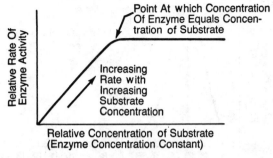

Figure 1.12 **Enzyme Activity Versus Concentration**

2.3. Cell Acidity (pH). The cell environment may be characterized by its level of acidity or alkalinity. To accurately measure this characteristic, scientists have devised methods of measuring the relative concentration of **hydrogen ion** (H^+), which is given off by acids in solution with water. The **pH scale** has been devised to indicate the relative hydrogen-ion concentration. This scale runs from 0 to 14, 0 being extremely acidic and 14 being extremely alkaline. A pH of 7 indicates a neutral condition (neither acidic nor alkaline). A pH less than 7 indicates an acidic condition; a pH greater than 7, an alkaline condition.

Most enzymes seem to work best at or near a pH of 7 (neutral). Despite this fact, there are many enzymes whose optimum pH's are extremely acidic (stomach enzymes) or extremely basic (intestinal enzymes). In general, however, as the cell's acidity increases (i.e., pH decreases below 7), most enzymes show a steady decrease in activity. The same phenomenon may be observed as the cell's alkalinity increases (i.e., pH rises above 7). Each enzyme has its own unique range of pH tolerance. Figure 1.13 illustrates the effect of pH on enzyme activity.

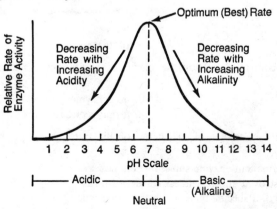

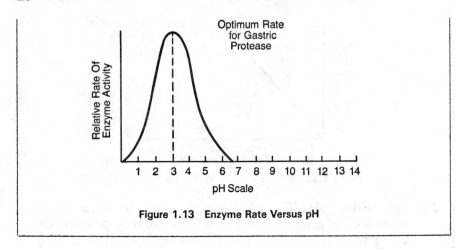

Figure 1.13   Enzyme Rate Versus pH

QUESTION SET 1.7

1. Which is characteristic of an enzyme?
   1. It is an inorganic catalyst.
   2. It is destroyed after each chemical reaction.
   3. It provides energy for any chemical reaction.
   4. It regulates the rate of a specific chemical reaction.

2. An enzyme-substrate complex may result from the interaction of molecules of
   1. glucose and lipase
   2. fat and amylase
   3. sucrose and maltase
   4. protein and protease

3. The diagram below represents an enzyme-catalyzed reaction. Which substance is represented by the letter *X*?

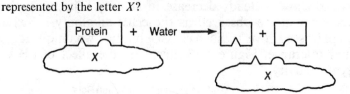

   1. maltase
   2. sucrase
   3. lipase
   4. protease

4. In an enzyme-controlled reaction, the role of certain vitamins such as niacin is to act as
   1. an enzyme
   2. a substrate
   3. a coenzyme
   4. a polypeptide

5. The "lock-and-key" model of enzyme action illustrates that a particular enzyme molecule
   1. forms a permanent enzyme-substrate complex
   2. may be destroyed and resynthesized several times
   3. interacts with a specific type of substrate molecule
   4. reacts at identical rates under all environmental conditions

BC **6.** Which graph below best illustrates the pattern of enzyme action rates when a specific substrate is slowly added to a system with a fixed enzyme concentration?

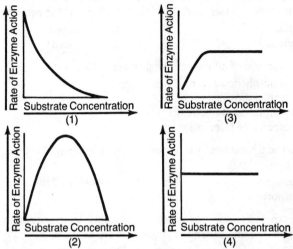

BC   Base your answers to questions 7 through 10 on the graph below, which represents the rate of action of four different enzymes, (*A, B, C, D*) at varying temperatures.

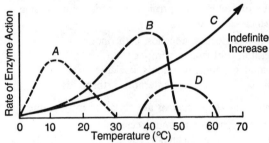

7. For which enzyme is the maximum rate of enzyme action achieved at 10°C?

1. *A*	**3.** *C*
2. *B*	**4.** *D*

8. Which enzyme begins its action at the highest temperature?

1. *A*	**3.** *C*
2. *B*	**4.** *D*

9. Which curve most closely illustrates the effect of temperature on human enzymes?

1. *A*	**3.** *C*
2. *B*	**4.** *D*

10. Which curve most likely represents incorrect data for the rate of enzyme action?

1. *A*	**3.** *C*
2. *B*	**4.** *D*

11. Which pH value indicates an acidic condition?

1. 8	**3.** 14
2. 10	**4.** 4

UNIT 1 REVIEW

1. The term that represents all the chemical activities that occur in an organism is
   1. synthesis
   2. regulation
   3. metabolism
   4. homeostasis

2. A characteristic of all known living things is that they
   1. use atmospheric oxygen
   2. use carbon dioxide
   3. carry on metabolic activities
   4. are capable of locomotion

3. The chemical bond energy in organic nutrients is changed to a usable form for living things by
   1. digestion
   2. transport
   3. photosynthesis
   4. respiration

4. Blue-green algae lack a membrane separating their nuclear material from their cytoplasm. On this basis these organisms are classified as
   1. fungi
   2. protists
   3. monerans
   4. plants

5. Which group of organisms is arranged in order of increasing complexity according to a modern classification system?
   1. protozoa, coelentrates, annelids, arthropods
   2. coelenterates, protozoa, arthropods, birds
   3. protozoa, arthropods, mammals, coelenterates
   4. amphibians, protozoa, coelenterates, arthropods

6. *Homo erectus* and *Homo sapiens* are classified in the same
   1. kingdom, phylum, genus, and species
   2. kingdom, phylum, and genus
   3. phylum, genus, and species
   4. genus and species

7. The diameter of the field of vision of a compound light microscope is 1.5 millimeters. This may also be expressed as
   1. 15 micrometers
   2. 150 micrometers
   3. 1,500 micrometers
   4. 15,000 micrometers

8. Some scientists disagree on whether viruses are alive. A major reason for this disagreement is that viruses
   1. cannot manufacture food
   2. are not composed of units of structure known as cells
   3. do not contain nucleic acids
   4. do not contain the element carbon

9. A student using a compound microscope measured several red blood cells and found that the average cell length was 0.008 millimeter. What is the average length in micrometers of a single red blood cell?
   1. 0.8
   2. 8
   3. 80
   4. 800

10. To collect mitochondria from cells and study their structure in fine detail, which instruments would a scientist most likely use?
    1. microdissection apparatus and compound light microscope
    2. ultracentrifuge and electron microscope
    3. microdissection apparatus and dissecting microscope
    4. ultracentrifuge and compound light microscope

11. Which organelle is primarily concerned with the conversion of the potential energy of organic compounds into a form suitable for immediate use by the cell?
    1. mitochondria   3. ribosomes
    2. centrosomes   4. vacuoles

Base your answers to questions 12 and 13 on the information below and on your knowledge of biology.

A student observed a green plant cell under the low-power objective of her microscope and noted the movement of organelles as shown in diagram A. She then added three drops of a 10% salt solution to the slide, waited a few minutes, and observed the cell as shown in diagram B.

A   B

12. The organelles observed were most likely
    1. centrosomes   3. mitochondria
    2. chloroplasts   4. ribosomes

13. The appearance of the clumped material as shown in diagram B is due to
    1. loss of water from the cell
    2. loss of salt from the cell
    3. addition of water into the cell
    4. addition of salt into the cell

---

14. Which is the principal inorganic solvent in cells?
    1. salt   3. alcohol
    2. water   4. carbon dioxide

Base your answers to questions 15 and 16 on the diagram below of a portion of a cell and on your knowledge of biology.

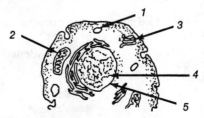

15. In which organelle would water and dissolved materials be stored?
    1. *I*   3. *3*
    2. *2*   4. *5*

16. Most of the enzymes involved in aerobic cellular respiration are located in the organelle labeled
    1. *I*   3. *3*
    2. *2*   4. *4*

---

17. Which element is present in living cells and in *all* organic compounds?
    1. potassium   3. nitrogen
    2. sulfur   4. carbon

18. Which substance is an inorganic compound that is necessary for most of the chemical reactions that take place in living cells?
    1. glucose   2. water
    2. starch   4. amino acid

19. If a specific carbohydrate molecule contains 10 hydrogen atoms, that same molecule will most probably contain
    1. 1 nitrogen atom   3. 5 oxygen atoms
    2. 10 nitrogen atoms   4. 20 oxygen atoms

20. A biochemist was given a sample of an unknown organic compound and asked to determine the class of organic compounds to which it belonged. The chart below represents the results of the biochemist's analysis of the sample.

Element	Number of Atoms per Molecule
C	12
H	22
O	11
S	0
N	0
P	0

Based on these results, to which class of organic compounds did this sample belong?
    1. lipid   3. salt
    2. protein   4. carbohydrate

21. Which substance is an enzyme?
    1. thymine   3. bile
    2. maltase   4. colchicine

22. An enzyme that acts as a catalyst in the digestion of certain carbohydrates is

1. maltase
2. lipase
3. protease
4. ATP-ase

23. An organic compound that has hydrogen and oxygen in a 2 : 1 ratio belongs to the group of compounds known as

1. lipids
2. fatty acids
3. proteins
4. carbohydrates

24. Examples of polymers that contain repeating units known as nucleotides are

1. hemoglobin and maltase
2. starch and glycogen
3. fats and oils
4. DNA and RNA

BC Base your answers to questions 25 through 28 on the structural formula below and on your knowledge of biology:

25. The structural formula represents a molecule of

1. glucose
2. glycerol
3. maltase
4. alanine

26. Which high-molecular-weight substances are made up of repeating units of these molecules?

1. starch and cellulose
2. hemoglobin and protease
3. fats and oils
4. polypeptides and nucleic acids

27. The complete aerobic oxidation of this compound produces

1. amino acids and urea
2. carbon dioxide and water
3. ethyl alcohol and carbon dioxide
4. glycerol and fatty acids

28. The process by which two or more of these molecules are bonded together in a muscle cell is known as

1. enzymatic hydrolysis
2. anaerobic respiration
3. dehydration synthesis
4. carbon fixation

BC Base your answers to questions 29 through 32 on the diagram below, which illustrates a biochemical reaction, and on your knowledge of biology.

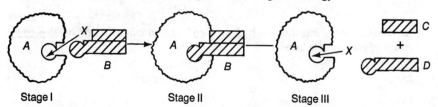

29. Which letter indicates a substrate molecule in this reaction?
    1. *A*   3. *C*
    2. *B*   4. *D*

30. Stage II in the diagram most accurately represents
    1. an inorganic catalyst
    2. a denatured enzyme
    3. a vitamin
    4. an enzyme-substrate complex

31. The area labeled *X* is known as
    1. an atomic nucleus
    2. an active site
    3. a pH indicator
    4. a temperature regulator

32. Which substance is needed in order for this biochemical reaction to occur?
    1. water   3. table salt
    2. iodine   4. Benedict's solution

---

33. Which structural formulas of organic compounds below represent carbohydrates.
BC

(A)

(B)

(C)

(D)

    1. *A* and *B*   3. *C* and *D*
    2. *B* and *C*   4. *A* and *D*

HOW DO LIVING THINGS MAINTAIN THEIR EXISTENCE?

KEY IDEAS	This unit deals with the way in which living things function so that they are able to survive from day to day. The concepts stressed in this unit include the processes that operate within each life function and the particular adaptations that enable representative plants and animals to perform these life functions efficiently.

KEY OBJECTIVES

Upon completion of this unit, you should be able to:

☐ List the major life functions carried on by living things, and describe how they contribute to the maintenance of steady state in the organism.

☐ Identify the diverse adaptations present in living things for carrying out the basic life functions, and compare these adaptations among selected organisms.

☐ Describe the way in which certain biochemical reactions correspond with physiological activities observable in living things.

☐ Describe the ways in which the adaptations of living things complement their life functions, allowing organisms to exist successfully in their environments.

I. THE LIFE FUNCTION OF NUTRITION

Nutrition is the life function by which organisms obtain materials needed for energy, growth and repair, and other life functions. As part of this process, these materials are converted to a simplified form that can be used by the cell. Some organisms manufacture their own food; they are known as "autotrophs." Organisms that do not manufacture their own food are called "heterotrophs."

A. Autotrophic Nutrition

Autotrophic nutrition includes the activities by which green plants and certain protists and monerans manufacture complex organic (food) molecules from simple inorganic molecules. The organisms capable of performing this activity are known as **autotrophs** ("self-feeders"). The principal modes of autotrophic nutrition are **photosynthesis** and **chemosynthesis.**

1. Process, Results, and Significance of Photosynthesis

1.1. The Photosynthetic Process. In photosynthesis, the energy of light is trapped and converted into the chemical bond energy of organic compounds such as sugar. Here are the important aspects of the process of photosynthesis:

- **Chlorophyll**, a green pigment found in the plant cell's **chloroplasts**, absorbs light energy, which strikes the plant from its environment.
- Normal sunlight appears white, but actually contains the types of light energy that produce all colors. The red and blue colors in this light are those absorbed and converted most readily by chlorophyll. Green is absorbed the least; instead, it is reflected.
- In the presence of the proper enzymes, the atoms of carbon dioxide (CO_2) and water (H_2O) are rearranged to form the more complex organic molecule glucose ($C_6H_{12}O_6$). A by-product of this process is molecular (atmospheric) oxygen (O_2).
- A simplified chemical equation illustrating the essential elements of photosynthesis is shown in Figure 2.1.

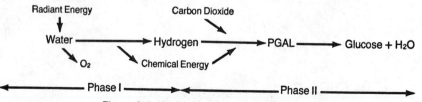

Figure 2.1   Photosynthesis Reaction Equation.

- A more detailed study of the photosynthetic process reveals that it actually consists of two separate processes, each characterized by its own set of chemical reactions.

BC

a. **Photochemical (light) reactions.** Stacked layers of chlorophyll-containing membranes within the chloroplast, known as **grana**, contain the enzymes that catalyze this process. The light energy absorbed by the chlorophyll and other pigments on these grana is used to split water molecules into their component elements, hydrogen and oxygen. This process is known as **photolysis** ("splitting with light") and involves the production of an energy-carrying molecule known as ATP. (See Section III for a more detailed description of ATP.) Atoms of oxygen recombine to form molecular (atmospheric) oxygen, which is released as a gas (the radioactive isotope **oxygen-18** has been used to trace this part of the process). The hydrogen released is transferred to the next phase of the process by a hydrogen-carrier compound.

b. **Carbon-fixation (dark) reactions.** A second set of photosynthetic reactions combines the released hydrogen atoms with the atoms making up carbon dioxide. An intermediate product of this reaction is a carbohydrate-like, three-carbon compound known as **phosphoglyceraldehyde (PGAL).** This compound may be used by the cell to synthesize several other compounds, such as glucose (the radioactive isotope **carbon-14** has been used to trace this part of the process). The name "carbon-fixation reactions" is derived from the fact that carbon atoms are "fixed in place" in a stable form as a result of this process. This set of reactions occurs in the **stroma** of the chloroplast; the stroma lie between the grana.

QUESTION SET 2.1

1. A bean plant is classified as an autotroph because it
   1. uses enzymes
   2. uses oxygen
   3. converts inorganic materials into organic nutrients
   4. absorbs organic nutrients from the soil

2. In most plants the process of photosynthesis occurs most rapidly when the plants are exposed to equal intensities of
   1. green and red light
   2. blue and red light
   3. yellow and orange light
   4. green and yellow light

3. Which organisms add more oxygen to the atmosphere than they remove?
   1. grasshoppers
   2. bread molds
   3. corn plants
   4. mushrooms

4. At optimum light intensity, which atmospheric gas most directly influences the rate of photosynthesis?
   1. nitrogen
   2. oxygen
   3. carbon dioxide
   4. hydrogen

5. In plants, the molecular oxygen concentration of a leaf cell usually increases during the process of
   1. aerobic respiration
   2. photosynthesis
   3. transpiration
   4. capillary action

6. Which word equation represents the process of photosynthesis?
   1. carbon dioxide + water → glucose + oxygen + water
   2. glucose → alcohol + carbon dioxide
   3. maltose + water → glucose + glucose
   4. glucose + oxygen → carbon dioxide + water

7. The raw materials used by green plants for photosynthesis are
   1. oxygen and water
   2. oxygen and glucose
   3. carbon dioxide and water
   4. carbon dioxide and glucose

8. An important function of chlorophyll molecules during photosynthesis is
   1. absorbing and storing water in root cells
   2. converting water into carbon dioxide
   3. absorbing certain wavelengths of light energy
   4. converting chemical bond energy to light energy

9. One bean plant is illuminated with green light, and another bean plant of similar size and leaf area is illuminated with blue light. If all other conditions are identical, how will the photosynthetic rates of the plants most probably compare?
   1. Neither plant will carry on photosynthesis.
   2. Photosynthesis will occur at the same rate in both plants.
   3. The plant under green light will carry on photosynthesis at a greater rate than the one under blue light.
   4. The plant under blue light will carry on photosynthesis at a greater rate than the one under green light.

10. During the photochemical reactions that occur during photosynthesis, light energy is used primarily to
    1. chemically combine carbon and oxygen
    2. chemically combine hydrogen and carbon dioxide
    3. split water into hydrogen and oxygen
    4. split maltose into glucose molecules

11. A product of the carbon-fixation reactions of photosynthesis is

BC **1.** phosphoglyceraldehyde   **3.** gaseous oxygen
    2. lactic acid   **4.** carbon dioxide

1.2. Results of Photosynthesis. The compound glucose, which results from the process of photosynthesis, may be utilized in many ways by the cell. Some of these ways include:

- Direct use as a source of energy in the life process of cellular respiration, which is carried on by all living things, plant as well as animal.

- Conversion, for purposes of storage, into more complex forms of carbohydrate, such as starch, by the process of dehydration synthesis. Such storage carbohydrates must be reconverted to simpler molecules by **intracellular digestion** (digestion inside the cell) before they can be utilized in metabolic processes or transported to other sites in the organism.

- Conversion into other types of metabolic compounds, such as proteins, lipids and nucleic acids.

1.3. Significance of Photosynthesis. Photosynthesis is perhaps the single most significant biochemical reaction carried on by living things. Significant results of the photosynthetic reactions include the following:

- Carbohydrates manufactured as a result of this process are used by virtually all living things as a source of cellular energy trapped from sunlight. This energy is released in a controlled fashion in the process of cellular respiration.

- Oxygen gas released as a by-product of the photosynthetic reactions is the principal source of atmospheric oxygen, required by a vast majority of living things for cellular respiration. Reductions in photosynthetic rate can have disastrous effects on such life forms by reducing the concentration of oxygen in the air.

2. Adaptations for Photosynthesis

2.1. Algae are unicellular or simple multicellular organisms capable of performing photosynthesis. Their simple structure allows them to absorb the raw materials of photosynthesis directly from their environment and to carry out the chemical processes required without specialized adaptations other than the chloroplast and other cellular organelles. A major portion of the photosynthetic activity on earth is performed by marine and freshwater algae.

2.2. Complex multicellular terrestrial plants have organs specialized to perform photosynthetic activities. The principal photosynthetic organ of such plants is the **leaf**. The characteristics of most leaves that make them efficient organs for photosynthesis include:

- A broad, flat shape that makes them efficient absorbers of light energy, while allowing heat to be given off readily to the environment.
- A waxy coating (**cuticle**) over the outer cell layers (**epidermis**), which acts to limit the amount of water loss from the leaf, as well as limiting the entry of disease organisms into the leaf. A certain amount of mechanical protection is also derived from these adaptations, preventing damage to the delicate inner structure of the leaf.
- Small openings in the bottom surface of the leaf, known as **stomates**, whose size is regulated by specialized pairs of **guard cells**. These small openings allow the entry of carbon dioxide into the inner leaf structure and the release of oxygen and water vapor from the leaf to its environment.
- A layer of cells (**palisade layer**) that contains large concentrations of chloroplasts. This layer lies just underneath the upper epidermis and is the principal tissue involved in the photosynthetic process.
- A loosely packed layer of cells (**spongy layer**) that allows free movement of gases to the photosynthetic cells of the leaf.
- Conducting tissues continuous with those of the stem bring water to the leaf tissues and allow manufactured glucose to be carried to other parts of the plant. These conducting tissues are located in the **veins** of the leaf.

A cross section of a leaf structure is shown in Figure 2.2.

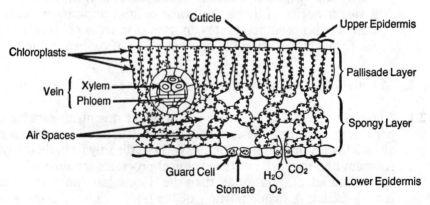

Figure 2.2 Leaf Structure — Cross Section

QUESTION SET 2.2

1. Starch stored in plants as a result of dehydration synthesis of glucose can be used for the plant's activities only after it has been
   1. digested intracellularly
   2. incorporated into cell walls
   3. converted into polypeptides
   4. excreted through stomates

2. Which process is the source of most of the oxygen in the atmosphere?
   1. aerobic respiration   **3.** transpiration
   2. fermentation   **4.** photosynthesis

Base your answers to questions 3 and 4 on the chemical equation below and on your knowledge of biology.

$$C_6H_{12}O_6 + C_6H_{12}O_6 \xrightarrow{\;X\;} C_{12}H_{22}O_{11} + H_2O$$

3. The process represented by the equation is known as
   1. fermentation
   2. hydrolysis
   3. aerobic oxidation
   4. dehydration synthesis

4. The substance represented by X is most likely
   1. vitamin B   **3.** RNA
   2. maltase   **3.** DNA

5. The broad, flat structure of the leaves of most plants makes them well adapted to
   1. store large quantities of starch for later use as food
   2. absorb sunlight used in the photosynthetic reactions
   3. support the upper portions of the plant
   4. carry out the process of chemical regulation

6. The size of stomate openings is regulated by the
   1. palisade cells   **3.** guard cells
   2. spongy cells   **4.** xylem cells

7. The openings on the lower surface of some leaves, which allow for the exchange of gases, are called
   1. stomates   **3.** lenticels
   2. guard cells   **4.** vascular bundles

Base your answers to questions 8 through 10 on the diagram below, which shows a leaf cross section, and on your knowledge of biology.

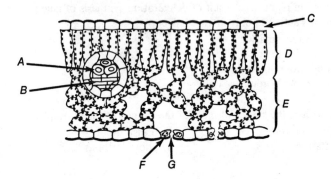

8. Which letter indicates the principal region of food manufacture?
   1. *E*   3. *C*
   2. *B*   4. *D*

9. Which letter indicates the area where carbon dioxide passes out of the leaf?
   1. *A*   3. *C*
   2. *G*   4. *D*

10. Which letter indicates a structure that regulates the size of a stomate?
    1. *A*   3. *F*
    2. *B*   4. *G*

B. Heterotrophic Nutrition

Heterotrophic nutrition includes the activities by which organisms unable to manufacture organic molecules of their own obtain and process preformed organic molecules from their environment. The organisms capable of performing this activity are known as **heterotrophs** ("different-feeders"), and include bacteria, fungi, protozoans, and animals.

1. Processes for Heterotrophic Nutrition

1.1. **Ingestion** includes the mechanisms by which organisms take in food from their environment. This food may be simple or complex in structure.

1.2. **Digestion** includes the mechanisms by which foods are broken down into simpler substances by the organism. This process may occur inside the cell (**intracellular digestion**) or outside the cell (**extracellular digestion**). If extracellular, the simple end products of the digestive process must be absorbed into the cell before they can be utilized.

- **Mechanical digestion** includes any processes by which foods are broken down physically, thus increasing their surface area for efficient enzyme action. Such processes include chewing, tearing, grinding, or cutting.
- **Chemical digestion** includes any processes by which foods are broken down chemically, thus reducing them to their simplest molecular forms. The chemical process involved is known as **enzymatic hydrolysis**. Hydrolysis ("splitting with water") involves replacing existing chemical bonds within a complex molecule with the atoms comprising a water molecule, thereby breaking the complex molecule apart at those points and producing simpler molecules. (See Unit 3 for further details on this process.) The end products of chemical digestion are as follows:

Complex Food Molecules	Molecular End Products
Carbohydrates	Simple sugars
Lipids	Fatty acids and glycerol
Proteins	Amino acids

The end products of chemical digestion are the same in all organisms.

1.3. Egestion includes the mechanisms by which the indigestible components of food materials are eliminated from the body. These food materials may remain undigested if specific enzymes needed to hydrolyze them are not present or if the time in which the foods and enzymes are in contact is not sufficient.

2. Adaptations for Heterotrophic Nutrition

2.1. Fungi are simple organisms that live, grow, and reproduce in the food on which they are found. Fungi penetrate this food with tiny, rootlike filaments known as **rhizoids**. The rhizoids secrete hydrolytic enzymes that digest (hydrolyze) the complex food molecules and then absorb the simple molecular end products. Fungi are a class of organisms whose digestion is extracellular.

2.2. Protozoans are simple unicellular organisms lacking specialized organs for nutrition. Nevertheless, they are known to have certain specializations of their organellar structure, which differ among different groups of protozoans:

- **Amebas** ingest their food by means of cytoplasmic projections known as **pseudopods**. The pseudopods surround and engulf food particles in a process known as **phagocytosis**. Digestion in amebas is intracellular within specialized food vacuoles.

- **Paramecia** ingest their food through a fixed opening in their plasma membrane known as the **oral groove**. Food particles are moved toward the oral groove by the action of **cilia**, which cover the outer cell membrane. Once the food particle has been moved into the oral groove, it is encased in a food vacuole to be digested intracellularly after merging with a lysosome in the cytoplasm. Egestion is accomplished through a fixed opening known as the **anal pore**.

- In all protozoans, the end products of digestion are absorbed directly into the cytoplasm from the food vacuole.

Figure 2.3 shows the anatomical features of two typical protozoans, the ameba and paramecium.

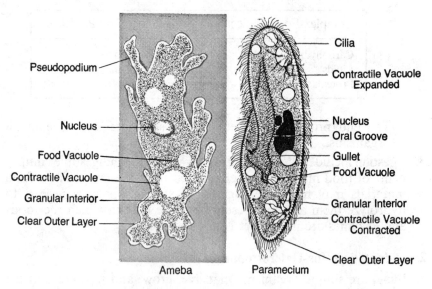

Ameba   Paramecium

Figure 2.3   Protozoa   Courtesy of Keuffel & Esser Rockaway, New Jersey

2.3. Animals

- The **hydra** belongs to Phylum Ceolenterata, and as such is classified as being among the simplest organisms in the Animal Kingdom (see Unit 1 of this book for the major characteristics of ceolenterates). Hydras ingest their food through a **mouth** opening after grasping it with stinging **tentacles**. Once inside the hydra's **digestive cavity**, the food particles are broken down by digestive enzymes secreted by specialized cells in the cavity wall (extracellular digestion). The molecular end products of this process may then be absorbed by the cells of the organism. The hydra also has specialized engulfing cells in its digestive cavity that can take in and digest small food particles too large for direct absorption by the cells (intracellular digestion). Egestion of undigested materials is accomplished through the mouth opening. It is sometimes said that the hydra's digestive cavity is a **two-way digestive tube**.

Figure 2.4 shows the anatomical features of the hydra, a typical ceolenterate.

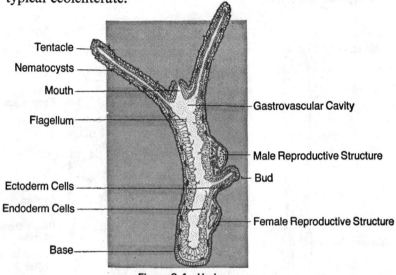

Tentacle
Nematocysts
Mouth
Flagellum
Gastrovascular Cavity
Male Reproductive Structure
Bud
Ectoderm Cells
Endoderm Cells
Female Reproductive Structure
Base

Figure 2.4   Hydra

Courtesy of Keuffel & Esser
Rockaway, New Jersey

- The **earthworm** belongs to Phylum Annelida, the "segmented worms" (see Unit 1 for the major characteristics of the annelids). The earthworm's digestive tract is tubular, with openings at each end. The earthworm's **mouth** is used for ingestion of organic materials found in its soil environment. As this food passes through the digestive tube, it is digested gradually in specialized areas of the tube, as follows:

a. **Pharynx**—a muscular organ for temporary storage of the food/soil mixture before processing.

b. **Esophagus**—the portion of the food tube that transports the food/soil mixture from the pharynx to the crop.

c. **Crop**—a thin-walled enlargement of the digestive tube used for storage of the food/soil mixture before its entry into the gizzard.

d. **Gizzard**—a muscular organ in which the ingested food is mechanically digested before its entry into the intestine.

e. **Intestine**—a long tube through which the food passes as it is digested chemically by enzymes secreted from the intestine lining.

Undigested materials are egested at the posterior end of the digestive tube through the **anus**. Food moves in one direction through this tube; for this reason, this design is sometimes referred to as a **one-way digestive tube**.

Figure 2.5 shows the anatomical features of the earthworm, a typical annelid.

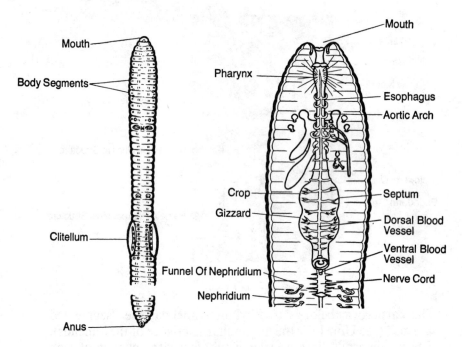

Figure 2.5   Earthworm

Courtesy of Keuffel & Esser
Rockaway, New Jersey

- The **grasshopper** is a member of Phylum Arthropoda, the "jointed-legged" animals (see Unit 1 for the major characteristics of the arthropods). Despite the great differences in appearance between earthworms and grasshoppers, their digestive systems are quite similar. Each of the digestive organs listed above for the earthworm is present also in the digestive system of the grasshopper. In addition, the grasshopper's digestive tract contains the following:

a. **Mouth parts**—specialized parts in the mouth that are used to chew food materials and mix them with salivary enzymes.
b. **Salivary glands**—glands that produce and secrete **saliva**, which contains digestive enzymes.
c. **Gastric caeca**—digestive glands that secrete hydrolytic (digestive) enzymes into the grasshopper's intestine to aid in the breakdown of food materials.
d. **Rectum**—a storage area for food wastes and metabolic wastes before their elimination through the anus.

Figure 2.6 shows the anatomical features of the grasshopper, a typical arthropod.

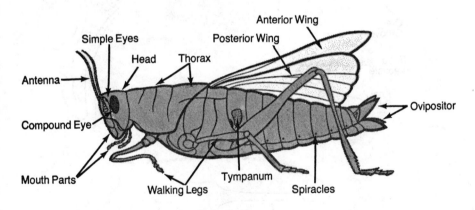

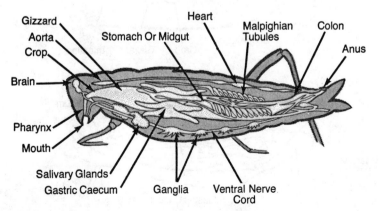

Figure 2.6   Grasshopper

Courtesy of Keuffel & Esser
Rockaway, New Jersey

- **Human beings** are members of Phylum Chordata, and as such are among the most complex of animal forms (see Unit 1 for the major characteristics of the chordates). In many ways the human being resembles the earthworm and the grasshopper in the general plan of its digestive tract. As in these organisms, the digestive tube is a "one-way" tract consisting of and surrounded by specialized organs of digestion. See Unit 3 of this book for a more detailed study of the structure and function of the human digestive system.

The adaptation for nutrition in various organisms are shown in Figure 2.7.

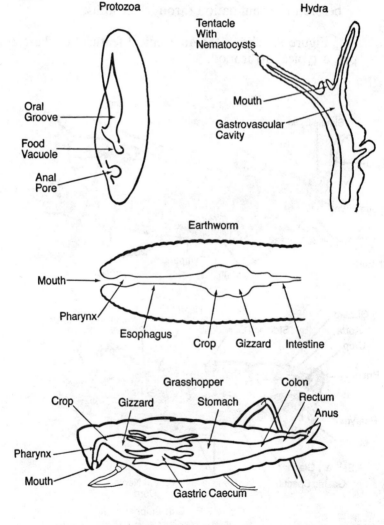

Figure 2.7 Adaptations For Nutrition

QUESTION SET 2.3

1. Animals that ingest organic materials are classified as
   1. heterotrophic
   2. photosynthetic
   3. autotrophic
   4. chemosynthetic

2. The process of heterotrophic nutrition occurs in
   1. algae
   2. bean plants
   3. grasshoppers
   4. mosses

3. Which function is performed by both an ameba and a maple tree?
   1. chlorophyll synthesis
   2. locomotion
   3. auxin secretion
   4. intracellular digestion

4. Nutrients are reduced to soluble form within the food vacuoles of the
   1. grasshopper and earthworm
   2. ameba and paramecium
   3. earthworm and human being
   4. hydra and grasshopper

5. A common characteristic of animals and fungi is their ability to carry on
   1. heterotrophic nutrition
   2. alcoholic fermentation
   3. auxin production
   4. transport through vascular tissue

6. During chemical digestion, large food molecules are broken down to smaller food molecules by the process of
   1. synthesis
   2. absorption
   3. hydrolysis
   4. excretion

7. A characteristic of heterotrophic organisms is that they
   1. obtain preformed organic molecules from other organisms
   2. need to live in a sunny environment
   3. are sessile for most of their lives
   4. use energy to manufacture organic compounds from inorganic compounds

8. Two organisms that possess a one-way digestive tract are the
   1. earthworm and grasshopper
   2. earthworm and hydra
   3. grasshopper and ameba
   4. hydra and ameba

9. Which is an example of enzyme-controlled intracellular digestion?
   1. An ameba digests a microorganism within its food vacuole.
   2. A human being digests food mechanically within its stomach.
   3. A grasshopper digests a piece of grass within its intestine.
   4. An earthworm grinds food within its gizzard.

10. The grinding action on food that occurs in the gizzard of a grasshopper is an example of which nutritional process?
    1. ingestion
    2. egestion
    3. chemical digestion
    4. mechanical digestion

11. The process of egestion can best be illustrated by the
    1. secretion of ear wax by a human being
    2. synthesis of urine by a mammal
    3. absorption of digested nutrients by a fungus
    4. elimination of undigested food by a frog

12. In a rabbit, which process would be most directly affected by a poison that destroyed some of the hydrolytic enzymes?
    1. photosynthesis
    2. ingestion
    3. selective absorption
    4. chemical digestion

13. In which organism is digestion accomplished mainly by the action of intracellular enzymes?
    1. paramecium
    2. human being
    3. earthworm
    4. grasshopper

14. The engulfing process utlilized by an ameba to ingest its food is known as
    1. hydrolysis
    2. synthesis
    3. cyclosis
    4. phagocytosis

II. THE LIFE FUNCTION OF TRANSPORT

Transport is the life function by which organisms absorb and distribute the materials necessary to maintain life.

A. Process

1. Absorption

Absorption is the first stage of transport. It is the process by which cells take dissolved materials through the plasma membrane into the cell interior. These materials may include the molecular end products of digestion, dissolved gases, salts, and other materials necessary for the continued health of the cell.

1.1. The plasma (cell) membrane is the key structure in the process of cellular absorption. Its structure and chemical composition are responsible for selectively regulating the passage of materials into and out of the cell.

The **fluid-mosaic model** (see Figure 2.8) is the currently accepted model of the structure of the cell membrane (although research is ongoing in this area). This model shows the principal component of the plasma membrane to be a double layer of lipid molecules with many embedded proteins. The nature of this structure allows the passage of many small, soluble molecules, such as monosaccharides, amino acids, and dissolved gases, while preventing the passage of larger molecules (e.g., starches and proteins). Current research indicates that size may be only one factor affecting the passage of materials through the cell membrane. Evidently, chemical factors are also important.

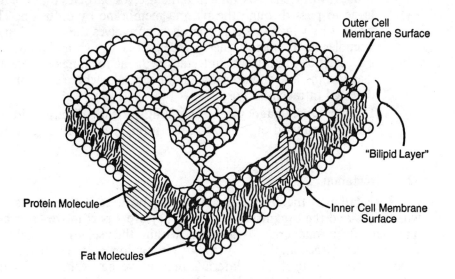

Outer Cell
Membrane Surface

"Bilipid Layer"

Protein Molecule

Inner Cell Membrane
Surface

Fat Molecules

Figure 2.8   Fluid-Mosaic Model

1.2. The functions of the plasma membrane in the various types of transport are as follows:

- **Passive transport.** All particles in the environment of the cell contain energy of motion (kinetic energy). As these particles collide with the plasma membrane, they may pass through by slipping between the lipid molecules making up the membrane. Because no cellular energy is expended in this absorption process, it is known as "passive" transport.

    Diffusion is a form of passive transport in which dissolved solids and gases pass through the plasma membrane from a region of *higher* relative concentration to a region of *lower* relative concentration. The diffusion of water molecules into or out of a cell (through the plasma membrane) is known as **osmosis**.

- **Active transport.** Some cells are capable of expending cellular energy in order to "pump" materials into themselves from regions of *low* relative concentration into the *higher* relative concentration of the cell interior. Because cellular energy must be expended to accomplish this process, it is known as "active" transport. The plasma membrane may be involved in active transport through the use of special **carrier proteins** embedded in the membrane's lipid layers. These carrier proteins take advantage of the specific chemical nature of desired components by attaching and "carrying" them through the membrane to the cytoplasm.

- **Pinocytosis.** Cells may take into themselves particles that are too large to pass through the plasma membrane by diffusion. The process by which this is accomplished involves the formation of vacuoles at the membrane surface and is known as "pinocytosis." Particles and large molecules near the forming vacuole are "sucked in" and enclosed within the vacuole, where they may be digested intracellularly.

- **Phagocytosis.** Phagocytosis (engulfing), already discussed in this unit as a method of protozoan nutrition, is also a form of cellular absorption.

2. Circulation

Circulation is the second stage of transport. It is the process by which cells and the organisms of which they are a part move absorbed materials from one area to another within themselves. Circulation within a cell is accomplished through circulation of the cell fluid, a process known as **cyclosis**. Materials may move between cells by the process of diffusion. In complex multicellular organisms, circulation is aided by the presence of **vascular tissues..** Vascular tissues are specialized, tubular, conducting tissues.

B. Adaptations for Transport

1. Plants

1.1. **Bryophytes** (mosses) are simple multicellular plants that lack organized vascular tissues for conducting materials within their bodies. These plants must depend on diffusion for intercellular conduction of water and other materials. Bryophytes also lack true roots, stems, and leaves, found commonly in higher plants.

1.2. **Tracheophytes** (vascular plants) contain tissues specialized for conducting materials throughout their bodies. Such tissues resemble tubes and are known as **vascular tissues.**

- **Xylem** is one of the most common types of vascular tissue found in plants. Xylem cells may interconnect to form long tubes that may reach from the roots, through the stem, all the way to the leaf. The principal function of xylem is to conduct water and dissolved minerals from the roots of a plant upward to the leaves, where it may be used in the process of photosynthesis.

  Upward movement of water in the xylem is accomplished by several mechanisms, the most important of which is **transpirational pull**. As water evaporates (or transpires) from leaf stomates, a pulling force is exerted on the column of water in the xylem tube. This force pulls more water up the xylem behind it. **Cohesive** forces that attract water molecule to water molecule help to maintain the water column intact, allowing a continuous flow of water from the roots to the leaf.

- **Phloem** is also a common vascular tissue found in plants. The principal function of phloem in plants is the downward (and upward) conduction of dissolved foods, such as sugar, from storage areas in the plant to other areas.

1.3. Plant vascular tissues are arranged into specialized organs as follows:

- **Roots** are specialized to absorb water and dissolved minerals from the soil surrounding them. Specialized cells, known as **root hairs**, found in the outer covering (**epidermis**) of the root, extend several millimeters into the soil and perform most of the water absorption. The thousands of root hairs found on the root increase the effective area of the root's absorbing surface. Absorption into the root is accomplished primarily by diffusion and osmosis, although active transport may also be involved. The secondary functions of roots may include anchorage, food storage, and reproduction.

- **Stems** contain vascular tissues that are continuous with those of the root. Water and dissolved minerals absorbed through the roots are conducted upward via xylem tissues. In most plants, organic foods dissolved in water are conducted downward through the stem's phloem tissues to the roots.

- **Leaves** have already been discussed as the principal food-producing organs of the plant. The **veins** of the leaf contain xylem and phloem that are continuous with those of the stem. Water and dissolved minerals carried via the xylem of the veins to leaf tissues may be used in the process of photosynthesis. A large amount of unused water evaporates from the leaf stomates in the process known as **transpiration**. Glucose produced in photosynthesis may then be carried, dissolved in water, via the phloem to other parts of the plant for consumption or storage.

2. Animals

2.1. The **hydra's** simple, two-cell-thick construction allows it to absorb needed materials readily from its environment and to circulate them within its body without specialized organs. To make molecular end products of digestion available to all cells in the hydra, special **flagella** are used to circulate the fluid contents of the digestive cavity. Cell-to-cell transfer of materials is accomplished by simple diffusion.

2.2 The **earthworm's** structure is more complex than that of the hydra, and many of its cells are not in direct contact with the earthworm's environment. The earthworm contains specialized absorbing surfaces: its moist **skin** absorbs oxygen; its intestine lining, with its infolded structure, absorbs digested foods. Once absorbed, these materials are circulated throughout the earthworm's body by means of a specialized system of blood vessels. The earthworm's blood, like that of human beings, contains a red, oxygen-carrying pigment known as **hemoglobin**, which flows suspended in the blood fluid. Hemoglobin is capable of carrying and releasing oxygen to the cells efficiently. Molecular end products of digestion are carried dissolved in the liquid portion of the blood. Because the earthworm's blood vessels are continuous and do not allow the blood to escape to the surrounding tissues, the circulatory pattern is known as a **closed transport system**. The blood is moved within this system by a group of five muscular pumps known as **aortic arches**.

2.3. The **grasshopper**, like the earthworm, uses its intestine lining to absorb digested foods. The circulatory fluid ("blood") flows freely through the tissues of the grasshopper and is not enclosed in blood vessels. This fluid dissolves and carries food molecules to all cells of the grasshopper's body. Oxygen is *not* distributed by the grasshopper's blood. The blood is pumped forward in the organism by a tubular pump sometimes called the dorsal aorta. This design, in which the blood is not always enclosed in blood vessels, is known as an **open transport system**.

2.4. **Human beings** have specialized areas for the absorption of needed materials. The human circulatory system is a closed transport system made up of specialized organs, including blood vessels and a muscular pump. The blood fluid contains hemoglobin and is used to transport dissolved gases, liquids, and solids. See Unit 3 of this book for a more detailed study of the structure and function of the human transport system.

The adaptation for transport in various organisms are seen in Figure 2.9.

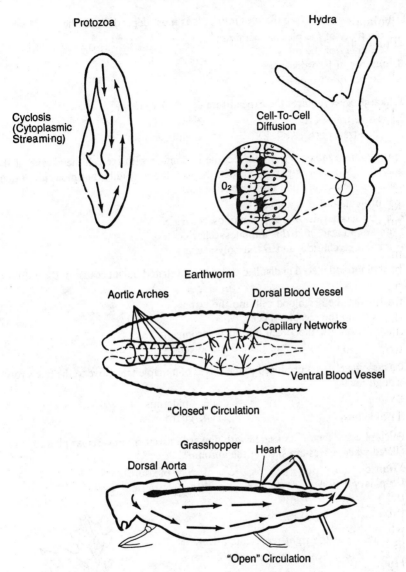

Figure 2.9 **Adaptations For Transport**

QUESTION SET 2.4

1. According to the fluid-mosaic model, a cell membrane is described as a structure composed of a
   1. protein layer in which large lipids are found
   2. carbohydrate structure in a "sea of lipids"
   3. double lipid layer in which proteins float
   4. phospholipid layer divided by carbohydrates

2. The net movement of molecules into cells is most dependent on the
   1. selectivity of the plasma membrane
   2. selectivity of the cell wall
   3. number of lysosomes
   4. number of chromosomes

3. Which process requires the expenditure of cellular energy?
   1. passive transport
   2. active transport
   3. osmosis
   4. diffusion

4. Carbohydrate molecules *A* and *B* come in contact with the cell membrane of the same cell. Molecule *A* passes through the membrane readily, but molecule *B* does not. It is most likely that molecule *A* is
   1. a protein, and *B* is a lipid
   2. a polysaccharide, and *B* is a monosaccharide
   3. an amino acid, and *B* is a monosaccharide
   4. a monosaccharide, and *B* is a polysaccharide

5. A red blood cell placed in distilled water will swell and burst because of the diffusion of
   1. salt from the red blood cell into the water
   2. water into the red blood cell
   3. water from the blood cell into its environment
   4. salts from the water into the red blood cell

6. The dissolved sugars produced in the leaves of a maple tree move to the tree's roots through the
   1. xylem
   2. phloem
   3. epidermis
   4. guard cells

7. Which physical factor associated with upward movement in vascular plants is most affected when leaves are shed in the autumn?
   1. transpirational pull
   2. capillary action
   3. root pressure
   4. active transport

8. The diagram at the right represents a leaf. The structure indicated by the arrow contains
   1. lenticels
   2. pollen grains
   3. xylem cells
   4. pistils

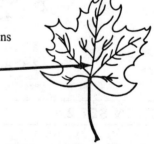

9. All of the following are functions of the roots of vascular plants except
   1. absorption of water and minerals from the soil
   2. conversion of phosphoglyceraldehyde into glucose
   3. conduction of water and minerals to the stem
   4. anchorage of the plant to the soil

10. The principal function of xylem tissue is the
    1. excretion of water from the respiratory reactions
    2. absorption of light for the photosynthetic reactions
    3. upward conduction of water and dissolved minerals
    4. downward conduction of water and dissolved glucose

11. Transport of essential materials within protozoa occurs primarily by
    1. cyclosis and diffusion
    2. a closed circulatory system
    3. pinocytosis and hydrolysis
    4. an open circulatory system

12. How are nutrients transported from the blood of an earthworm to the muscle cells of its body wall?
    1. as a result of blood flowing directly into muscle cells
    2. as a result of diffusion through capillary walls
    3. through the pores at the ends of nephridia
    4. through the skin from the outside environment

13. Which organism has a specialized organ system for transport?
    1. ameba
    2. hydra
    3. grasshopper
    4. paramecium

14. Which organism has an internal, closed circulatory system that brings materials from the external environment into contact with its cells?
    1. ameba
    2. paramecium
    3. hydra
    4. earthworm

III. THE LIFE FUNCTION OF RESPIRATION

Respiration is the life process by which organisms convert the chemical energy stored in foods to a form that may more easily be used by cells. As part of this process, respiratory gases are usually exchanged between the organism and its environment.

A. Process

1. Cellular Respiration

Cellular respiration is a biochemical process that includes the reactions used by cells to release energy from organic molecules such as glucose. The energy released as a result of these reactions is temporarily stored in the bonds of molecules of **adenosine triphosphate (ATP)**. This energy may be released for use in cell processes when ATP is hydrolyzed (converted) to adenosine diphosphate (ADP) and phosphate (P). Like other reactions that occur in the cell, the reactions of respiration are controlled by enzymes. The following equation illustrates the reactions involving the formation of ATP:

$$H_2O + ATP \xrightarrow{\text{ATP-ase}} ADP + P + energy$$

The two forms of cellular respiration carried on by living things are known as "aerobic" ("with oxygen") and "anaerobic" ("without oxygen").

1.1. Anaerobic respiration is a form of cellular respiration carried on by certain cells in the absence of molecular oxygen. This process operates by means of chemical reactions catalyzed by enzymes located in the cytoplasm of the cell. Organisms typically employing this form of cellular respiration as their primary mode of energy production include certain bacteria and fungi (yeasts). Human beings have learned to take advantage of this process by using these organisms in the manufacture of such foods as cheese, buttermilk, and yogurt. These organisms also play an important role in the baking, wine-making, and brewing industries. There are two principal types of anaerobic respiration:

- **Alcoholic fermentation**, so called because one of the major waste chemicals produced in this process is ethyl alcohol.

$$\text{Glucose} \xrightarrow{\text{enzymes}} 2 \text{ ethyl alcohol} + 2\,CO_2 + 2\,\text{ATP}$$

- **Lactic acid fermentation**, which produces the end product chemical lactic acid.

$$\text{Glucose} \xrightarrow{\text{enzymes}} 2 \text{ lactic acid} + 2\,\text{ATP}$$

Both types of anaerobic respiration are considered relatively inefficient, since the end products of both contain considerable amounts of energy that are not effectively released for use by the cell.

BC

The reactions of anaerobic respiration are actually a series of reactions that progressively convert glucose to end products and energy. A more detailed study of the process reveals that glucose molecules are first converted to two molecules of an intermediate compound known as **pyruvic acid**. To begin this conversion, the energy stored in two molecules of ATP is released. However, the fact that four molecules of ATP are *formed* in this process makes possible the net gain of two ATP. The following equation illustrates this process:

$$\text{Glucose} + 2\,\text{ATP} \xrightarrow{\text{enzymes}} 2 \text{ pyruvic acid} + 4\,\text{ATP}$$

2 lactic acid   2 alcohol + 2 CO_2

1.2 **Aerobic respiration** is a form of cellular respiration carried on by certain cells in the presence of molecular oxygen. This process operates by means of chemical reactions catalyzed by enzymes located primarily in the **mitochondria** of the cell. As in anaerobic respiration, the principal "product" of aerobic respiration is the energy released from organic molecules for use by the cell in other processes. Unlike anaerobic respiration, however, aerobic respiration uses molecular oxygen to release substantially more energy per glucose molecule metabolized. For this reason, aerobic respiration is considered to be a relatively more efficient form of respiration than anaerobic respiration. In general, the process of respiration may be illustrated as follows:

$$\text{Glucose} + \text{oxygen} \xrightarrow{\text{enzymes}} \text{water} + \text{carbon dioxide} + \text{ATP}$$

$$C_6H_{12}O_6 + 6\,O_2 \xrightarrow{\text{enzymes}} 6\,H_2O + 6\,CO_2 + 36\,\text{ATP}$$

BC

Like anaerobic respiration, aerobic respiration is characterized by a series of enzyme-catalyzed reactions that progressively convert glucose to end products and energy. These reactions are conceptually divided into two phases, known as the "anaerobic phase" and the "aerobic phase."

- In the **anaerobic phase**, glucose molecules are converted to two molecules of pyruvic acid, the process being initiated through release of the energy stored in two molecules of ATP. Four molecules of ATP are formed in this conversion, resulting in a net gain of two molecules of ATP. A summary equation of this phase is as follows:

$$\text{Glucose} + 2\,\text{ATP} \xrightarrow{\text{enzymes}} 2\,\text{pyruvic acid} + 4\,\text{ATP}$$

- In the **aerobic phase**, the cell uses molecular oxygen to further break down pyruvic acid to form the waste end products water and carbon dioxide. In this part of the process, an additional 34 molecules of ATP are formed. A summary equation of this phase is as follows:

$$2\,\text{Pyruvic acid} + 6\,O_2 \xrightarrow{\text{enzymes}} 6\,H_2O + 6\,CO_2 + 34\,\text{ATP}$$

The net gain of ATP molecules in aerobic respiration is 36, making it 18 times more efficient than anaerobic respiration.

2. Gas Exchange

Gas exchange between the environment and the cell interior involves the inward diffusion of molecular oxygen into the cell, as well as the outward diffusion of gaseous carbon dioxide from the cell interior to the environment, via the plasma membrane.

B. Adaptations for Gas Exchange

The chemical aspects of respiration are the same for all organisms, since all organisms require energy for their life processes. However, a wide variety of mechanisms for gas exchange have evolved over time and are observable in the following representative organisms:

1. Monera, Protista, Fungi

These simple organisms absorb oxygen and release carbon dioxide directly through the moist outer membrane of the cell. This gas exchange is accomplished by diffusion.

2. Plants

Plants are complex multicellular organisms and as such contain many cells that are not in direct contact with their environment. In addition, most living cells close to the surface are enclosed within specialized protective structures that limit gas movement. Adaptations that aid gas exchange in plants include: (1) the **stomates**, which allow gases to reach the inner cells of the leaf; (2) the **lenticels**, which allow gases to reach the inner cells of the stem; and (3) **root hairs** and **epidermis**, which permit gas exchange between the root tissues and the environment. Cells beyond the reach of these oxygen-providing adaptations, such as those in the inner wood of trees, may die of oxygen starvation.

3. Animals

3.1. The **hydra** has a simple structure that allows all of its cells to be in direct contact with its watery environment. This structure permits dissolved oxygen to diffuse directly into the cells for use in the respiratory process.

3.2. The **earthworm** contains many cells that are not in direct contact with the environment. The earthworm uses its moist **skin** as a respiratory surface through which it absorbs oxygen and releases carbon dioxide. The moistness of the skin surface is maintained by means of **mucus** produced by special glands embedded in the skin, as well as by behavioral adaptations that keep the earthworm in contact with moist soil. To transport the absorbed oxygen to all cells, the earthworm utilizes the blood fluid (containing hemoglobin) and the closed circulatory system described in Section II.

3.3. The **grasshopper** lives in a dry terrestrial environment that would quickly dehydrate exposed respiratory membranes. The grasshopper's impermeable outer covering (**exoskeleton**) prevents such dehydration. However, this same covering also prevents the exchange of respiratory gases between the environment and the grasshopper's internal cells. A system of tubes, known as **tracheal tubes**, conducts oxygen to the moist inner tissues for use in respiration. Small pores, the **spiracles**, connect the tracheal tubes with the environment.

3.4. **Human** respiration is aided by a system of respiratory tubes and an extensive internal surface for gas exchange. The blood's hemoglobin carries oxygen to all cells. See unit 3 of this book for a more detailed study of the structure and function of the human respiratory system.

The adaptations for respiration in various organisms are shown in Figure 2.10.

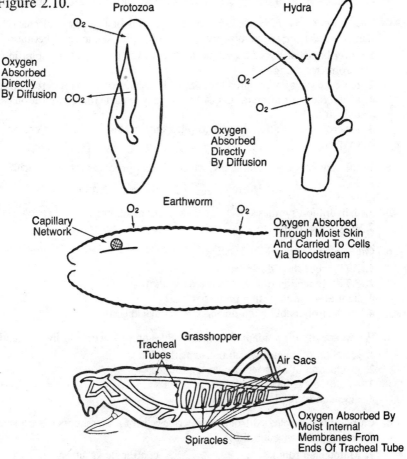

Figure 2.10   Adaptations For Respiration

QUESTION SET 2.5

1. The site of aerobic cellular respiration is the
   1. nucleus
   2. ribosome
   3. chromosome
   4. mitochondrion

2. Organisms make energy readily available by transferring the chemical bond energy of organic molecules to
   1. mineral salts
   2. adenosine triphosphate
   3. light energy
   4. nitrogenous wastes

3. Which of the following processes releases the greatest amount of energy?
   1. the oxidation of one glucose molecule to lactic acid molecules
   2. the oxidation of one glucose molecule to carbon dioxide and water molecules
   3. the conversion of two glucose molecules to a maltose molecule
   4. the conversion of one glucose molecule to alcohol and carbon dioxide molecules

4. The energy released from the anaerobic respiration of a glucose molecule is less than that released from the aerobic respiration of a glucose molecule because
   1. fewer bonds of the glucose molecule are broken in anaerobic respiration than in aerobic respiration
   2. more enzymes are required for anaerobic respiration than for aerobic respiration
   3. anaerobic respiration occurs 24 hours a day while aerobic respiration can occur only at night.
   4. anaerobic respiration requires oxygen but aerobic respiration does not require oxygen

5. In the summary equation below, which process produces the lactic acid?

$$\text{glucose} \xrightarrow{\text{enzymes}} 2 \text{ lactic acid} + 2 \text{ ATP}$$

   1. dehydration synthesis
   2. enzymatic hydrolysis
   3. fermentation
   4. aerobic respiration

6. Green plants use molecular oxygen for
   1. ATP production during anaerobic respiration
   2. ATP production during aerobic respiration
   3. light absorption during photosynthesis
   4. the hydrolysis of starch during intracellular digestion

7. Which activity is an adaptation that enables an earthworm to live on land?
   1. secretion of mucus, which moistens the skin
   2. production of nitrogenous waste products
   3. oxidation of glucose with the aid of enzymes
   4. digestion of food sequentially in a one-way tract

8. Carbon dioxide released from the interior cells of a grasshopper is transported to the atmosphere through the
   1. Malpighian tubules
   2. tracheae
   3. contractile vacuoles
   4. lungs

9. Vigorous activity of human voluntary muscle tissues may result in the production of lactic acid. Insufficient amounts of which gas would result in the buildup of lactic acid in muscle cells?

   1. carbon dioxide   **3.** oxygen
   2. nitrogen   **4.** hydrogen

10. In human beings, anaerobic respiration of glucose is a less efficient energy-releasing system than aerobic respiration of glucose. One reason is that in anaerobic respiration

    1. lactic acid contains much unreleased potential energy
    2. water contains much released potential energy
    3. oxygen serves as the final hydrogen acceptor
    4. chlorophyll is hydrolyzed into PGAL molecules

11. Which organism carries on gas exchange in a terrestrial environment?

    1. ameba   **3.** hydra
    2. shark   **4.** grasshopper

IV. THE LIFE FUNCTION OF EXCRETION

Excretion is the life process by which metabolic (cellular) wastes are removed from the cells of an organism and released to the environment.

A. Process

1. Products Resulting from Metabolism

Products resulting from metabolic processes may be "useful end products" or "waste end products." Waste products may be toxic to the cell if they are allowed to accumulate. These waste end products may include:

- **Carbon dioxide**, which results from aerobic respiration.
- **Water**, which results from numerous processes, including aerobic respiration and dehydration synthesis.
- **Mineral salts**, which result from several of the cell's metabolic processes.
- **Nitrogenous wastes**, which result from protein/amino acid metabolism; examples include **ammonia** (extremely toxic), **urea** (moderately toxic), and **uric acid** (relatively nontoxic).

2. Results of Excretory Activities

These results vary from process to process. In animals, the toxic wastes that result from metabolic processes are usually removed, whereas in plants they are stored in special vacuoles. In both plants and animals, nontoxic wastes may be recycled as useful reactants in other life processes.

B. Adaptations for Excretion

1. Protists

Protists, as a rule, lack specialized excretory structures. Instead, they depend on the plasma membrane's ability to selectively regulate the passage of materials out of the cell. Small dissolved molecules, such as carbon dioxide, salts, and water, diffuse readily through this membrane without the expenditure of cell energy. Large metabolic wastes may be stored in special vacuoles or removed by active transport. Algae, like complex plants, recycle much of the waste that results from metabolic processes such as respiration and photosynthesis.

2. Plants

Plants are well adapted to recycle the wastes of certain processes into the reactants of other processes. Most important is the recycling of carbon dioxide gas and water (waste products of aerobic respiration) into the process of photosynthesis; at the same time, the plant recycles oxygen gas (a waste product of photosynthesis) into the process of aerobic respiration.

During the day, more oxygen gas is produced than can be used by the plant. At these times the excess oxygen gas is released through stomates to the atmosphere. At night, more carbon dioxide gas is produced than the plant can use, and then the excess is released by diffusion from stomates, lenticels, and the root epidermis to the surrounding environment.

Some wastes, because of their molecular size or toxicity, are not released at all, but are stored within the plant in specialized vacuoles or in dead tissues, where they cannot harm living cells. Such wastes may include organic acids.

3. Animals

3.1. The **hydra's** simple multicellular structure permits an excretory function similar to that of the protists. Most wastes are released directly from the cells to the surrounding environment by diffusion.

3.2. The **earthworm's** more complex structure requires a specialized excretory system. The moist **skin** of the earthworm serves as a respiratory surface for the release of carbon dioxide gas from the blood. Nitrogenous wastes, salts, and water are collected in specialized organs known as **nephridia**, a pair of which is located in each of the earthworm's segments. Once concentrated in the nephridia, these wastes are released to the terrestrial environment through small pores in the earthworm's skin.

3.3. The **grasshopper** rids its body of carbon dioxide gas through the same system of **tracheal tubes** that serves to bring oxygen gas to the tissues. The carbon dioxide is released to the atmosphere through the **spiracles**. Nitrogenous wastes (in the form of **uric acid**), salts, and small amounts of water are concentrated in a specialized organ known as the **Malpighian tubules** for elimination through the grasshopper's digestive tract. These wastes are mixed with undigested food matter in the intestine and are egested in a semisolid form along with the fecal material. The grasshopper reabsorbs most of the "waste" water before it can be expelled, thereby aiding water conservation in the body of the insect. The fact that nitrogenous wastes are converted into uric acid, a relatively insoluble compound, is considered a further water-conserving adaptation.

3.4. **Human** excretion is aided by the presence of specialized organs in the form of respiratory surfaces, glands, and an extensive blood-filtering system similar to that of the earthworm. See Unit 3 of this book for a more detailed study of the structure and function of the human excretory system.

The adaptation for excretion in various organisms are shown in Figure 2.11.

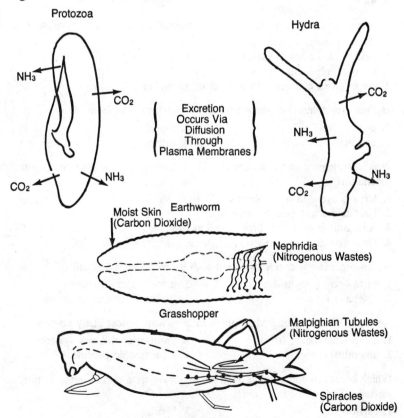

Figure 2.11 Adaptations For Excretion

QUESTION SET 2.6

1. Metabolic wastes of animals most likely include
   1. water, carbon dioxide, oxygen, and salts
   2. carbon dioxide, nitrogenous compounds, water, and salts
   3. hormones, water, salts, and carbon dioxide
   4. glucose, carbon dioxide, nitrogenous compounds, and water

2. Nitrogenous waste products are produced from the complete metabolism of
   1. water   3. starches
   2. sugars   4. proteins

3. In plants, the process by which water vapor passes through stomates and lenticels is known as
   1. cyclosis   3. transpiration
   2. respiration   4. hydrolysis

4. In plants, which waste materials may be stored in vacuoles rather than being released to the environment?
   1. starch   3. respiratory gases
   2. water   4. organic acids

5. A hydra can function without any organized excretion system because its cells
   1. do not produce wastes
   2. change all wastes to useful substances
   3. remove only solid wastes
   4. are in direct contact with a water environment

6. Hydras excrete most of their nitrogenous wastes in the form of
   1. urea   3. ammonia
   2. uric acid   4. nitrates

7. Which statement best describes the excretion of nitrogenous wastes from paramecia?
   1. Urea is excreted by nephrons.
   2. Uric acid is excreted by nephridia.
   3. Ammonia is excreted through cell membranes.
   4. Urea and uric acid are excreted through Malpighian tubules.

8. Which organism is correctly paired with its excretory structure?
   1. earthworm—nephridium   3. grasshopper—nephron
   2. ameba—skin   4. hydra—Malpighian tubule

9. An organism containing Malpighian tubules would most likely possess
   1. a four-chambered heart   3. an open circulatory system
   2. an endoskeleton   4. a contractile vacuole

10. Which organism contains an open circulatory system and excretes its nitrogenous waste in the form of uric acid?
    1. earthworm   3. hydra
    2. grasshopper   4. ameba

V. THE LIFE FUNCTION OF REGULATION

Regulation is the life function by which organisms control and coordinate their other life functions to maintain life. Regulation includes both nerve control (animals only) and chemical control (all organisms).

A. Nerve Control

1. Definition of Terms

Definitions of the elements important in the process of nerve regulation are essential to a full discussion of the process itself.

1.1. Stimulus. Any change that occurs in the environment of an organism may be considered a stimulus. Such stimuli may be external (outside the organism's body) or internal (within the body of the organism). Examples of stimuli include light, sound, and chemical stimuli.

1.2. Response. A reaction that an organism has to a specific stimulus is its response to that stimulus. Responses may include physical movements or glandular secretions.

1.3. Neuron. A cell specialized for the transmission of nerve impulses from place to place in the body is a neuron. Its parts include:

- **Dendrites**—fibers that serve to detect a stimulus (when the neuron is part of a receptor) and direct an impulse toward the cell body (cyton).
- **Cyton**—the main cell body, containing most of the cytoplasm, the nucleus, and other organelles. It is here that the nerve impulse is generated.
- **Axon**—an elongated portion of the neuron that carries the nerve impulses from the cyton toward the terminal branches.
- **Terminal branches**—fibers leading from the axon of the nerve cell that reach toward the dendrites of adjacent neurons and secrete neurotransmitters.

A vertebrate neuron is diagramed in Figure 2.12.

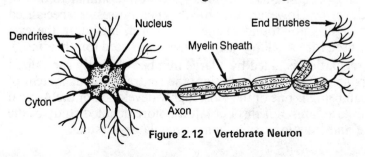

Dendrites

Nucleus

End Brushes

Myelin Sheath

Cyton

Axon

Figure 2.12 Vertebrate Neuron

1.4. Synapse. A gap that occurs between adjacent neurons, or between a neuron and an effector, is a synapse. Nerve impulses are prevented from being transmitted between neurons across this gap unless a special chemical (neurotransmitter) is secreted.

1.5. Neurotransmitters. These are special chemicals, produced and secreted by the neuron's terminal branches, which serve to carry the nerve impulse from one neuron to the next. An example of such a neurotransmitter is **acetylcholine**.

1.6. Impulse. This term refers to the electrochemical change in the surface of a nerve cell that transmits information from receptor organs to the central nervous system and back to effectors.

1.7. Receptors. These are structures specialized to receive stimuli from the environment of an organism. Receptor organs are composed of neurons specialized for the detection of specific stimuli, such as light, vibration, or touch. Examples of receptors include the eye, the ear, and the skin.

1.8. Effectors. These are structures specialized to produce responses by an organism. Effector organs may be muscles (producing physical movement) or glands (producing chemical secretions).

2. Adaptations for Nerve Control in Animals

2.1. The **hydra** lacks specialized receptors and a complex central nervous system. However, an arrangement of modified neurons, known as a **nerve net**, is specialized to receive environmental stimuli and transmit impulses to other parts of the hydra's body. Impulses may travel over these neurons in either direction, and responses are extremely general as a result.

2.2. The **earthworm's** nervous system consists of a primitive, two-lobed **brain** (really fused ganglia) and a **nerve cord** located on the ventral side of its body. **Ganglia** (bunched nerve cells) located in each of the body segments serves to connect the nerve cord with the **peripheral nerves**, which branch out to remote areas of the earthworm's body. This design aids the process of regulation, since the nerve impulses are directed in specific pathways from receptor organs to the central nervous system and back to effector organs.

2.3. The **grasshopper** contains a nervous system quite similar to that of the earthworm. However, the grasshopper possesses specialized receptor organs not found in the earthworm, including the **eye**, **antenna** (used for "smelling"), and **tympanum** (used for "hearing").

2.4. **Human** regulation is aided by a large number of highly specialized receptors and effectors. In addition, the central nervous system of the human being is one of the most highly developed of the Animal Kingdom. See Unit 3 of this book for a more detailed study of the structure and function of the human regulatory system.

Adaptations for regulation in various organisms are shown in Figure 2.13.

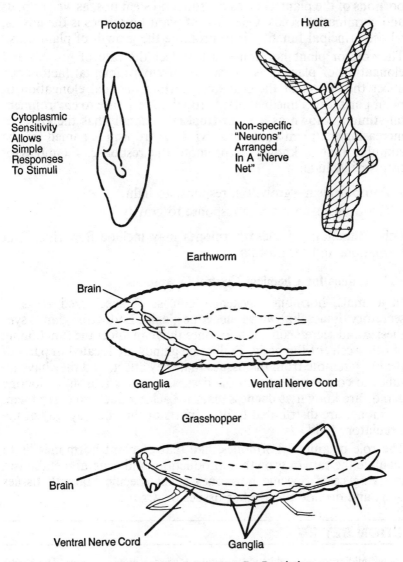

Figure 2.13   Adaptations For Regulation

B. Chemical Control

Like nervous control, chemical control results in the coordination of life functions important to the survival of the organism. In both plants and animals chemical control mechanisms depend on the presence of chemicals known as **hormones**, which can affect the operation of living cells.

1. Adaptations for Chemical Control in Plants

1.1. Plant hormones are chemicals produced by cells in the growing portions of the plant, such as the root and stem tips, as well as buds and germinating seeds. One class of plant hormones is the **auxins,** whose principal function is to promote the growth of plant cells.

1.2. The role of plant hormones is to affect the rate of growth and elongation of plant cells. A variety of environmental factors can affect the action of these auxins, causing unequal elongation of plant parts. Such unequal plant growth in response to environmental stimuli, known as a plant **tropism,** generally has the effect of increasing the plant's rate of survival. A tropism is named for the stimulus that is known to promote the response. Examples of tropisms include:

- Phototropism—growth in response to light.
- Geotropism—growth in response to gravity.

Other functions of these hormones may include flowering, fruit formation, and seed formation.

2. Adaptations for Chemical Control in Animals

2.1. In animals, hormone "systems" consist of specialized areas of secretory tissues, known as **endocrine glands.** Endocrine glands synthesize and secrete animal hormones that can affect the functioning of other cells of the body. Endocrine glands are located in parts of the body remote from the tissues that they affect, and they have no tubes to connect them to these tissues. For this reason, endocrine glands are known as **ductless glands.** Lacking ducts to carry them, hormones are distributed to other parts of the body by way of the circulatory fluid.

2.2. The role of animal hormones, like that of plant hormones, is to regulate the rate of growth, reproduction, and general metabolism of living cells. Specific hormones affect specific "target" tissues only, and do not affect other tissues or organs.

QUESTION SET 2.7

1. A hawk gliding over a field suddenly dives toward a moving rabbit. The hawk's reaction to the rabbit is known as a

  1. stimulus   **3.** response

  2. synapse   **4.** tropism

2. What is the role of sensory organs in the body?

  1. the transmission of impulses directly to effectors

  2. the detection of environmental stimuli

  3. the conduction of impulses from the spinal cord

  4. the interpretation of impulses from motor neurons

3. Neurotransmitters, such as acetylcholine, are initially detected by which part of a neuron?
   1. dendrite
   2. nucleus
   3. terminal branch
   4. mitochondrion

4. Animal cells that are specialized for conducting electrochemical impulses are known as
   1. neurons
   2. synapses
   3. nephrons
   4. neurotransmitters

5. Impulses are transmitted from the cyton to the terminal branches of a neuron along the membranes of the
   1. dendrite
   2. axon
   3. nucleus
   4. mitochondrion

6. When a tentacle of a hydra is touched with a needle, the entire body responds as a result of impulses traveling to all cells through the
   1. central nervous system
   2. spinal cord
   3. posterior brain
   4. nerve net

7. The central nervous system of a grasshopper is most similar in structure to the central nervous system of
   1. an earthworm
   2. a human being
   3. an ameba
   4. a hydra

8. The two systems that directly control homeostasis in most animals are the
   1. nervous and endocrine systems
   2. endocrine and excretory systems
   3. nervous and locomotive systems
   4. excretory and locomotive systems

9. Which animal has a ventral nerve cord?
   1. grasshopper
   2. ameba
   3. hydra
   4. human being

10. An auxin that is produced in the root tips of an apple tree may function as a
    1. food producer
    2. water carrier
    3. growth regulator
    4. light absorber

11. The left side of the stem of a young bean plant has a greater concentration of auxin than the right side has. The stem will most likely
    1. bend to the right
    2. bend to the left
    3. grow straight upward
    4. stop growing

12. Removing the tip of the stem of a young plant will most directly interfere with the production of
    1. sugars
    2. auxins
    3. carbon dioxide
    4. oxygen

13. In animals, hormones are transported throughout the organism by the
    1. lysosomes
    2. synapses
    3. circulatory system
    4. excretory system

14. The hormones released during insect metamorphosis serve as growth regulators. Which substance present in the stem of a bean plant has a similar function?

    1. DNA   **3.** ATP

    2. auxin   **4.** ribose

VI. THE LIFE FUNCTION OF LOCOMOTION

Locomotion is the life process by which organisms move from place to place within their environment.

A. Advantages

Locomotion provides certain advantages to the organism able to utilize it. The principal advantages to an organism of the process of locomotion include:

- Increased opportunities to locate food.
- Increased ability to seek shelter.
- Increased ability to avoid predators.
- Increased ability to move away from toxic waste.
- Increased opportunities to locate mates.

B. Adaptations for Locomotion

1. Protists

Protists have evolved a variety of structural adaptations to aid the process of locomotion. These adaptations include:

- **Flagella**—long, whiplike structures found in certain algae and protozoans.
- **Cilia**—short, hairlike projections of the cytoplasm common to paramecia and related protozoans.
- **Pseudopods**—cytoplasmic extensions resembling "feet," which enable amebas to move from place to place.

2. Animals

2.1. The **hydra** moves very little in the adult stage; for this reason, it is said to be **sessile**. However, hydras have been observed to move occasionally by utilizing their contractile fibers to perform a "somersaulting" motion.

2.2. The **earthworm** contains specialized **muscle** tissues that are capable of producing movement of the earthworm's body. These muscles are able to provide lengthening and shortening motions, as well as expansion and contraction of body width. To promote this movement short bristles, known as **setae**, provide anchorage of the earthworm's body against the soil in which it lives.

2.3. The **grasshopper** also contains muscle tissues, but these muscles are more highly specialized than those of the earthworm in order to provide control of the grasshopper's jointed appendages (limbs). The grasshopper's external skeleton (**exoskeleton**) is composed of a polysaccharide known as **chitin**, which provides anchorage for muscles, levers for bodily movement, and protection of internal tissues from dehydration and mechanical damage.

2.4. **Human** movement is accomplished through the interaction of an internal, jointed skeleton and attached muscles. See Unit 3 of this book for a more detailed study of the structure and function of the human locomotor system.

Adaptations for locomotion in various organisms are shown in Figure 2.14.

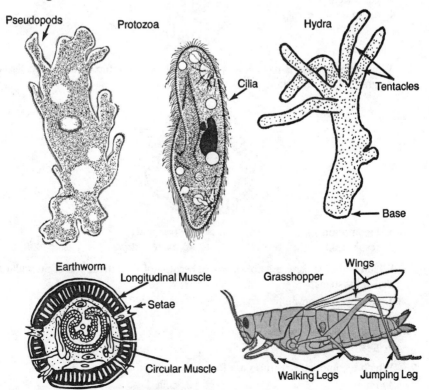

Figure 2.14   Adaptations For Locomotion

QUESTION SET 2.8

1. Locomotion does *not* increase an animal's opportunity to
   1. obtain food
   2. reproduce
   3. escape from predators
   4. transmit impulses

2. Which organism is essentially a sessile animal?
   1. ameba
   2. grasshopper
   3. earthworm
   4. hydra

3. Which organism moves by the interaction of muscles and chitinous appendages?
   1. hydra
   2. paramecium
   3. grasshopper
   4. human being

4. A student viewing a sample of pond water through a microscope observed and drew a protist as shown at the right. Which organism has locomotive structures most similar to this organism?

   1. earthworm
   2. ameba
   3. paramecium
   4. hydra

5. Which life function is most closely associated with structure *A* in the cross-section diagram of an earthworm below?

   1. locomotion
   2. respiration
   3. transport
   4. excretion

6. Which two organisms are able to move because of the interaction of muscular and skeletal systems?
   1. earthworm and human being
   2. grasshopper and hydra
   3. hydra and earthworm
   4. grasshopper and human being

7. Which locomotive structures are found in some protozoa?
   1. muscles
   2. tentacles
   3. cilia
   4. setae

UNIT 2 REVIEW

1. A characteristic of animals that makes them similar to heterotrophic plants is that animals
    1. obtain preformed organic molecules from other organisms
    2. need to live in a sunny environment
    3. are sessile for most of their lives
    4. use energy to manufacture organic compounds from inorganic compounds

2. The summary word equation shown below represents a set of reactions occurring in photosynthesis.

$$\text{water} \xrightarrow[\text{light}]{\text{chlorophyll}} \text{hydrogen} + \text{oxygen}$$

These reactions are known as
    1. carbon fixation reactions
    2. photochemical reactions
    3. dark reactions
    4. fermentation reactions

3. In the diagram below, the plant was exposed to several different colors of light. If all the light intensities were the same, under which color of light would oxygen be produced at the *lowest* rate?

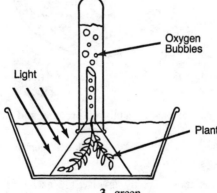

    1. red   3. green
    2. blue   4. white

4. In a maple tree, the enzymatic hydrolysis of starches, lipids, and proteins occurs
    1. extracellularly only
    2. intracellularly only
    3. both extracellularly and intracellularly
    4. neither extracellularly nor intracellularly

5. Nitrogenous compounds may be used by plants in the synthesis of
    1. glucose   3. proteins
    2. waxes   4. starch

6. On the basis of their pattern of nutrition, most animals are classified as
    1. autotrophic   3. photosynthetic
    2. heterotrophic   4. phagocytic

7. The process by which digestive enzymes catalyze the breakdown of larger molecules to smaller molecules with the addition of water is known as
   1. synthesis
   2. pinocytosis
   3. hydrolysis
   4. photosynthesis

8. In an ameba, the structures formed as a result of the ingestion of yeast cells are known as
   1. food vacuoles
   2. contractile vacuoles
   3. cell walls
   4. buds

9. A fruit fly is classified as a heterotroph, rather than as an autotroph, because it is unable to
   1. transport needed materials through the body
   2. release energy from organic molecules
   3. manufacture its own food
   4. divide its cells mitotically

10. The diagram below illustrates phases of a specific life activity being carried on by a cell. Which process is occurring at phase 3?

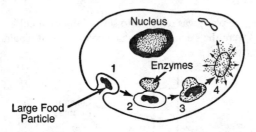

   1. intracellular digestion
   2. extracellular digestion
   3. ingestion
   4. excretion

11. Which structure, composed mainly of proteins and lipids, aids in maintaining homeostasis in a cell?
   1. chromosome
   2. centrosome
   3. cell membrane
   4. nucleolus

12. Which processes most directly involve the exchange of gases through moist membranes?
   1. digestion and locomotion
   2. reproduction and growth
   3. transport and regulation
   4. respiration and excretion

13. Which process requires cellular energy to move molecules across the cell membrane from a region of lower concentration to a region of higher concentration?
   1. active transport
   2. diffusion
   3. osmosis
   4. hydrolysis

14. In vascular plants, the absorption of water from the soil into root hairs depends principally on the presence of a
   1. phototropic response by the root hairs
   2. geotropic response by the conducting tissue
   3. higher concentration of water in the soil than in the root hairs
   4. higher concentration of water in the root hairs than in the soil

15. Which part of a plant is specialized for anchorage and absorption?
   1. leaf   **3.** stem
   2. flower   **4.** root

16. In plants, transpirational pull is a major factor in the transport of
   1. water   **3.** auxin
   2. food   **4.** chlorophyll

17. Ferns may grow several feet in height, whereas mosses seldom grow more than a few inches tall. What structural adaptation enables fern plants to grow taller than moss plants?
   1. Ferns contain chlorophyll molecules, but mosses do not.
   2. Ferns are heterotrophic, but mosses are not.
   3. Ferns contain xylem and phloem, but mosses do not.
   4. Ferns are aerobic, but mosses are not.

18. Which structures transport food downward in a geranium plant?
   1. phloem vessels   **3.** root hairs
   2. xylem tubes   **4.** chloroplasts

19. A student filled a bag of dialysis tubing with a milky-white starch solution and placed the bag in a beaker of iodine-water, as shown in the diagram below. An hour later, the student observed that the starch solution had turned blue-black (positive test for starch). What is the most probable explanation for the change?

   1. The iodine diffused into the bag.
   2. The starch was changed to sugar.
   3. The iodine was changed to starch.
   4. The starch diffused out of the bag.

20. There is a higher concentration of mineral salts within the body of a paramecium than in its external water environment. This higher concentration is maintained as a result of the action of
   1. pinocytosis   **3.** CO_2
   2. cyclosis   **4.** ATP

21. In the diagram of root cells shown below, in which direction would the net flow of water be the greatest as a result of osmosis?
   1. *A* to *C*
   2. *A* to *B*
   3. *B* to *C*
   4. *C* to *B*

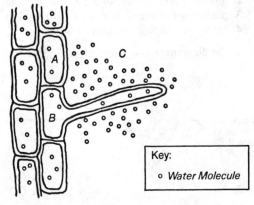

22. The diagram below shows the method of entry of a molecule too large to diffuse through the plasma membrane of a cell.

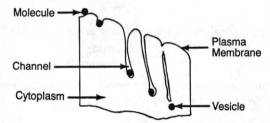

The process represented in the diagram above is known as

1. homeostasis
2. osmosis
3. pinocytosis
4. cyclosis

23. An increase in the concentration of ATP in a muscle cell is a direct result of which life function?

1. respiration
2. reproduction
3. digestion
4. excretion

24. In animal cells, the energy to convert ADP to ATP comes directly from

1. hormones
2. sunlight
3. organic molecules
4. inorganic molecules

25. Respiratory enzymes are present in

1. animal cells, but not in plant cells
2. plant cells, but not in animal cells
3. neither animal nor plant cells
4. both animal and plant cells

26. In a laboratory culture of yeast, it may be concluded that fermentation has occurred if chemical tests indicate the production of

1. carbon dioxide and water
2. PGAL and nitrates
3. oxygen and ATP
4. ethyl alcohol and carbon dioxide

27. Which summary equation represents a process that releases the greatest amount of energy from a molecule of glucose?

1. glucose $\rightarrow$ 2 lactic acid
2. glucose $\rightarrow$ 2 ethyl alcohol + 2 carbon dioxide
3. glucose + 6 oxygen $\rightarrow$ 6 water + 6 carbon dioxide
4. glucose + fructose $\rightarrow$ sucrose + water

28. In the diagram below, what gas is probably present in fermentation tube B?

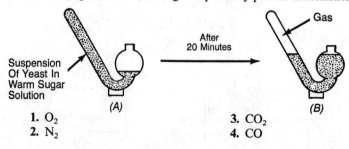

1. O_2
2. N_2
3. CO_2
4. CO

29. When does the process of cellular respiration occur in bean plants?
    1. in the daytime only
    2. at night only
    3. both in the daytime and at night
    4. only when photosynthesis stops

30. What do the tracheal tubes of the grasshopper and the air spaces of a geranium leaf have in common?
    1. They regulate the flow of urea into and out of the organism.
    2. They are the major sites for the ingestion of nutrients.
    3. They contain enzymes that convert light energy to chemical bond energy.
    4. They are surrounded by moist internal surfaces where gas exchange occurs.

31. The respiratory system of an earthworm utilizes the skin as an external gas-exchange surface. What additional system is used to carry gases to moist internal body tissues?
    1. digestive   3. nervous
    2. circulatory   4. endocrine

32. Nitrogenous wastes result from the metabolism of
    1. amino acids   3. fatty acids
    2. glucose molecules   4. water molecules

33. The principal structure for excretion in protozoans is the
    1. cell membrane   3. lysosome
    2. food vacuole   4. nucleus

34. Which organism contains a closed transport system and excretes nitrogenous wastes through nephridia?
    1. hydra   3. grasshopper
    2. earthworm   4. human being

35. A ventral nerve cord and a closed circulatory system are characteristics of
    1. a hydra   3. a human being
    2. an earthworm   4. a grasshopper

36. In the diagram of two neurons shown below, at which point would a substance that interferes with the action of a neurotransmitter be most effective?

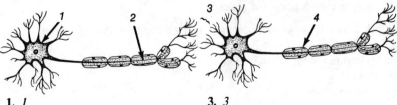

    1. *1*   3. *3*
    2. *2*   4. *4*

37. Tropisms in plants most directly result from the
    1. unequal distribution of auxins
    2. transpirational pull of water
    3. transmission of impulses by acetylcholine
    4. excitation of chlorophyll molecules by light

38. Geranium leaves grow in positions that permit the optimum use of light as a result of

1. phototropic responses
2. capillary action
3. transpirational pull
4. symbiotic relationships

39. The structure and function of the nervous system of the earthworm are most similar to those of the

1. ameba
2. hydra
3. grasshopper
4. paramecium

40. The diagram below represents four stages of development in an arthropod. Which substances most directly regulate the process illustrated in the diagram?

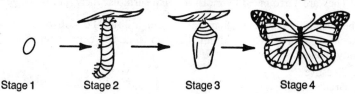

Stage 1   Stage 2   Stage 3   Stage 4

1. auxins
2. hormones
3. vitamins
4. minerals

41. A chemical injected into a tadpole caused the tadpole to undergo rapid metamorphosis into a frog. This chemical was most probably

1. an enzyme
2. a neurohumor
3. a hormone
4. a blood protein

42. Cilia in protozoans perform the same life function as the

1. setae in earthworms
2. food vacuoles in protozoans
3. nephridia in earthworms
4. tracheal tubes in grasshoppers

43. An earthworm that has partially entered its burrow can be surprisingly difficult to pull from the ground. This is due primarily to the earthworm's

1. chitinous outer covering and legs
2. bristlelike setae and muscles
3. powerful ventral suckers and claws
4. grasping mouth parts and scales

44. Which organ of the grasshopper is correctly matched with its function?

1. chitinous appendage—locomotion
2. Malpighian tubule—respiration
3. trachea—digestion
4. gonad—excretion

45. Given the summary equation:

BC

$$\text{glucose} \xrightarrow{\text{enzymes}} 2\,\text{ethyl alcohol} + 2\,\text{carbon dioxide} + 2\,\text{ATP}$$

This equation represents a form of respiration carried on by some types of

1. algae and maple trees
2. paramecia and hydras
3. human beings and grasshoppers
4. bacteria and yeasts

BC Base your answers to questions 46 through 50 on the word equation below, which represents a summary of the two major sets of reactions occurring during photosynthesis.

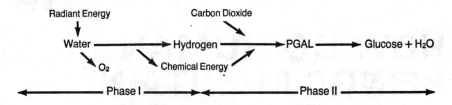

46. Which form of energy is absorbed by green plants during phase I?
    1. heat energy
    2. light energy
    3. nuclear energy
    4. chemical energy

47. The oxygen present in the water molecule in phase I is
    1. released as molecular oxygen
    2. released as chemical energy
    3. incorporated into PGAL
    4. incorporated into glucose

48. Phase II is often referred to as
    1. oxidation
    2. hydrolysis
    3. carbon fixation
    4. aerobic respiration

49. A three-carbon sugar formed during phase II is
    1. carbon dioxide
    2. glucose
    3. ATP
    4. PGAL

50. The reactions in phase I occur in the
    1. grana
    2. stoma
    3. Golgi apparatus
    4. cell wall

BC Base your answers to questions 51 and 53 on the equation below, which represents a process that occurs in both plants and animals, and on your knowledge of biology.

$$\text{Glucose} + \text{oxygen} \xrightarrow{\text{enzymes}} \text{water} + \text{carbon dioxide} + 36\ \text{ATP}$$

51. Within which organelles are most of the 36 ATP molecules produced?
    1. ribosomes
    2. endoplasmic reticula
    3. nuclei
    4. mitochondria

52. In animal cells, much of the carbon dioxide produced is
    1. used for energy
    2. converted to sugar
    3. excreted as waste
    4. stored in vacuoles

53. On a sunny day, much of the carbon dioxide produced by a green plant may be
    1. used for fermentation
    2. used for photosynthesis
    3. stored in vacuoles
    4. converted to oxygen gas

HOW DO HUMAN BEINGS MAINTAIN THEIR EXISTENCE?

KEY IDEAS	This unit explores the structure and function of the human body. All life functions discussed in Units 1 and 2 are explored here in detail for the human being, outlining the adaptations employed by humans to carry out these life functions efficiently.

KEY OBJECTIVES

Upon completion of this unit, you should be able to:

☐ Describe the similarities between human beings and other animal groups in terms of their structure and function.

☐ Determine the relationship between physiological function and proper nutrition.

☐ List the major organs and organ systems of the human body, and describe their roles in the maintenance of homeostatic balance.

☐ Describe the interdependence of the body's organ systems in the maintenance of life.

☐ Define some of the malfunctions of the human body, and describe their relationships to the life functions that they affect and their impact on the maintenance of homeostasis.

I. THE LIFE FUNCTION OF NUTRITION

Nutrition is the life function by which human beings obtain materials needed for energy, growth and repair, and other life functions. As part of this process, these materials are converted to a simplified form that can be used by the cell. Nutritional requirements of humans are similar to those of other heterotrophic animals. A diet that includes the proper balance of carbohydrate, protein, lipid, roughage, vitamins, and minerals is essential to the physical health of the human body. These requirements are known to vary, however, with the age, sex, and physical activity of the individual.

A. Functional Organization for Nutrition

In general, the human digestive tract resembles the tracts of the earthworm and the grasshopper in that it is a one-way tube in which food materials are progressively converted into molecular end products. The parts of this tube are as follows:

1. Oral Cavity

The human **mouth (oral) cavity** is used for the ingestion of food materials. The **teeth** and **tongue** help to manipulate and break the food down mechanically. This process increases the surface area of the food, thereby aiding the action of salivary enzymes.

The **salivary glands** produce and secrete an enzyme-containing fluid known as **saliva** into the oral cavity. The enzymes in saliva are responsible for acting on complex carbohydrates (e.g., starch), converting them into disaccharides such as maltose.

HP

Carbohydrates are an important source of energy for the body. They should constitute approximately 50 percent of the diet. Certain undigestible carbohydrates, such as cellulose, pass through the digestive tract without being affected by the body's digestive enzymes. Such materials, grouped into a class of nutrients called **roughage**, are found commonly in whole grain foods, garden vegetables, fruits, and other vegetable matter. Roughage is important in maintaining regularity of egestion and is increasingly identified as essential in preventing certain disorders of the digestive tract.

---

2. Esophagus

The **esophagus** is a short tube (about a foot in length) that connects the oral cavity to the stomach. The swallowing action initiated at the back of the oral cavity is continued in the esophagus as a wave of muscular contraction known as **peristalsis**. This peristaltic action moves the chewed food to the stomach for further digestive action.

3. Stomach

The **stomach** is a muscular organ whose main function is to liquefy and further digest food materials. The lining of the stomach contains digestive glands that secrete digestive enzymes and hydrochloric acid. Gastric (stomach) enzymes are specific for proteins, which they break down into short amino acid chains known as **dipeptides**. Unlike most of the body's enzymes, the enzymes of the stomach work best in an acidic (low pH) condition. This acidic condition is provided by the hydrochloric acid secreted by the stomach's digestive glands.

4. Small Intestine

Liquefied and partially digested food enters the **small intestine** from the stomach. Enzymes secreted by intestinal glands complete the digestive process. These enzymes include **protease** (to digest proteins and dipeptides), **lipase** (to digest fats and oils), and **disaccharidase** such as maltase (to digest maltose). Fluids produced and/or stored in **accessory organs**, such as those listed below, contain enzymes and other substances that aid digestion.

4.1. The **liver** manufactures **bile**, which is secreted via a duct to the gall bladder.

4.2. The **gallbladder** stores bile and releases it slowly into the small intestine. The function of bile is to liquefy fats in a process known as **emulsification**. This process increases the surface area of the fat and allows fat-digesting enzymes to work more efficiently.

4.3. The **pancreas** produces several enzymes that act on all major food groups, breaking them down chemically into molecular end products. These enzymes, which include protease, lipase, and **amylase** (to digest starch), travel to the small intestine by way of a special duct.

In addition to the digestive process carried on within the small intestine, the lining of the small intestine acts as the principal surface for the absorption of the molecular end products of digestion. To facilitate this process, the lining contains millions of microscopic projections known as **villi**. The villi contain capillaries and small extensions of the lymphatic system known as **lacteals**, which receive the dissolved nutrients and conduct them throughout the body. The lacteals absorb fatty acids and glycerol; the capillaries absorb monosaccharides and amino acids.

HP

The liver temporarily stores monosaccharides in the form of **glycogen** (animal starch) for later distribution through the body. Glycogen must be reconverted into glucose before it can be transported by way of the bloodstream.

Amino acids are also stored in the liver for later use in the process of protein synthesis. Approximately 20 different kinds of amino acids are used in the manufacture of proteins in the human body. Some of these amino acids can be produced within the body. Eight amino acids, however, can be obtained only by ingesting them as part of the diet. These eight are known as **essential amino acids**, since they must be obtained in the diet in order to ensure human health. When the cell produces new proteins, it must have all the amino acids necessary to manufacture these proteins. If any are missing, the manufacturing process is disrupted and the amino acids are broken down into simpler components in a process known as **deamination**. A diet that supplies **complete proteins** (those containing all eight "essential amino acids") helps to ensure that protein synthesis will occur in the most efficient manner possible. Complete proteins are generally found in foods that contain animal protein.

---

5. Large Intestine

The **large intestine** receives food materials that have passed through the entire digestive tract, but have not been affected by the digestive process. These materials are normally in a liquid state when they pass from the small to the large intestine. The large intestine reabsorbs much of the water in this waste matter and solidifies it into **feces**. The feces are forced out of the **anus** by strong peristaltic contractions in the process of egestion.

Figure 3.1 shows a diagram of the human digestive tract.

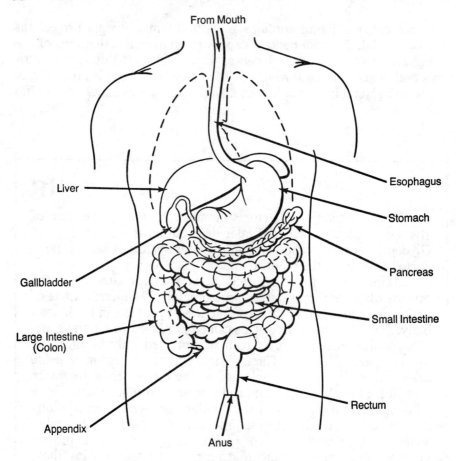

Figure 3.1 Digestive Tract

B. Mechanism for Chemical Digestion

The enzyme-controlled reaction that takes place in chemical digestion is **hydrolysis** ("splitting by adding water"). In this reaction, large,
insoluble molecules are converted into small, soluble molecules. To accomplish this, the cell adds the atoms that constitute water to the structure of the complex molecule, thereby enabling it to be converted into
the simpler, soluble form. This reaction may be accomplished with each
of the major types of food, as illustrated by the following equations:

$$\text{Maltose} + \text{water} \xrightarrow{\text{maltase}} 2 \text{ glucose}$$

$$\text{Proteins} + \text{water} \xrightarrow{\text{protease}} \text{amino acids}$$

$$\text{Lipids} + \text{water} \xrightarrow{\text{lipase}} 3 \text{ fatty acids} + 1 \text{ glycerol}$$

The hydrolytic process is further illustrated by the structural formulas
and associated equations shown in Figure 3.2.

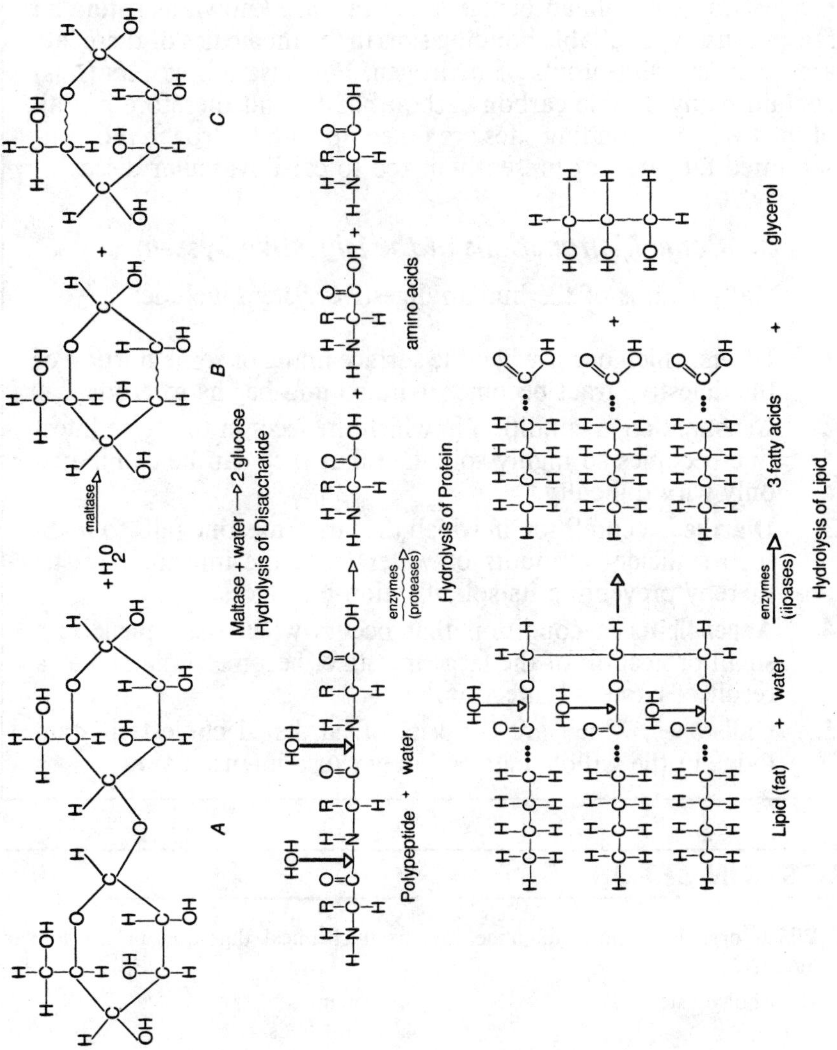

Figure 3.2 Hydrolysis

HP

Fats and oils provide the greatest source of nutritional energy per gram of any of the major food groups. Animal fats, however, have increasingly been linked to **cardiovascular** (heart and circulatory) diseases in human beings. Such fats are known as **saturated fats** because all available bonding sites in the molecules of these fats are taken up with atoms of hydrogen. **Polyunsaturated fats** (oils) contain many double carbon-carbon bonds, and therefore not all of the available bonding sites are taken up with hydrogen. Polyunsaturated fats are not presently linked to cardiovascular disease.

C. Some Malfunctions of the Digestive System

Malfunctions of the human digestive system include:

1.   **Ulcers**, which occur when the surface lining of some portion of the digestive tract becomes irritated and begins to erode.
2.   **Constipation**, a condition in which the feces in the large intestine becomes so highly solidified that they can be eliminated only with difficulty.
3.   **Diarrhea**, a condition in which the large intestine fails to reabsorb sufficient amounts of water from the intestinal waste, thereby preventing its solidification into feces.
4.   **Appendicitis**, a condition that occurs when the **appendix**, a small projection of the large intestine, becomes inflamed as a result of bacterial infection.
5.   **Gallstones**, which are deposits of hardened cholesterol that lodge in the gallbladder and may cause intense pain.

QUESTION SET 3.1

1. Which organic compounds undergo partial chemical digestion in the human mouth?
   1. carbohydrates
   2. fats
   3. proteins
   4. amino acids

2. Which structures aid in the mechanical digestion of foods in human beings?
   1. pancreas and salivary glands
   2. teeth and tongue
   3. small and large intestines
   4. intestinal and gastric glands

3. Which are produced as a result of the mechanical digestion of a piece of meat?
   1. amino acids
   2. fatty acids
   3. smaller meat particles
   4. larger glycerol molecules

4. Which organ forms part of the human gastrointestinal tract?
   1. trachea
   2. esophagus
   3. diaphragm
   4. aorta

5. After a person's stomach was surgically removed, the chemical digestion of ingested protein would probably begin in the
   1. mouth
   2. small intestine
   3. large intestine
   4. liver

6. The intestinal fold of the earthworm and the villi of the human small intestine function primarily to
   1. increase the surface area for absorption of digested nutrients
   2. excrete metabolic wastes
   3. circulate blood
   4. force the movement of food in one direction through the digestive tract

7. The removal of a human gallbladder interferes most directly with the
   1. production of gastric juice
   2. production of saliva
   3. storage of pancreatic juice
   4. storage of bile

8. Which process is represented by the diagram below?

   1. emulsification
   2. excretion
   3. absorption
   4. peristalsis

9. Into which parts of the human digestive system are digestive enzymes secreted?
   1. mouth, esophagus, stomach
   2. stomach, small intestine, large intestine
   3. mouth, stomach, small intestine
   4. esophagus, stomach, large intestine

10. The complete hydrolysis of carbohydrates results directly in the production of
    1. starch molecules
    2. urea molecules
    3. carbon dioxide
    4. simple sugars

11. Small lymphatic vessels that extend into the villi are
    1. veins
    2. lacteals
    3. nodes
    4. capillaries

12. The chemical process carried on in "chemical digestion" is known as
    1. cyclosis
    2. hydrolysis
    3. dehydration synthesis
    4. carbon fixation

13. Within the liver cells of animals, glycogen is most directly converted to glucose by the process of
    1. enzymatic hydrolysis
    2. cellular respiration
    3. active transport
    4. molecular adhesion

14. In human beings, glucose is stored in the liver chiefly as a polysaccharide known as

    HP
    1. fat
    2. protein
    3. maltose
    4. glycogen

15. Carbohydrate foods that add undigestible bulk to the contents of the human intestine are known collectively as

    HP
    1. roughage
    2. enzymes
    3. vitamins
    4. emulsifiers

16. Compared to the other major types of nutrients, fats provide a greater quantity of

    HP
    1. nitrogen per gram digested
    2. salts per gram metabolized
    3. energy per gram oxidized
    4. roughage per gram ingested

17. Under certain conditions, hardened deposits of cholesterol may form in the structures that store and transport bile. These hardened deposits are known as

    HP
    1. ulcers
    2. gallstones
    3. roughage
    4. glands

18. A person who consumes large amounts of saturated fats may increase his/her chances of developing

    HP
    1. meningitis
    2. hemophilia
    3. viral pneumonia
    4. cardiovascular disease

II. THE LIFE FUNCTION OF TRANSPORT

Transport is the life function by which human beings absorb and distribute the materials necessary to maintain life.

A. Functional Organization for Transport

In general, the human transport system resembles that of the earthworm in that it is made up of a series of blood vessels which carry a blood fluid containing hemoglobin.

1. Transport Media

1.1. Blood. Blood is a complex tissue (group of similar cells) suspended in a fluid medium for easy transport through blood vessels. The blood is responsible for carrying essential materials throughout the body, thereby helping it to maintain homeostasis. The following components make up the blood tissue:

- **Plasma.** The principal component of plasma is water. Carried in this water are dissolved salts, nutrients, gases, and molecular wastes. Also transported in the plasma are hormones and a large variety of manufactured proteins, such as antibodies, clotting proteins, and enzymes. The plasma also suspends the cellular fraction of the blood.

- **Red blood cells.** The most abundant cell type in the blood fluid is the red blood cell. These small (about 8 micrometers in diameter), dish-shaped cells lack nuclei and cannot reproduce. Red blood cells contain **hemoglobin,** a red oxygen-carrying pigment that makes the blood an efficient medium for transporting oxygen to all body tissues. Hemoglobin that is bound to oxygen is known as **oxyhemoglobin.**
- **White blood cells.** Of the many different types of white blood cells that exist, two principal types are described below. Both are important in the control of disease.
  a. **Phagocytes** engulf and destroy bacteria that enter the bloodstream through breaks in the skin surface. These phagocytic white blood cells gather in large numbers at sites of bacterial infection in the body.
  b. **Lymphocytes** produce **antibodies** specifically designed to fight off particular types of foreign proteins, known as **antigens,** that enter the bloodstream by various routes.
- **Platelets.** These are small, noncellular components of the blood that contain chemicals important to the clotting process.

HP

It is thought that the platelets in the area of a cut break open and release these chemicals into the bloodstream, initiating the formation of a **clot** through a series of enzyme-controlled reactions. A blood clot forms as a mass of protein fibers that trap lymph and red blood cells and eventually harden into a "cap" protecting the damaged area.

1.2. Intercellular Fluid (ICF) and **Lymph.** All cells in the body are bathed in a fluid rich in salts and various components important to the homeostatic balance of the cell. This fluid is known as the "intercellular fluid" or "ICF." The ICF drains from the tissues within lymphatic vessels. When the ICF enters these lymphatic vessels, it is known as "lymph."

2. Transport Vessels

2.1. Arteries are relatively thick-walled blood vessels that contain muscle tissues. These muscles enable the artery to maintain blood flow via rhythmic contractions known as the **pulse**. Arteries are always involved with conduction of the blood away from the heart toward the body tissues.

2.2. **Capillaries** are microscopic blood vessels whose walls are only one cell in thickness. Capillaries branch from the ends of small arteries and carry blood rich in oxygen and nutrients to all tissues in the body. Dissolved materials are readily exchanged between the blood and the body tissues through the thin walls of the capillaries by diffusion.

2.3. **Veins** are relatively thin-walled blood vessels that lack muscular tissues. Veins contain one-way valves that aid the forward movement of blood by preventing its "backflow" within the vein. Veins are always involved with conduction of the blood back toward the heart from the body tissues.

2.4. **Lymph vessels** form a branching series of microscopic vessels containing lymph. The lymph vessels carry lymph to and from the body tissues, where it bathes the cells and is known as "ICF." To aid the movement of the lymph, the lymph vessels contain valves similar to those found in veins. Lymph vessels become enlarged and are gathered in masses known as **lymph nodes** at specific parts of the body. These nodes contain phagocytic white blood cells that attack and destroy bacteria in the lymph.

The movement of lymph within the lymph vessels is sometimes referred to as the **lymphatic circulation**.

Circulatory patterns are shown in Figure 3.3.

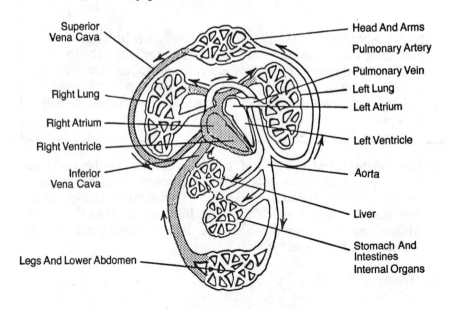

Figure 3.3 Circulatory Patterns

3. Transport Mechanisms

3.1. Structure. The structure of the **heart** permits efficient movement of blood throughout the body within blood vessels. The heart is a muscular pump with four chambers. Two of these chambers, the **atria**, receive blood from veins leading from body organs. Two other chambers, the **ventricles**, have thick, muscular walls that contract to force blood out under great pressure through arteries to other organs.

Figure 3.4 shows circulation within the heart.

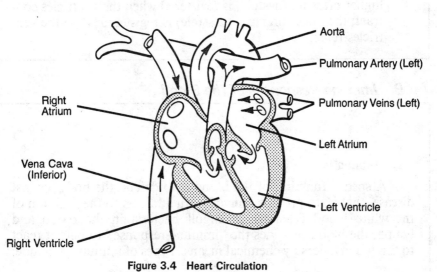

Figure 3.4 Heart Circulation

HP

3.2. Function. The circulatory function can be traced from the entry of **deoxygenated** blood into the right atrium of the heart from the **vena cavae**. This blood then passes through a one-way valve to the right ventricle, where strong muscular contractions force it out through the **pulmonary arteries** to the lungs. A valve in the pulmonary artery prevents the backflow of blood. In the lungs, the blood passes through capillaries, where gas exchange occurs. The **oxygenated** blood then returns via the **pulmonary veins** to the heart, entering the left atrium. From the left atrium, the blood passes through another one-way valve on its way to the left ventricle. Contraction of the left ventricle sends the oxygenated blood out of the heart to the body organs by way of the **aorta**. As in the pulmonary artery, a valve prevents the blood from flowing from the aorta back into the heart.

The circulation of blood through the lungs is known as the **pulmonary circulation**. The circulation of blood through the body organs is known as the **systemic circulation**. The movement of blood through the blood vessels serving the heart muscle is known as the **coronary circulation**.

Blood pressure from both the pumping action of the heart and the contractions of the muscular artery walls is responsible for maintaining the flow of blood through the arteries and capillaries. During blood pressure testing, the higher pressure (**systole**) is registered when the ventricles contract; the lower pressure (**diastole**) is registered when the ventricles relax.

B. Immune Responses of the Blood

HP

1. Immunity

A special function of the blood is to defend the body against disease, that is, to provide **immunity**. In addition to the function of the white blood cells, which engulf bacteria in the blood and lymph, the blood provides the "immune response" to help it react to foreign invaders by chemical means. Types of immunity include:

1.1. **Active immunity**, which results when the body produces specific **antibodies** in response to the presence of invading organisms or their products (**antigens**). The antibodies neutralize the antigens, making them harmless to the body. Antibody production may be stimulated either by actual infection by the disease organisms or by **vaccination** (inoculation with dead or weakened disease organisms). Active immunity is long-lasting.

1.2. **Passive immunity**, which may be acquired by the injection of preformed antibodies into the body. These antibodies may be produced by human beings or animals, and have the same effect on the antigens as do antibodies produced within the body. The immunity to disease produced by this method is usually only temporary.

2. Allergies

Allergies are the result of the body's reaction to the chemical composition of such materials as pollen, dust, animal dander, insect saliva (injected during insect bites), foods, drugs, molds, and many other substances. Sensitive people produce antibodies as though the chemicals on the materials were the antigens of disease-causing organisms. An additional response to the antigens is the production of **histamines**, which may cause irritation and swelling of mucous membranes, a symptom typical of allergic reactions.

3. Applications of Research into the Immune Response

3.1. Blood Typing. Over 50 different blood antigens are common to human blood. Of these, the antigens of the ABO blood group are most commonly checked for compatibility when a blood transfusion is performed. These antigens are known as antigens A and B and are responsible for producing the blood types A, B, AB, and O according to the following schedule:

Blood Antigen on RBC	Blood Type	Antibody in Plasma
A	A	Anti-B
B	B	Anti-A
A and B	AB	Neither Anti-A nor Anti-B
Neither A nor B	O	Both Anti-A and Anti-B

3.2. Organ/Tissue Transplants. Organ and tissue transplants, including blood transfusions, can be safely accomplished only if the antigen types of the donor and the recipient are the same or very similar. If there is not a close match between the recipient's and the donor's antigens, the recipient's body will react to the transplanted tissues as though they were foreign, disease-causing organisms by producing antibodies. The result of such a reaction is the rejection of the transplanted tissue or organ.

C. Some Malfunctions of the Transport System

1. Cardiovascular Diseases

Cardiovascular disorders are associated with the heart and blood vessels. They include:

1.1. High blood pressure, which may result from narrowing of the arteries caused by the buildup of fatty deposits on their walls. This condition can be aggravated by factors such as stress, smoking, diet, aging, and hereditary factors. Untreated, high blood pressure can cause serious damage to the heart and blood vessels and may result in death.

1.2. Coronary thrombosis, commonly known as "heart attack," which results from the blockage of the **coronary artery** leading to and feeding the heart muscle itself. This condition starves the heart muscle of oxygen and can cause severe damage to the heart if left untreated.

1.3. Angina pectoris, also known as "chest pain," which results from the gradual deterioration of coronary circulation caused by fatty deposit buildup on the walls of the coronary artery. The restricted blood flow deprives the heart muscle of a sufficient oxygen supply. Angina is usually felt as an intense pain through the chest and shoulders, radiating down the arms.

2. Blood Disorders

These problems are associated with the blood tissue. They include:

2.1. Anemia, a condition in which the blood's ability to carry oxygen is impaired. This condition may be caused by the body's reduced ability to produce red blood cells or hemoglobin, as a result of dietary or hereditary factors.

2.2. Leukemia, a disease of the bone marrow characterized by the uncontrolled production of nonfunctional white blood cells. Leukemia is a form of cancer.

QUESTION SET 3.2

1. Which part of human blood is primarily responsible for transporting nutrients, hormones, and wastes?

   1. red blood cell   **3.** white blood cell
   2. plasma   **4.** platelet

2. In the human body, which blood components engulf foreign bacteria?

   1. red blood cells   **3.** antibodies
   2. white blood cells   **4.** platelets

3. In human beings, a function of intercellular fluid is to

   1. produce red blood cells
   2. serve as a transport medium
   3. produce white blood cells
   4. serve as a filter for uric acid

4. Which of the following human transport vessels has thick, muscular walls for forcing blood through the body?

   1. capillaries   **3.** lymph vessels
   2. arteries   **4.** veins

5. Oxygenated blood leaves the heart and is transported to other parts of the body through
    1. lymph vessels   **3.** capillaries
    2. veins   **4.** arteries

6. In human beings, the exchange of materials between blood and intercellular fluid directly involves blood vessels known as
    1. capillaries   **3.** venules
    2. arterioles   **4.** arteries

7. The synthesis of a clot is controlled directly by
[HP]
    1. enzymes present in the blood
    2. hormones secreted by the pituitary gland
    3. neurons in the medulla of the brain
    4. phagocytic cells within the intercellular fluid

8. The accumulation of specific antibodies in the plasma, due to the introduction of an antigen, is characteristic of
[HP]
    1. an immune response   **3.** a coronary thrombosis
    2. angina pectoris   **4.** cerebral palsy

9. Blood serum is tested in a laboratory and is found to contain the antibodies anti-A and anti-B. The blood type of the donor of this blood is
[HP]
    1. A   **3.** AB
    2. B   **4.** O

10. A small boy catches chicken pox from his older sister. Upon recovery, he is immune to future infections with this disease. The type of immunity he has acquired is known as
[HP]
    1. passive   **3.** active
    2. antibiotic   **4.** allergic

[HP] Base your answers to questions 11 through 13 on the diagram below and on your knowledge of biology. The diagram represents the human heart, and the direction of blood flow is indicated by arrows.

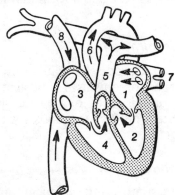

11. The aorta is represented by number
    1. *1*   **3.** *8*
    2. *6*   **4.** *4*

12. Deoxygenated blood returns to the heart through the structure represented by number

    1. *8*   **3.** *3*
    2. *7*   **4.** *5*

13. The chamber that pumps blood to all parts of the body except the lungs is represented by number

    1. *1*   **3.** *3*
    2. *2*   **4.** *4*

HP For questions 14 through 16 choose from the list below the human disorder that best matches the description in the question.

Human disorders

(1) Anemia
(2) Coronary thrombosis
(3) Angina pectoris
(4) Leukemia

14. A human cardiovascular disorder characterized by a complete blockage of the artery feeding the heart muscle.

15. A human blood disorder characterized by impaired ability of the blood to transport oxygen because of a reduction in the concentration of hemoglobin.

16. A disease of the bone marrow resulting in the uncontrolled production of nonfunctional white blood cells.

III. THE LIFE FUNCTION OF RESPIRATION

Respiration is the life process by which human beings convert the chemical energy stored in foods to a form that may more easily be used by cells.

A. Cellular Respiration

In human beings, the process of cellular respiration is essentially the same as in other aerobic organisms. Lactic acid fermentation is carried on by cells under conditions of oxygen deprivation as a temporary survival measure. (See Unit 2 of this book for a more detailed description of this process.)

B. Gas Exchange

In human beings, a moist internal membrane is maintained for the exchange of gases between the body and the external environment. This membrane and the structures responsible for conducting air to and from the membrane constitute the human respiratory system.

1. Functional Organization for Gas Exchange

1.1. The **nasal cavity** is composed of a series of channels through which outside air is admitted to the body interior. The cavity is lined by ciliated cells capable of producing mucus. Air passing through the cavity is moistened, warmed, and filtered.

1.2. The **pharynx** is the area in the back of the oral cavity where the nasal cavity joins it. In the pharynx, a flap of tissue, the **epiglottis**, covers the open end of the trachea to prevent food from entering the respiratory tubes.

1.3. The **trachea** is a cartilage-ringed tube used to conduct air from the pharynx deeper into the respiratory system. The cartilage rings maintain the open condition of the trachea at all times. The trachea is lined with ciliated tissues whose function is to sweep dust particles up and out of the trachea so that they can be swallowed or expelled. A variety of atmospheric pollutants, including cigarette smoke, can inhibit the action of these cilia.

1.4. Two **bronchi** branch from the end of the trachea and lead to the two lungs. Like the trachea, the bronchi are ringed with cartilage and lined with ciliated mucous membrane.

1.5. The **bronchioles** are highly branched tubules that subdivide from the ends of the bronchi and become progressively smaller as they pass deeper into the lung. The bronchioles lack cartilage rings.

1.6. The **alveoli**, tiny "air sacs," are found at the ends of each of the bronchioles. These alveoli are lined with cells that constitute the actual respiratory surface in the lung. Hence they are the functional unit of the lung. Capillaries surround the alveoli to carry away the oxygen absorbed in the moist lining.

1.7. The **lung** is composed of the bronchi, bronchioles, alveoli, and their supporting tissues.

Adaptations for human respiration are diagramed in Figure 3.5.

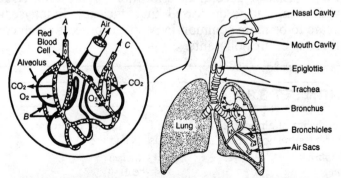

Deoxygenated Blood Enters Through Capillaries (A) Branching From The Pulmonay Artery. As The Blood Passes Through The Capillary Network (B) Surrounding The Alveoli, Oxygen Diffuses Into The Blood And Carbon Dioxide Diffuses Out Of The Blood. Oxygenated Blood Then Returns To The Heart Via Capillaries (C) Leading To The Pulmonary Veins.

Figure 3.5 Adaptations For Human Respiration

2.   Mechanisms for Gas Exchange

2.1. **Breathing** is a mechanical process used to move air into the lungs as efficiently as possible. This process involves muscular movements of the diaphragm and rib cage, which raise and lower pressures within the chest cavity. As the pressure is reduced, air is forced into the lungs by atmospheric pressure; as the pressure is increased, the process is reversed, allowing gases to be expelled. The breathing rate is regulated by the nervous system as it monitors the concentration of carbon dioxide dissolved in the blood.

2.2. Gas exchange within the alveoli is accomplished by simple diffusion. Oxygen in the bloodstream combines with hemoglobin to form **oxyhemoglobin**. In a reversal of the process that formed the oxyhemoglobin, oxygen is released to the cells, where it is used in the chemical process of cellular respiration. At the cells, carbon dioxide formed as a waste product of cellular respiration diffuses into the blood and is carried in the plasma in the form of the **bicarbonate ion**. Carbon dioxide and water vapor are released from the blood into the alveoli for removal from the body.

HP

C.   Some Malfunctions of the Respiratory System

Malfunctions of the human respiratory system include:

1.   **Bronchitis**, an inflammation of the bronchial tubes caused by infectious agents such as bacteria.

2.   **Asthma**, a condition in which the bronchial tubes become constricted because of tissue swelling brought on by an allergic reaction.

3.   **Emphysema**, a general deterioration of the lung structure, which can lead to decreased lung efficiency. Emphysema appears to be more common in people exposed to high concentrations of air pollutants.

QUESTION SET 3.3

1. The exchange of air between the human body and the environment is a result of rhythmic contractions of the rib cage muscles and the

   1. diaphragm   3. trachea
   2. lung   4. heart

2. The immediate source of the energy used by human muscle cells for contraction is

   1. DNA   3. ATP
   2. RNA   4. ADP

3. Which condition would most directly result in the production of lactic acid by some cells of the human body?
1. an excess of nitrogen in the atmosphere
2. an insufficient amount of nitrogen in the atmosphere
3. an excess of oxygen reaching the cells
4. an insufficient amount of oxygen reaching the cells

4. Which sequence most accurately represents the network of passageways that normally permits air to flow from the external environment to the human lungs?
1. nasal cavity, bronchioles, trachea, pharynx, bronchi
2. nasal cavity, pharynx, trachea, bronchi, bronchioles
3. nasal cavity, trachea, pharynx, bronchi, bronchioles
4. nasal cavity, bronchi, bronchioles, pharynx, trachea

5. The human trachea is prevented from collapsing by the presence of
1. respiratory cilia
2. smooth muscle
3. cartilage rings
4. striated muscle

6. Oxygen molecules absorbed by moist respiratory surfaces in human beings diffuse immediately into
1. endocrine glands
2. blood capillaries
3. external tubules
4. skin pores

7. Most carbon dioxide is carried in the plasma in the form of
1. hydrogen ions
2. bicarbonate ions
3. lactic acid
4. oxyhemoglobin

8. In the diagram below, which structure is indicated by the arrow?

1. bronchus
2. nasal cavity
3. trachea
4. lung

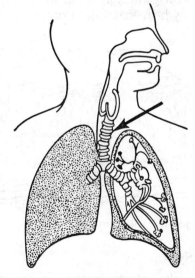

9. A disorder of the human respiratory system characterized by degeneration of the alveoli, resulting is decreased lung capacity, is known as
1. bronchitis
2. asthma
3. emphysema
4. anemia

IV. THE LIFE FUNCTION OF EXCRETION

Excretion is the life process by which human beings remove metabolic (cellular) wastes from their cells and release them to the environment.

A. Functional Organization for Excretion

1. The Lungs

The **lungs** are responsible for the excretion of carbon dioxide and water, which result from the process of aerobic respiration. These materials diffuse out of the blood into the alveoli and are expelled in the breathing process.

2. The Liver

The **liver's** role in excretion includes the recycling of worn-out red blood cells and the production of urea, which results from amino acid deamination.

3. The Sweat Glands

The **sweat glands** of the skin (see Figure 3.6) play a role in excretion by removing water, salts, and urea from the blood and excreting them as **perspiration**. Heat, a by-product of metabolism, is also removed from the tissues by the sweat glands; as water in the perspiration evaporates, it carries with it body heat that would otherwise overheat the cells.

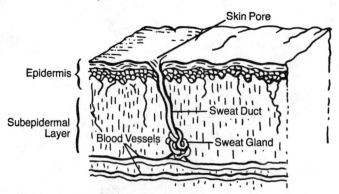

Figure 3.6 Skin Structure

4. The Urinary System

4.1. The **kidneys** in the **urinary system** are major excretory organs in human beings. Through their activities, the chemical composition of the blood is regulated, and thereby the chemical composition of the body tissues.

Two arteries that branch off the aorta carry blood to the kidneys for filtering. These arteries quickly branch into capillary networks, each known as a **glomerulus**. The glomerulus is nested inside a cup-shaped structure, **Bowman's capsule**, where many soluble blood components (including water, salts, urea, and soluble nutrients) are absorbed from the blood by diffusion. Bowman's capsule is part of a larger structure, known as the **nephron** (see Figure 3.7), which is the functional unit of the kidney. In a looped portion of the nephron, active transport is used to reabsorb most of the soluble nutrient molecules, certain mineral ions, and some water. These reabsorbed components are returned to the blood flowing out of the kidney via veins to the vena cava. The concentrated mixture of waste materials remaining in the nephron, including water, salts, and urea, is known as **urine**.

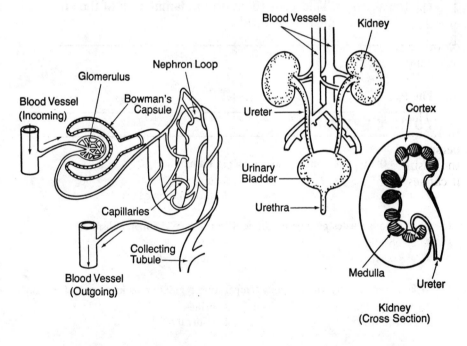

Figure 3.7 Nephron (Magnified)   **Figure 3.8 Urinary System**

4.2. The **ureters**, two small tubes, conduct urine from the kidneys through the lower abdomen to the urinary bladder.

4.3. The **urinary bladder** collects urine from the ureters and stores it for periodic excretion from the body.

4.4. The **urethra** is a small tube that leads from the urinary bladder to the outside of the body. Urine is released to the environment through the urethra.

Figure 3.8 shows a diagram of the urinary system.

HP

B. Some Malfunctions of the Excretory System

Malfunctions of the human excretory system include:

1. **Kidney failure**; this general term refers to any condition that results in the malfunctioning of the kidney or its functional unit, the nephron. Kidney failure can be brought on by physical trauma, chemical imbalance, high blood pressure, or dietary factors. It is potentially serious, since without the kidneys the body has no way of filtering many wastes out of the blood.

2. **Gout**, which is caused by uric acid deposition in the joints of the skeleton. This leads to painful inflammation of the affected joints, similar to arthritis.

QUESTION SET 3.4

1. What is the principal nitrogenous waste in human beings?
   1. salt
   2. urea
   3. uric acid
   4. carbon dioxide

2. Nitrogenous wastes may be produced as a result of the metabolism of
   1. glucose
   2. glycogen
   3. fatty acids
   4. amino acids

3. Which human excretion has body temperature regulation as its primary function?
   1. perspiration
   2. carbon dioxide
   3. urine
   4. mineral salt

4. Wastes that may be excreted from human liver cells include
   1. water, oxygen, and mineral salts
   2. water, carbon dioxide, and urea
   3. hormones, urea, and carbon dioxide
   4. hormones, oxygen, and water

5. Compared to blood entering the human kidney, blood leaving the kidney normally contains a lower concentration of
   1. red cells
   2. proteins
   3, white cells
   4. urea

6. In the diagram below, which structure is the principal site of the filtration and reabsorption processes that occur during the formation of urine?

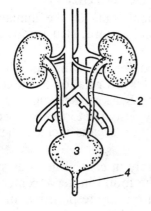

1. *1*   **3.** *3*
2. *2*   **4.** *4*

7. Which process reduces the concentration of urea in human blood?
   1. excretion   **3.** digestion
   2. egestion   **4.** synthesis

8. In human beings, the filtrate of the nephrons is stored in the
   1. glomerulus   **3.** gallbladder
   2. alveolus   **4.** urinary bladder

9. In addition to water, the principal components of urine are
   1. amino acids and fatty acids
   2. urea and salts
   3. ammonia and bile
   4. hydrochloric acid and bases

V. THE LIFE FUNCTION OF REGULATION

Regulation is the life function by which the human body controls and coordinates its other life functions to maintain existence. Regulation includes both nerve control and chemical (endocrine) control.

The nervous system and the endocrine system are similar in that both are responsible for helping to regulate the body's activities. To do this, both systems utilize chemical "messengers" that affect other living tissues. However, the systems are different in that they work at different rates and over different time periods. Nervous responses are very rapid in the rate of response and are of very short duration; endocrine responses require more time than nervous responses to be initiated, but their effect is longer lasting.

A. Nervous System

1. Functional Organization of the Nervous System

1.1. Neurons are the basic functional units of the human nervous system. These neurons are found in three structurally and functionally different types:

1.1.1. Sensory neurons are specialized for receiving stimuli from the environment and transmitting this information to the central nervous system for interpretation. Sensory neurons are normally concentrated in organs specially designed to receive specific types of stimuli. These sensory receptor organs include the eye, ear, nose, tongue, and skin.

1.1.2. Interneurons are located primarily in central nervous system organs, although they may also be found in nerve centers in other parts of the body. The interneurons are responsible for interpreting sensory impulses brought to them by the sensory neurons. Interneurons are also known to transmit "commands" to the motor neurons leading back to the body's effector organs.

1.1.3. Motor neurons carry impulses from the "command centers" in the central nervous system to effector organs (muscles or glands), where an appropriate response is initiated.

Figure 3.9 shows the interaction of a sensory neuron, an interneuron, and a motor neuron in a reflex arc (see 1.3.2).

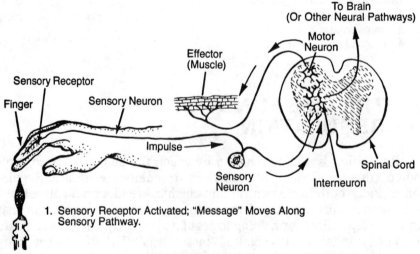

1. Sensory Receptor Activated; "Message" Moves Along Sensory Pathway.

Figure 3.9   Reflex Arc

1.2. Nerves are bundles of neurons that may contain a single type of neuron (sensory or motor nerves) or two separate types (mixed nerves). A fatty sheath protects the neurons in a nerve from coming into contact with each other and "short-circuiting" the impulses they carry. Nerves are specialized for conducting impulses over comparatively long distances at high rates of speed.

1.3. The **central nervous system** consists of the brain and the spinal cord.

1.3.1. The **brain** is a large organ composed of a mass of interneurons located within the cranial cavity. The brain is one of the most highly specialized parts of the human body and is responsible for regulating everything from the simplest to the most complex human activity. The brain is subdivided into three major regions, each responsible for a specific type of human behavior:

- In human beings, the **cerebrum** is the largest portion of the brain. It regulates conscious thought, memory, sense interpretation, reasoning, and other voluntary activities. All learned behavior, including **conditioned behavior** (habit), is centered in the cerebrum. Learned behavior depends on the cerebrum's ability to establish reflexlike nerve pathways that permit rapid, automatic responses to particular stimuli. These nerve pathways are established through repetition of the desired response.
- The **cerebellum**, located at the rear of the cranium, is responsible for coordinating muscular activities and helping the body to remain in physical balance relative to its surroundings.
- The **medulla**, at the base of the brain, regulates the automatic, rhythmic processes of the body. These processes include regulation of heartbeat, breathing rate, blood pressure, and peristaltic activity.

The central nervous system is diagramed in Figure 3.10.

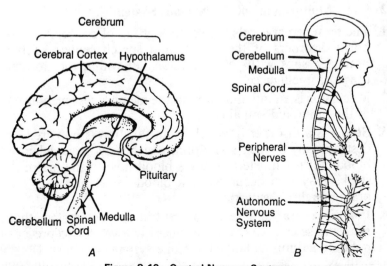

Figure 3.10   Central Nervous System

1.3.2. The **spinal cord** is continuous with the brain in the area of the medulla. It extends downward from the base of the brain along the dorsal surface of the body and is encased within the bony vertebral column, which protects it from mechanical damage. The major functions of the spinal cord include connecting the brain to the peripheral nerves and coordinating the **reflex** response. Reflexes are simple, inborn, involuntary patterns of behavior that permit immediate, unthinking response to potentially dangerous situations. This rapid response is made possible by the **reflex arc** (see Figure 3.9). The reflex arc is a series of three neurons working together: a sensory neuron (which receives a stimulus in a receptor), an interneuron in the spinal cord (which interprets the sensory impulse), and a motor neuron (which sends an impulse to an effector). Reflex actions are so rapid and automatic that many are completed before our cerebrum makes us aware of the stimuli initiating them.

1.4. The **peripheral nervous system** consists of all the nerves of all types that branch through the body from the central nervous system. The peripheral nervous system is organized into two divisions:

HP

1.4.1. The **somatic nervous system** includes the nerves that control the actions of the voluntary skeletal muscles.

1.4.2. The **autonomic nervous system** consists of nerves regulating automatic functions such as the actions of glands and involuntary muscle.

2. Some Malfunctions of the Nervous System

2.1. The term **cerebral palsy** refers to a group of congenital (inborn) disorders of the muscle and speech centers of the brain. Persons with this condition fall within the normal range of intelligence.

2.2. **Meningitis** is an inflammation of the membranes (meninges) of the brain and spinal cord.

2.3. **Stroke** is a disorder in which brain function is impaired or destroyed by oxygen starvation, which may occur when a blood clot in an artery restricts blood flow to the brain. A stroke may also occur when a blood vessel bursts in the brain (**cerebral hemorrhage**).

2.4. **Polio** is a viral disease of the central nervous system that results in severe paralysis. Vaccines are available to prevent polio by aiding the body's immune system.

B. Endocrine System

The endocrine system consists of a number of discrete glands located at various points in the body. These glands lack ducts to carry their secretions to the "target" tissues and are therefore known as "ductless" glands. The endocrine glands deliver their secreted hormones throughout the body by way of the bloodstream.

HP

1. Functional Organization of the Endocrine System

1.1. The **hypothalamus** is a small gland located within the brain. It produces secretions that affect the operation of the pituitary gland.

1.2. The **pituitary gland**, which is located under the brain, is sometimes referred to as the "master gland" because of the large number of hormones it produces, many of which control other endocrine glands. These hormones include:

- **Growth-stimulating hormone**, which affects the growth of the long bones of the body.
- **Thyroid-stimulating hormone (TSH)**, which stimulates the thyroid gland to produce its principal hormone, thyroxin.
- **Follicle-stimulating hormone (FSH)**, which affects the ovary in females, stimulating the maturation and release of human eggs on a monthly basis (see Unit 4 of this book for additional information).

1.3. The **thyroid gland** is located in the neck region, surrounding the trachea. Its hormone is **thyroxin**, which regulates general metabolic rate. Thyroxin production depends on a dietary supply of iodine, found in many seafoods.

1.4. The **parathyroid glands** are embedded within the thyroid gland. The hormone of the parathyroids is **parathormone**, which regulates the metabolism of calcium in the body. Proper calcium levels are essential for nerve function and blood clotting, as well as bone and tooth structure.

1.5. The **adrenal glands**, located on the kidneys within the abdomen, consist of two separate regions, the **cortex** (outer region) and **medulla** (inner region). Each region produces its own hormones as follows:

- The adrenal cortex secretes steroid hormones that (1) regulate water balance and blood pressure by controlling the reabsorption of sodium salt into the blood from kidney tubules, and (2) stimulate the conversion of complex stored foods, such as fat and protein, into glucose.

- The adrenal medulla secretes the hormone **adrenalin**, which increases the rates of metabolism, heartbeat, and breathing, and stimulates the conversion of glycogen into glucose. These activities have the effect of readying the body for "fight or flight" from danger.

1.6. The **islets of Langerhans**, small groups of glandular tissue scattered throughout the pancreas, produce the hormones **insulin** and **glucagon**. These two hormones have opposite effects on the storage of sugar in the liver and muscles of the body. Insulin promotes the storage of excess blood glucose as glycogen, thus lowering blood sugar levels; glucagon stimulates the conversion of glycogen back into glucose, thus raising blood sugar levels.

1.7. The **gonads** ("sex glands") differ in men and women. In males, the **testes** produce the hormone **testosterone**, important in the promotion of male secondary sex characteristics. In females, the **ovaries** secrete **estrogen**, which influences various aspects of the female secondary sex characteristics.

The locations of the endocrine glands (and of major nonendocrine organs) are shown in Figure 3.11.

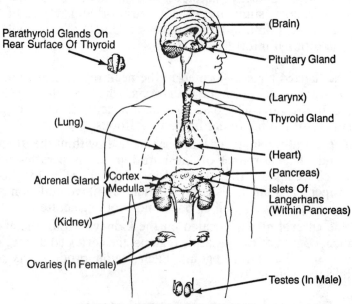

Parathyroid Glands On
Rear Surface Of Thyroid

(Brain)

Pituitary Gland

(Larynx)

Thyroid Gland

(Lung)

(Heart)

(Pancreas)

Adrenal Gland {Cortex Medulla

Islets Of
Langerhans
(Within Pancreas)

(Kidney)

Ovaries (In Female)

Testes (In Male)

Locations Of Major Nonendocrine Organs
Are Indicated By Dotted Outline.

Figure 3.11 Endocrine Glands

2. Negative Feedback

The mechanism that controls many aspects of endocrine regulation is **negative feedback**. This mechanism operates on the principle that the effects of a particular hormone may inhibit the further production of that hormone, but may stimulate the production of another hormone. A good example of negative feedback is the "insulin-glucagon" feedback loop, in which elevated sugar levels stimulate the production of insulin (which stores glucose in the liver), while inhibiting the production of glucagon. As sugar levels drop, glucagon production increases, with the effect that glucose is released from the liver, while insulin production is inhibited.

3. Some Malfunctions of the Endocrine System

3.1. Goiter is a condition resulting in enlargement of the thyroid gland. Goiter is caused when the thyroid is unable to produce adequate supplies of thyroxin, and may be traced to iodine deficiencies in the diet.

3.2. Diabetes is a disorder characterized by inability of the body to remove sugar from the blood and store it as glycogen in the liver. Diabetes is associated with insufficient insulin production by the body.

QUESTION SET 3.5

1. The correct sequence for the pathway of an impulse in a simple reflex arc is
  1. effector → motor neuron → sensory neuron → interneuron → receptor
  2. receptor → interneuron → sensory neuron → motor neuron → effector
  3. receptor → sensory neuron → interneuron → motor neuron → effector
  4. effector → sensory neuron → receptor → interneuron → motor neuron

2. Impulses from the spinal cord to the muscle fibers in the human leg are transmitted through structures known as
  1. sensory neurons   **3.** motor neurons
  2. interneurons   **4.** connective neurons

3. In human beings, which structure is primarily responsible for maintaining balance and coordinating motor activities?
  1. spinal cord   **3.** medulla
  2. cerebrum   **4.** cerebellum

4. Neurons are to neurotransmitters as endocrine glands are to
  1. hormones   **3.** nucleic acids
  2. vitamins   **4.** enzymes

5. The breathing rate of a human being increases after rapid exercise. The part of the nervous system that controls this involuntary action is the
   1. cerebrum
   2. cerebellum
   3. medulla
   4. spinal cord

6. The somatic nervous system contains nerves that run from the central nervous system to the
   1. muscles of the skeleton
   2. heart
   3. smooth muscles of the gastrointestinal tract
   4. endocrine glands

7. Which condition results from brain cell damage due to blocked or burst blood vessels and is characterized by impaired speech or motor patterns?
   1. polio
   2. meningitis
   3. stroke
   4. diabetes

8. A viral disease of the central nervous system that may result in permanent paralysis or death if untreated is
   1. cerebral palsy
   2. stroke
   3. meningitis
   4. polio

9. Which structure secretes the substance that it produces directly into the bloodstream?
   1. gallbladder
   2. salivary gland
   3. adrenal gland
   4. skin

10. The maintenance of proper blood sugar level involves the storage of excess sugar in the
    1. salivary glands
    2. stomach
    3. pancreas
    4. liver

11. The hormones TSH and FSH are products of the
    1. pituitary gland
    2. hypothalamus
    3. gonads
    4. thyroid gland

12. Which is regulated by thyroxin?
    1. the rate of metabolism in the body
    2. the condition of cartilage in the joints of bones
    3. ACTH production
    4. calcium production

13. Which hormone lowers blood sugar levels by increasing the rate of entry of glucose into the cells?
    1. follicle-stimulating hormone
    2. insulin
    3. parathormone
    4. adrenalin

14. The body normally responds to low concentrations of sugar in the blood by secreting
    1. glucagon
    2. estrogen
    3. insulin
    4. testosterone

15. Calcium metabolism, necessary for normal nerve conduction, is regulated through
HP   the action of the hormone

1. glucagon   3. testosterone
2. parathormone   4. thyroxin

16. A human disorder characterized by a deficiency in the output of insulin by the
HP   pancreas is

1. goiter   3. gout
2. diabetes   4. emphysema

VI. THE LIFE FUNCTION OF LOCOMOTION

Locomotion is the life process by which human beings move from place to place within their environment.

A. Functional Organization for Locomotion

1. Bones

Bones provide mechanical support and protection for the body and its internal organs. These bones are arranged into an internal skeleton (**endoskeleton**; see Figure 3.12), which gives human beings their characteristic body shape. Other functions of the bones include anchorage sites for muscles, leverage for movement, and production of blood cells in the bone marrow.

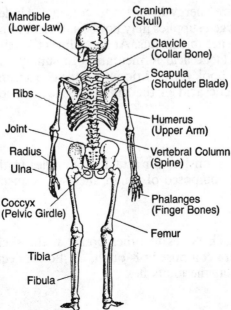

Figure 3.12   Skeletal System. *Knowledge Of Specific Bones Is Not Required.

2. Cartilage

Cartilage is a flexible, fibrous connective tissue that pads the joints between bones. Cartilage is found in many of the flexible parts of the body, including the outer ear and the nose. In embryos and very young children, cartilage is abundant, making up the major portion of the skeleton; in adults, it is much reduced, being found only at bone ends and in flexible portions of the body such as the trachea.

3. Muscles

Muscles in the human being are of three types: **visceral** (smooth), **cardiac** (striated, or "striped"), and **skeletal** (striated). These three types of muscle have different appearances and functions, although they have in common the mechanism of muscle contraction.

3.1. Visceral and cardiac muscles are involuntary muscles controlled primarily by the autonomic nervous system.

3.2. Skeletal muscles are voluntary and are those most directly involved in human locomotion. Skeletal muscles normally work in opposing pairs, each pulling a joint in a different direction. By coordinating these muscle pairs, the body is able to move skeletal levers to accomplish movement from place to place. In opposing pairs, muscles that extend joints are known as **extensors**, while muscles that flex joints are known as **flexors**.

Muscles require energy in order to contract. Under conditions of heavy exercise, oxygen supplies may not be sufficient to release energy to the muscle cells by aerobic means. At such times the muscle cell has the capacity to utilize lactic acid fermentation to supply additional energy (see Unit 2 of this book for additional details concerning this process). The buildup of lactic acid in the muscle tissues is known as "muscle fatigue."

4. Tendons

Tendons are responsible for attaching muscles to the bones of the skeleton. They are composed of tough, inelastic connective tissue.

5. Ligaments

Ligaments attach bones to other bones at the skeletal joints (see Figure 3.13) and are composed of tough, elastic connective tissues that are able to stretch as the joints flex.

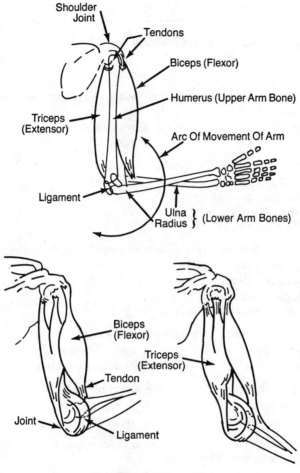

Figure 3.13   Elbow Joint

HP

B. Some Malfunctions of the Locomotor System

Malfunctions of the human locomotor system include:

1. **Arthritis,** a condition in which the joints of the skeleton become inflamed and swollen.
2. **Tendonitis,** a condition in which the junction between a tendon and its bone attachment becomes irritated and inflamed. It is frequently associated with physical stress such as that encountered in sports activities.

QUESTION SET 3.6

1. Which structure contains pairs of opposing skeletal muscles?
    1. stomach
    2. small intestine
    3. heart
    4. hand

2. Which type of connective tissue makes up the greatest proportion of the skeleton of a human embryo?
    1. ligaments
    2. cartilage
    3. tendons
    4. bone

3. Which is *not* a major function of cartilage tissues in a human adult?
    1. giving pliable support to body structures
    2. cushioning joint areas
    3. adding flexibility to joints
    4. providing skeletal levers

4. Which statement most accurately describes human skeletal muscle tissue?
    1. It is involuntary and striated.
    2. It is involuntary and lacks striations.
    3. It is voluntary and striated.
    4. It is voluntary and lacks striations.

5. Tendons are tough, inelastic fibrous cords that attach
    1. muscles to bones
    2. bones to bones
    3. nerves to muscles
    4. capillaries to nerves

6. At movable joints, bones are attached to each other by
    1. elastic ligaments
    2. cartilaginous tissues
    3. smooth muscles
    4. skeletal muscles

7. A human disorder of the locomotor system characterized by inflammation of the structures attaching muscles to bones is
    HP
    1. arthritis
    2. gout
    3. goiter
    4. tendonitis

UNIT 3 REVIEW

1. Which occurs as a result of the action of hydrolytic enzymes?
    1. Inorganic substances are converted directly to organic substances.
    2. Complex organic molecules are made more soluble.
    3. Glucose molecules are converted to starches.
    4. Glucose molecules are converted to maltase molecules.

2. The principal function of mechanical digestion is the
    1. hydrolysis of food molecules for storage in the liver
    2. production of more surface area for enzyme action
    3. synthesis of enzymes necessary for food absorption
    4. breakdown of large molecules to smaller ones by the addition of water

3. The most common food reserves in human beings are fat and

   1. hemoglobin   **3.** DNA
   2. maltose   **4.** glycogen

4. Phagocytosis and the production of antibodies are functions associated with specialized

   1. white blood cells   **3.** platelets
   2. red blood cells   **4.** neurons

5. Which is a characteristic of lymph nodes?

   1. They carry blood under great pressure.
   2. They move fluids by means of a muscular pump.
   3. They produce new red blood cells.
   4. They contain phagocytic cells.

6. The thin-walled vessels of the circulatory system where most oxygen and carbon dioxide are exchanged are

   1. alveoli   **3.** capillaries
   2. arteries   **4.** veins

7. Which compounds are produced in human muscle cells as a result of the oxidation of glucose in the absence of oxygen?

   1. lipase and water
   2. sucrase and carbon dioxide
   3. ethyl alcohol and ATP
   4. lactic acid and ATP

8. Which human excretory organ breaks down red blood cells and synthesizes urea?

   1. lung   **3.** skin
   2. kidney   **4.** liver

9. Water, salts, and urea are removed from the blood by the

   1. liver and lungs
   2. sweat glands and kidneys
   3. pancreas and adrenal glands
   4. kidneys and liver

10. Which human body system includes the lungs, liver, skin, and kidneys?

   1. respiratory   **3.** transport
   2. digestive   **4.** excretory

11. In the diagram below, which number indicates the region of the human brain that maintains balance and coordinates motor activities?

   1. *1*
   2. *2*
   3. *3*
   4. *4*

12. Which part of the human central nervous system is involved primarily with sensory interpretation and thinking?

1. spinal cord   **3.** cerebrum
2. medulla   **4.** cerebellum

13. The breathing rate of human beings is regulated principally by the concentration of

1. carbon dioxide in the blood
2. oxygen in the blood
3. platelets in the blood
4. white blood cells in the blood

14. Which is a correct route of an impulse in a reflex arc?

1. receptor → sensory neuron → interneuron → motor neuron → effector
2. effector → receptor → motor neuron → sensory neuron → interneuron
3. sensory neuron → effector → motor neuron → receptor → interneuron
4. motor neuron → sensory neuron → interneuron → effector

15. Which is an example of an effector?

1. a taste bud of the tongue
2. the auditory nerve of the ear
3. the retina of the eye
4. a muscle of the arm

16. In human beings, the elastic, flexible connective tissue found between the vertebrae is known as

1. chitin   **3.** bone
2. cartilage   **4.** tendons

17. In the human elbow joint, the bone of the upper arm is connected to the bones of the lower arm by flexible connective tissue known as

1. tendons   **3.** muscles
2. ligaments   **4.** neurons

HP For each phrase in questions 18 through 21, select the organ, *chosen from the drawing below*, which is most closely related to that phrase. (*A number may be used more than once or not at all.*)

18. Where feces are formed

19. Where protein digestion begins

20. Where lipid digestion is completed

21. Contains gastric glands

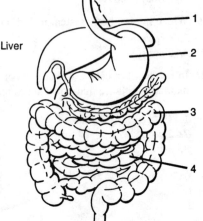

22. An erosion of the lining of the alimentary canal, generally associated with some
HP kind of irritant, is known as
  1. an ulcer   **3.** an allergy attack
  2. a gallstone   **4.** a coronary attack

HP Base your answers to questions 23 through 27 on your knowledge of biology and
on the graph below, which shows the extent to which carbohydrates, proteins, and fats
are chemically digested as food passes through the human digestive tract. The letters
represent sequential structures that make up the digestive tract.

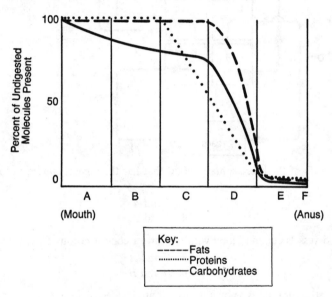

23. Proteins are digested in both
  1. *A* and *B*   **3.** *C* and *D*
  2. *B* and *C*   **4.** *A* and *C*

24. The organ represented by the letter *C* is most probably the
  1. esophagus   **3.** small intestine
  2. stomach   **4.** large intestine

25. Enzymes secreted by the pancreas enter the system at
  1. *E*   **3.** *C*
  2. *B*   **4.** *D*

26. The final products of digestion are absorbed almost entirely in
  1. *F*   **3.** *C*
  2. *B*   **4.** *D*

27. Water is removed from the undigested material in
  1. *A*   **3.** *E*
  2. *B*   **4.** *D*

HP Base your answers to questions 28 through 30 on the schematic diagram below of blood flow throughout the human body and on your knowledge of biology.

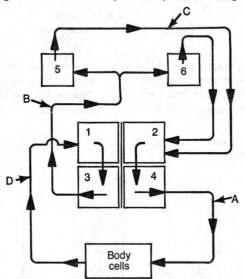

28. Which chambers of the heart contain blood that has the highest concentration of oxygen?

  1. *1* and *2*   **3.** *3* and *4*
  2. *2* and *4*   **4.** *1* and *3*

29. Which blood vessels contain blood with the lowest concentration of oxygen?

  1. *A* and *D*   **3.** *C* and *A*
  2. *B* and *C*   **4.** *D* and *B*

30. Microscopic structural units known as alveoli are located in structures

  1. *1* and *3*   **3.** *5* and *6*
  2. *B* and *C*   **4.** *D* and *A*

---

31. A complete blockage of a section of a main heart artery, resulting in damage to the HP heart muscle, is known as

  1. sickle-cell anemia
  2. high blood pressure
  3. angina pectoris
  4. coronary thrombosis

32. Which chamber of the human heart receives most of the blood returning from the HP brain?

  1. left ventricle   **3.** left atrium
  2. right ventricle   **4.** right atrium

33. Which structures in human blood contain enzyme molecules necessary for the HP clotting process?

  1. phagocytes   **3.** red blood cells
  2. lymphocytes   **4.** platelets

34. An organism develops active immunity as a result of
HP **1.** manufacturing its own antigens
  2. producing antibodies in response to a vaccination
  3. receiving an injection of antibodies produced by another organism
  4. receiving an injection of a dilute glucose solution

35. Which is an allergic reaction characterized by constriction of the bronchial tubes
HP and reduced airflow to the lungs?
  1. anemia   **3.** asthma
  2. arthritis   **4.** angina

36. In human beings, circulation to and from the lungs is known as
HP **1.** systemic circulation   **3.** coronary circulation
  2. pulmonary circulation   **4.** lymphatic circulation

37. Substances formed in the human body in response to foreign proteins entering the
HP blood are known as
  1. antigens   **3.** antibodies
  2. platelets   **4.** red blood cells

38. A student's blood pressure measures 116/70. The number 116, or systolic number,
HP refers to the amount of blood pressure exerted on the walls of the student's
  1. veins   **3.** capillaries
  2. lymph glands   **4.** arteries

39. A type of "heart attack" in which a narrowing of the coronary artery causes an
HP inadequate supply of oxygen to reach the heart muscle is known as
  1. anemia   **3.** angina pectoris
  2. leukemia   **4.** cerebral palsy

HP Base your answers to questions 40 through 42 on the graph below, which represents
relative blood pressure in human circulatory structures *A* through *D*.

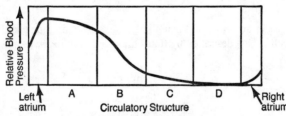

40. Blood pressure drops most drastically within structure
  1. *A*   **3.** *C*
  2. *B*   **4.** *D*

41. What prevents the backflow of blood in structure *D*?
  1. ganglia   **3.** pumps
  2. enzymes   **4.** valves

42. A heavy, muscular ventricle would be found as part of structure
  1. *A*   **3.** *C*
  2. *B*   **4.** *D*

HP Base your answers to questions 43 through 45 on the diagram below of a portion of the human nervous system and on your knowledge of biology.

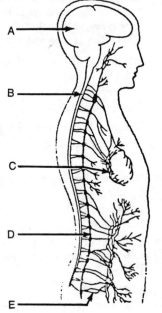

43. The structure labeled *B* would be most directly involved with the regulation of

1. conscious thought
2. reflex action
3. sense interpretation
4. reasoning ability

44. The human disorder known as meningitis would cause inflammation of which portions of the nervous system?

1. *A* and *B*
2. *B* and *C*
3. *C* and *D*
4. *D* and *E*

45. A stroke occurs in the area indicated by

1. *A*
2. *B*
3. *C*
4. *D*

46. In human beings, the center for regulating the amount of oxygen in the blood is situated in the

1. cerebrum
2. cerebellum
3. medulla
4. spinal cord.

For each phrase in questions 47 through 51, select the endocrine gland, *chosen from the list below*, which is best described by that phrase. (*A number may be used more than once or not at all.*)

Endocrine Gland

(1) Thyroid
(2) Adrenal
(3) Pancreas
(4) Pituitary
(5) Ovary
(6) Parathyroid

47. Is located on top of the kidney and influences the heartbeat and breathing rates

48. Produces FSH, which influences a phase of the human female menstrual cycle

49. Produces insulin, which helps to regulate the blood sugar levels

50. Produces thyroxin, which regulates the rate of metabolism

51. Produces estrogen, which influences the development of secondary sex characteristics

52. A person was admitted to the hospital with abnormally high blood sugar and an
HP abnormally high sugar content in his urine. Which gland most likely caused this
condition by secreting lower than normal amounts of its hormone?

   1. pancreas   **3.** salivary

   2. parathyroid   **4.** thyroid

53. Human ovaries do not function properly unless they receive a stimulating hormone
HP from the

   1. thyroid   **3.** adrenal gland

   2. pituitary gland   **4.** pancreas

54. Which gland does *not* secrete hormones?
HP   **1.** pituitary gland   **3.** sex gland

   2. thyroid gland   **4.** salivary gland

55. The reabsorption of sodium and chloride ions by the kidney tubules is regulated by
HP the

   1. ovaries   **3.** adrenal glands

   2. testes   **4.** parathyroid glands

56. Which of the following contains the greatest amount of skeletal muscle tissue?
HP   **1.** cerebrum   **3.** kidney

   2. small intestine   **4.** foot

BC Base your answers to questions 57 through 59 on the chemical equation below,
which represents a metabolic activity.

57. This metabolic activity is known as

   1. glucose oxidation

   2. enzymatic hydrolysis

   3. polysaccharide formation

   4. alcoholic fermentation

58. The chemical reaction represented can be performed by

   1. vertebrate animals only

   2. multicellular plants only

   3. protists only

   4. plants, animals, and protists

59. When substance *B* and substance *C* combine chemically and produce substance *A*
and water, this chemical process is known as

   1. anaerobic respiration   **3.** dehydration synthesis

   2. digestion   **4.** enzymatic hydrolysis

4

HOW DO LIVING THINGS REPRODUCE AND GROW?

<table>
<tr><td>KEY
IDEAS</td><td>This unit focuses on the process by which living things pro- duce new organisms of the same species. This is accomplished by means of the life process of reproduction. Concepts pre- sented in this unit include the types of reproduction and the mechanisms by which the cells divide. When reproduction is accomplished by a single parent organism, the reproductive process is known as asexual. If two parents must be involved in the production of offspring, the reproductive process is known as sexual. The basic mechanism of reproduction in- volves cell division, in which mature cells separate into two new cells with identical characteristics. Also presented in this unit are the processes controlling embryological development.</td></tr>
</table>

KEY OBJECTIVES

Upon completion of this unit, you should be able to:

☐ Compare and contrast the processes of mitosis and meiosis, and describe their respective roles in the reproductive process.

☐ Describe the process of fertilization and its role in maintaining the species chromosome number during sexual reproduction.

☐ Compare and contrast the processes of sexual and asexual reproduc- tion in terms of mechanism and result.

☐ Describe the adaptations for sexual reproduction in both plants and animals and the ways in which they aid the reproductive process.

☐ Compare animal reproductive processes that occur outside and in- side the mother's body in terms of the numbers of eggs and offspring produced and the ways in which they affect species survival.

☐ Compare plants and animals in terms of their patterns of embryonic development.

☐ Describe how the interaction of hormones is important to the pro- cess of sexual reproduction in human males and females.

I. REPRODUCTION BY ASEXUAL MEANS

Asexual reproduction involves the production of a new organism of a species from a cell or cells of a single parent organism. There is no fusion of cells or cell nuclei in asexual reproduction. Instead, there is a duplication of the nucleus, followed by the splitting of a single cell or group of cells. This results in the production of more cells with characteristics identical to those of the single parent organism.

A. Mitotic Cell Division

In Unit 1 we learned that the cell theory states that all cells arise from preexisting cells by cell division. This type of cell division is known as **mitotic cell division** and involves two distinct stages:

A-1. Mitosis

Mitosis is a precise duplication of the contents of the parent cell nucleus, followed by an orderly separation of those contents into two new, identicial nuclei.

A-2. Cytoplasmic Division

The separation of the two new nuclei into two new **daughter cells** as the cytoplasm of the parent cell divides is known as **cytoplasmic division.** The daughter cells contain an identical complement of chromosomes and so will share identical genetic characteristics. (See Unit 5 for a more detailed treatment of genetic theory.)

The events of mitotic cell division are as follows:

1. Mitosis

- **Replication** (exact self-duplication) of each chromosome strand in the nucleus of the parent cell, which results in the doubling of each chromosome strand to form **double-stranded chromosomes.** Each of the two strands of a double-stranded chromosome is known as a **chromatid.** Chromatids are chemically identical to each other and carry identical genetic information. These chromatids are held together by a **centromere.** Figure 4.1 shows a double-stranded chromosome.

Centromere   Chromatids

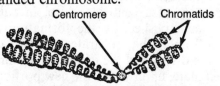

Figure 4.1   Double-Stranded Chromosome

- Disappearance of the membrane surrounding the nucleus. In this disappearance, the complex structure of the nuclear membrane disintegrates and becomes part of the cytoplasm.
- Appearance of a network of fibers (**spindle apparatus**), which attach to the double-stranded chromosomes at the centromere during the early stages of mitosis.
- Replication of the centromere of each double-stranded chromosome, followed by separation of the two chromatids of the double-stranded chromosomes to form single-stranded chromosomes.
- Migration of the single-stranded chromosomes along the spindle apparatus toward opposite ends of the cell. The chromatids in each pair separate in this stage of mitosis, allowing a full set of single-stranded chromosomes to migrate to each pole.
- Reformation of the nuclear membrane around the chromosomes grouped at the ends of the cell. These two **daughter nuclei** are identical to each other and to the parent nucleus in terms of the number and type of chromosomes, as well as in the particular genetic information found within these chromosomes. The significance of mitosis is the *exact duplication* of the parent nucleus, forming two identical daughter nuclei.

2. Cytoplasmic Division

- After the formation of the daughter nuclei, the cytoplasm of most (though not all) cells is divided into two roughly equal portions, each enclosing one of the daughter nuclei. These daughter cells have characteristics identical to each other and to the parent cell. The daughter cells will grow and eventually go through the cell cycle themselves.
- The mechanisms governing this cytoplasmic division are not fully understood, but are known to occur differently in different types of organisms. For example, in plants and animals, mitosis occurs in much the same way as described above. However, in animals the centriole is known to function in the formation of the spindle apparatus, whereas in plants the spindle forms without the presence of a centriole. Also, in animals cytoplasmic division is accomplished by the formation of a constriction ("pinched-in" area) that separates the two daughter cells. In plant cells, the division of the cytoplasm occurs after the formation of a **cell plate**, upon which the new portions of the cell walls form following mitosis.

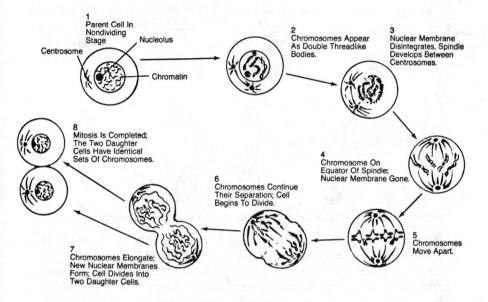

Figure 4.2   Mitosis

Normal mitotic cell division (see Figure 4.2) results in the production of new cells for growth and for the repair of damaged or worn-out body tissues. It is a process that is controlled within the cell itself, and that occurs countless times in every living thing without flaw. However, in some cells at some times, the mitotic process appears to break down and begins to occur so rapidly that insufficient time is available for normal replication and chromosome separation. This rapid, abnormal cell division is known as **cancer**. In a very short time cancer may produce a large number of such abnormal cells, which begin to crowd out the normal tissues, resulting in damage to these tissues and often in the death of the host organism.

B.   Types of Asexual Reproduction

All forms of asexual reproduction have in common the production of new organisms from a single parent organism. However, the process is carried on differently by different living things. Some of the types of asexual reproduction are as follows:

1.   **Binary fission** is carried on by many forms of unicellular organisms, including the paramecium and the ameba (see Figure 4.3). Bacteria are also known to reproduce in this manner. Binary fission is accomplished when a single cell undergoes mitosis followed by equal cytoplasmic division, forming two daughter cells having roughly the same size and shape and containing identical genetic information.

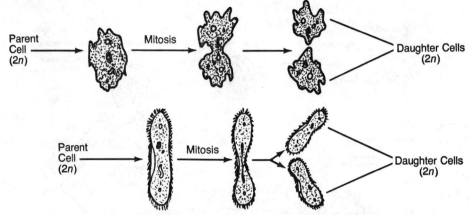

Figure 4.3 Binary Fission in Ameba (above) and Paramecium (below)

2. **Budding** may also occur in certain unicellular life forms such as yeast. Budding is accomplished when mitosis is followed by unequal cytoplasmic division, resulting in daughter cells of unequal size, but containing identical genetic information. The larger of the two cells may divide rapidly several more times, producing a chain or colony of daughter cells.

Certain simple multicellular organisms, including the hydra, carry on a form of asexual reproduction which may be called budding. In this process, new embryonic cells begin to form by mitosis within the mature tissues of the organism. As these embryonic tissues begin to differentiate (specialize), they take on the physical appearance of the parent organism. After a period of growth, the new organism, or "bud," identical to the parent organism, separates from the parent and begins to live an independent existence. In some species, such as the corals, the newly created offspring may remain in the same area and form a colony.

Budding in yeast and hydra is diagramed in Figure 4.4.

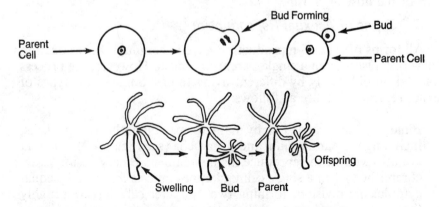

Figure 4.4 Budding in Yeast (above) and Hydra (below)

3. **Sporulation** involves the formation of specialized reproductive cells, known as **spores**, within the parent organism. Each spore contains a nucleus surrounded by cytoplasm. When these spores are released from the parent plant and land in an environment containing conditions favorable to their growth, they begin to undergo mitotic division. The spores of most species require moisture and warmth to germinate. The mitotic divisons result in the formation of a new multicellular organism genetically identical to the original parent organism. This type of sporulation is carried on by fungi such as mold (see Figure 4.5) and mushrooms.

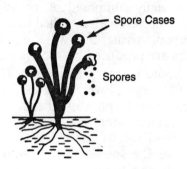

Figure 4.5   Sporulation in Bread Mold

4. **Regeneration** involves the production of one or more new organisms from the severed parts of a single parent organism. The pieces of cut-up planarian worms may frequently grow back lost tissues to produce as many as four or five new, identical planarian worms. Starfish cut into pieces that contain part of the central disc area may undergo regeneration that produces several new, genetically identical starfish. The term "regeneration" may also be used to refer to the production of new tissues to replace those lost or damaged by accident or disease.

    In both cases, invertebrate animals, with their less highly differentiated (specialized) tissues, are known to display a higher degree of regenerative ability than vertebrate animals. In other words, invertebrates both reproduce and repair damaged tissues by regeneration more readily than do vertebrates.

5. **Vegetative propagation** is a general term referring to any of several forms of asexual reproduction carried on by multicellular plants. An aspect common to all types of vegetative propagation is the production of new plant organisms from the leaves, stems, or roots ("vegetative" parts) of the parent plant, rather than from the flower. As in other forms of asexual reproduction, the new organisms produced by vegetative propagation are genetically identical to the parent plant.

Examples of vegetative propagation include:

- **Cuttings.** By removing a portion of the root, stem, or leaf of certain plants and placing it in an environment favorable to growth, it is possible to stimulate the plant to replace the missing parts of its body and to establish itself in the new environment as an independent organism. Coleus and geranium plants are both artificially reproduced in this way.

- **Bulbs.** Certain plants produce specialized asexual structures known as "bulbs," which bud from mature "parent" bulbs. A bulb is actually composed of a very short underground stem (with thick leaves) with attached root structure. Within the bulb are embryonic tissues that will eventually form leaves and flowers. Bulbs are produced by such plants as the onion and tulip.

- **Tubers.** Some plants produce specialized underground stems, known as "tubers," which are used for storage of excess food produced during photosynthesis. Tubers also contain the undifferentiated embryonic tissues necessary to produce roots, above-ground stems, and leaves, so can be used to grow new members of the species. The potato is a good example of a tuber-producing plant.

- **Runners.** Certain plants produce specialized asexual structures know as "runners" (see Figure 4.6). Runners are stemlike structures that grow from the main stem of the parent plant over the surface of the soil. When the runner reaches a spot favorable to growth of a new plant, it produces roots and leaves. Once established, the new plant can live independently of the parent plant. Poison ivy and strawberries can reproduce using runners.

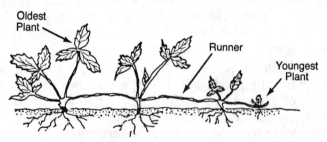

Figure 4.6 Vegetative Propagation: Runners

- **Grafting.** Commercial flower and fruit growers make use of grafting to produce large quantities of genetically identical plants for sale. In grafting, the embryonic tissues, known as **cambium** tissues, of a rootstock are attached to like tissues of an upper stem taken from the parent plant to be reproduced. Properly attached, the two sections grow together, and the new plant may be established in a favorable growing environment until mature. Fruits or flowers that grow from the grafted upper stem will be genetically identical to those produced by the original parent plant. Empire apples and seedless oranges are examples of fruits produced on trees that were grafted.

QUESTION SET 4.1

Base your answers to questions 1 and 2 on the diagram below, which represents a microscopic structure observed during cell division, and on your knowledge of biology.

1. Letter *A* indicates a
   1. nucleolus
   2. ribosome
   3. centriole
   4. centromere

2. Letter *B* indicates a
   1. centrosome
   2. spindle fiber
   3. chromatid
   4. cell plate

3. Each strand of a double-stranded chromosome is known as a
   1. centromere
   2. homologue
   3. chromatid
   4. tetrad

4. Normally, a complete set of chromosomes (2*n*) is passed on to each daughter cell as a result of
   1. reduction division
   2. mitotic cell division
   3. meiotic cell division
   4. nondisjunction

5. A cell with 16 chromosomes in its nucleus forms daughter cells each of which has 16 chromosomes in its nucleus. These nuclei were produced by the process of
   1. nondisjunction
   2. meiosis
   3. mitosis
   4. polyploidy

6. What is the result of normal chromosome replication?
   1. Lost or worn-out chromosomes are replaced.
   2. Each daughter cell is provided with twice as many chromosomes as the parent cell.
   3. The exact number of centrioles is provided for spindle fiber attachment.
   4. Two identical sets of chromosomes are produced.

7. Which event occurs in the cytoplasmic division of plant cells but *not* in the cytoplasmic division of animal cells?
   1. cell plate formation
   2. centromere replication
   3. chromosome replication
   4. centriole formation

8. After normal mitotic division, how many chromosomes does each new daughter cell contain as compared to the mother cell?
   1. the same number   3. half as many
   2. twice as many   4. four times as many

9.

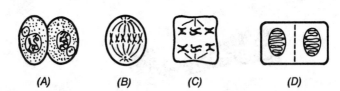

(A)   (B)   (C)   (D)

Which diagram shows the formation of a cell plate?
   1. *A*   3. *C*
   2. *B*   4. *D*

10. A human disorder characterized by rapid and abnormal cell division is known as
    1. sickle-cell anemia   3. Down's syndrome
    2. cancer   4. hemophilia

11. Within a species, the retention of the same genetic makeup in the offspring, generation after generation, is most likely the result of
    1. internal fertilization   3. asexual reproduction
    2. external fertilization   4. sexual reproduction

12. Which is a type of asexual reproduction that commonly occurs in many species of unicellular protists?
    1. external fertilization
    2. tissue regeneration
    3. binary fission
    4. vegetative propagation

13. Which organism undergoes a form of asexual reproduction characterized by an unequal division of its cytoplasm?
    1. earthworm   3. paramecium
    2. yeast   4. bean

14. Which is a form of vegetative reproduction used to propagate desirable varieties of plants quickly?
    1. hybridization   3. fertilization
    2. pollination   4. grafting

15. The diagram below illustrates which type of reproduction?

1. cleavage
2. fission
3. zygote formation
4. vegetative propagation

16. Cell *A* and cell *B* are both in the final stages of cell division. Which represent(s) budding?

Cell *A*   Cell *B*

1. cell *A* only
2. cell *B* only
3. both cell *A* and cell *B*
4. neither cell *A* nor cell *B*

17. Budding, spore formation, and vegetative propagation are methods of reproduction that involve the process of
1. mitosis   **3.** fusion of gametes
2. meiosis   **4.** crossing-over

18. A certain fruit tree is found to have desirable characteristics. These characteristics could be propagated most easily and quickly by
1. planting seeds produced by the tree
2. grafting cuttings taken from the tree
3. cross-pollinating with another tree
4. culturing highly differentiated cells of the tree

19. A fruit-grower found that one of his trees bore only seedless grapefruit. More trees of this kind could be produced by means of
1. bulbs   **3.** runners
2. grafts   **4.** special seeds

20. Compared to vertebrates, invertebrate animals exhibit a higher degree of regenerative ability because they
1. produce larger numbers of gametes
2. produce larger numbers of spindle fibers
3. possess more chromosomes in their nuclei
4. possess more undifferentiated cells

II. REPRODUCTION BY SEXUAL MEANS

Sexual reproduction involves the production of a new organism of a species from the fusion of specialized sex cells (**gametes**), which are produced in separate male and female sex organs. The fusion of the nuclei of these gametes is known as **fertilization**. The single cell that results from this fusion is a **zygote**. As in asexual reproduction, the zygote undergoes mitotic cell division to produce more cells genetically identical to it. This process is carried on somewhat differently in animals than in plants.

A. *Reproduction and Development in Animals*

1. Gametogenesis

Gametogenesis, the production of specialized sex cells, known as gametes, is the first stage of sexual reproduction.

Each body cell of an organism contains a number of chromosomes characteristic of the species. This number is known as the **diploid chromosome number**. The term "diploid" means literally the "2 number," and refers to the fact that the chromosomes in the nucleus are found in pairs with similar structure. These chromosome pairs are known as **homologous chromosomes**; they carry genes for the same traits (see Unit 5 for a more detailed treatment of this concept). The symbol "$2n$" is used to represent the diploid chromosome number.

In gametogenesis, body cells located in the sex organs (**gonads**) undergo a special type of cell division (**meiosis**) that results in the formation of sperm cells (in males) or egg cells (in females). In this process, the number of chromosomes is reduced by half.

In most animal species, sperm cells and egg cells are produced in separate (male or female) animals. In a few species, known as **hermaphrodites**, both male and female gonads are found in the same organism.

1.1. Meiosis is a special type of cell division that results in daughter cells with one-half the chromosome number of the "parent" cell. This number is known as the **monoploid chromosome number** and is represented by the symbol "n". In meiosis, homologous chromosome pairs separate and move to opposite poles of the cell before cell division. Because the number of chromosomes is "reduced" in meiosis, this type of cell division is frequently referred to as **reduction division**. Meiotic cell division occurs in two distinct and separate stages. Details of this process are as follows:

1.1.1. First Meiotic Division (see Figure 4.7)

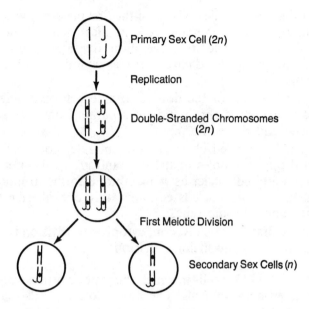

Figure 4.7   Meiosis — First Meiotic Division

- **Replication** of the cell's single-stranded chromosomes, forming double-stranded chromosomes, is the first event in the process of meiosis. As in mitosis, the chromatids resulting from this replication are chemical and genetic duplicates of the original chromosome strands.
- **Synapsis**, the second phase of meiosis, involves the close pairing of homologous double-stranded chromosomes, forming groupings of four chromatids. Such groupings are known as **tetrads**.
- Soon after synapsis, the tetrads become aligned on the spindle apparatus in preparation for the first meiotic division.
- **Disjunction** is the separation of homologous pairs into two groups as the cell enters the actual division phase. The result of disjunction is the formation of two monoploid sets of double-stranded chromosomes. These sets migrate along the spindle apparatus to opposite poles of the cell.
- Cytoplasmic division is the last step in the first meiotic division. As in mitosis, cytoplasmic division is accomplished by means of a "pinching in" of the plasma membrane, which separates the monoploid sets of double-stranded chromosomes into each of the two new cells.

1.1.2. Second Meiotic Division (see Figure 4.8)

- After the cytoplasmic division of the first meiotic division, each of the resulting monoploid cells readies itself for the second meiotic division. The first stage of this division is the alignment of the double-stranded chromosomes on a newly formed spindle apparatus.
- The centromeres of the double-stranded chromosomes replicate, allowing the chromatids to separate from each other as single-stranded chromosomes.
- The single-stranded chromosomes migrate along the spindle toward opposite poles of the cell, and group themselves into new monoploid nuclei as a membrane forms around them. Each monoploid nucleus contains only one of each type of homologous chromosome.
- As in the first meiotic division, cytoplasmic division is the final event in the second meiotic division.

The diploid ($2n$) cells in which the meiotic process begins are known as **primary sex cells**. The monoploid (n) cells that result from the meiotic process will mature into specialized reproductive cells known as **gametes**.

The exact pattern of separation of homologous pairs during the disjunction phase of meiosis is entirely random. This random distribution of homologous pairs is an important aspect in the production of variations within a species. (See Units 5 and 6 for additional discussion of the concept of variation within a species.)

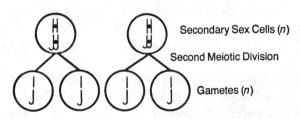

Secondary Sex Cells (n)

Second Meiotic Division

Gametes (n)

Figure 4.8 Meiosis — Second Meiotic Division

1.2. **Spermatogenesis** (see Figure 4.9) is a specific type of gametogenesis carried on in the male gonad, or **testis**. The kind of gamete produced by the testis is the **sperm** cell. In spermatogenesis, the meiotic process normally results in the production of four monoploid nuclei housed within individual cells. Each of these cells has the potential to mature into a functional, motile sperm cell containing its own monoploid nucleus.

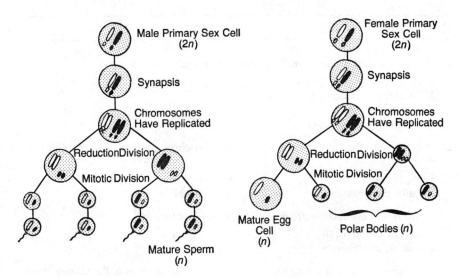

Figure 4.9   Spermatogenesis   Figure 4.10   Oogenesis

1.3.   **Oogenesis** (see Figure 4.10) is a specific type of gametogenesis carried on in the female gonad, or **ovary**. The gamete produced by the ovary is known as the **ovum** (egg cell). In oogenesis, the meiotic process normally results in the production of four monoploid nuclei housed within individual cells. However, only one of these cells has the potential to mature into a functional egg cell. The remaining cells, known as **polar bodies**, serve to separate homologous sets of chromosomes from each other to ensure a monoploid chromosome complement in the ovum. The polar bodies disintegrate after the meiotic process. The mature ovum contains food substances, including **yolk**, as well as the normal monoploid nucleus.

1.4.   Comparison of Meiosis to Mitotic Cell Division. Both mitosis and meiosis are forms of cell division. There are important similarities and differences between the two processes, which are summarized in the following chart:

	Mitosis	Meiosis
Associated with:	Asexual reproduction	Sexual reproduction
Parent cells	Diploid	Diploid

Daughter cells	Diploid
	Identical to
	  parent cell
	Two daughter cells
		Monoploid
		Specialized for
		  reproduction
		Four daughter cells

QUESTION SET 4.2

1. Meiotic cell division in animals is directly responsible for the
   1. formation of gametes
   2. fertilization of an egg
   3. growth of a cell
   4. production of muscle cells

2. Only one member of each pair of homologous chromosomes is normally found in a
   1. zygote
   2. multicellular embryo
   3. gamete
   4. cheek cell

3. Synapsis and disjunction are processes directly involved in
   1. mitotic cell division
   2. meiotic cell division
   3. fertilization
   4. fission

4. In a fruit fly in which the diploid number of chromosomes is 8, the chromosome number in each gamete is normally
   1. 16
   2. 2
   3. 8
   4. 4

5. Which is an example of sexual reproduction?
   1. zygote formation in bean plants
   2. budding in hydras
   3. spore formation in molds
   4. propagation of geraniums by cuttings

6. Homologous pairs of chromosomes are *not* normally found in
   1. zygotes
   2. body tissue cells
   3. gametes
   4. embryonic nerve cells

7. Male and female reproductive cells develop in specialized organs known as
   1. excretory glands
   2. lymph glands
   3. gametes
   4. gonads

8. The earthworm is classified as a hemaphrodite because it has the ability to
   1. produce both eggs and sperm
   2. produce eggs and develop without fertilization
   3. reproduce asexually
   4. reproduce by budding

9. A male fish produces gametes called
   1. egg cells
   2. sperm cells
   3. testes
   4. zygotes

10. A human zygote is produced from two gametes that are identical in
    1. size
    2. method of locomotion
    3. genetic composition
    4. chromosome number

11. Identical twins would result from

 1. the fertilization of two eggs
 2. eggs from the same ovary
 3. the same zygote
 4. the formation of different zygotes

12 The diagram below illustrates the important stages in the life cycle of an organism that reproduces sexually. Which processes result in the formation of cells with the monoploid number of chromosomes?

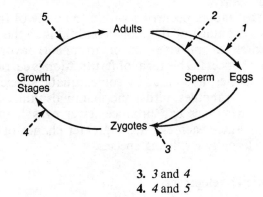

1. *1* and *2*	**3.** *3* and *4*
2. *2* and *3*	**4.** *4* and *5*

2. Fertilization

 Fertilization (see Figure 4.11) is defined as the fusion (union) of a sperm cell nucleus with an egg cell nucleus. Since both these nuclei are monoploid (n), the nucleus that results from this fusion will be diploid ($2n$). The fertilized egg cell that contains this newly formed diploid nucleus is known as a **zygote**, and is the first cell of a new living organism. The process of fertilization restores the diploid condition reduced by meiosis, since it brings together two homologous sets of chromosomes separated in the disjunction phase of meiosis. Fertilization is frequently classified according to the environment in which it occurs.

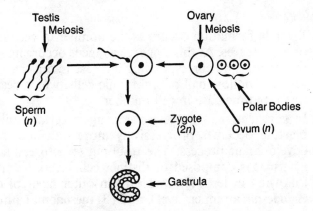

Figure 4.11 Fertilization and Early Development

2.1. External fertilization, as its name implies, occurs in the environment outside (external to) the body of the female parent. Because the fertilized egg is vulnerable to drying and mechanical damage, the only organisms that can utilize this form of fertilization are those whose natural habitats are either aquatic or moist terrestrial. Among vertebrate animal species, only fish and amphibians use external fertilization. To ensure species survival, these species normally produce large quantities of eggs, so that at least a few offspring will survive.

2.2. Internal fertilization, occurring within the body of the female parent, takes advantage of the moist conditions in the female reproductive tract to provide an environment favorable for the fertilization process. This form of fertilization is carried on by most terrestrial species, as well as by some aquatic species. Among the vertebrates, the reptiles, birds, and mammals utilize internal fertilization. Such species can produce fewer eggs than do fish and amphibians, since each egg has a greater chance of survival than the eggs externally fertilizing species.

3. Embryonic Development

3.1. Process. In **embryonic development**, after fertilization the zygote begins to undergo a series of rapid mitotic divisions. This process, known as **cleavage**, increases the number of cells in the growing cell mass. Little cell growth accompanies cleavage, resulting in a reduction of cell size with each division. A zygote undergoing cleavage is known as an **embryo**. As cleavage continues, distinct stages of development may be noted.

RD

3.1.1. Blastula formation occurs as the number of cells in the embryo increases. At this stage, cleavage is occurring so rapidly that there is no corresponding increase in cell size. The blastula is a hollow ball of embryonic cells, which results from this rapid increase in cell number.

3.1.2. In **gastrulation**, a process occurring in certain animals, the blastula becomes progressively more indented during the development process. The resulting structure, known as a **gastrula**, is composed of distinct cell layers formed in the following order: (1) **ectoderm** (an outer layer of cells), (2) **endoderm** (an inner layer), and (3) **mesoderm** (a middle cell layer).

3.1.3. Differentiation is a process by which embryonic cells become specialized to perform the various tasks of particular tissues throughout the body. As shown in the following chart, these tissues arise from the embryonic cell layers listed above:

Cell Layer	Tissues
Ectoderm	Outer skin, nervous system
Endoderm	Lining of digestive tract, lining of respiratory tract, portions of liver and pancreas
Mesoderm	Muscles, circulatory system, skeleton, excretory system, gonads

The stages in cleavage and differentiation are shown in Figure 4.12

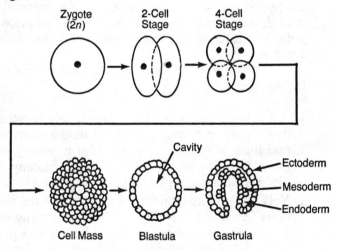

Figure 4.12   Cleavage and Differentiation

3.1.4. Growth is a process that involves increase in cell number and cell size. As differentiation continues, specialized tissues and organs develop, and the bodies of most multicellular organisms increase in size to accommodate this growth.

3.2. Site. Like fertilization, embryonic development may occur in a variety of sites, both inside and outside the mother's body.

3.2.1. External development may occur in both aquatic and terrestrial environments.

RD

a. Aquatic, or "water-based," development is common among fish and amphibians. The eggs of these species are generally thin-walled and poorly protected from mechanical damage. Since they are deposited in a moist environment, they are prevented from drying out. The water environment also provides cushioning from physical shock, as well as dissolved oxygen for respiration. The developing embryo obtains its nutrition from yolk located within the egg.

b. Terrestrial, or "land-based," development can be successfully carried on only by species whose eggs are protected by specialized membranes and other structures. Birds, reptiles, and a small number of mammal species have developed the adaptations necessary for external terrestrial development. As in external aquatic development, yolk is the primary food source for the developing embryo.

The adaptations found in the eggs of these species include the following:

- **Amnion**—a membrane that immediately surrounds the developing embryo and contains fluid known as **amniotic fluid.** Amniotic fluid provides a watery environment for the embryo, as well as cushioning it from physical shock and preventing its adhesion to the shell.

- **Yolk sac**—as its name implies, a membrane that contains the yolk food supply of the developing embryo. Blood vessels originating from the embryo penetrate the yolk sac and provide a transport route for digested yolk food to reach the embryo.

- **Allantois**—a membrane that provides a storage area for the nitrogenous wastes, principally uric acid, that are produced by the embryo during its development. The allantois also serves as a moist membrane for the absorption of dissolved oxygen during respiration.

- **Chorion**—an outer membrane that surrounds all other membranes in the terrestrial egg. The chorion is sometimes referred to as the "shell membrane," since it directly underlies the shell.

- **Shell**—a tough outer structure covering the chorion membrane of the egg. In reptiles the shell is leathery; in birds it is limey. The shell protects the embryo from mechanical damage.

    Figure 4.13 shows the development of a bird egg.

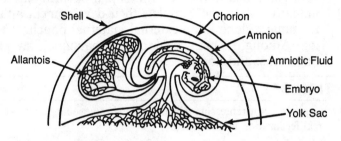

Figure 4.13   Bird Egg Development

3.2.2. Internal development, like external development, may occur in animal species living in both aquatic and terrestrial environments. Among the vertebrates, mammal species are those utilizing this form of development.

   a. Placental mammals are adapted to carry on internal development. After internal fertilization, the zygote begins this development in a specialized organ known as the **uterus**. The uterus provides an environment favorable for the development process.

   The eggs of mammal species contain little yolk, but have relatively long development periods. Mammalian embryos derive the food necessary to support this lengthy development through the **placenta**, a structure providing connection with the mother's bloodstream. The placenta is composed of tissues produced by both the mother and the embryo. Nutrients, respiratory gases, and wastes are passed between the maternal and embryonic bloodstreams in the placenta. There is no direct connection between the bloodstreams of mother and embryo in the placenta. Instead, all exchanges of materials occur by diffusion or active transport across membranes.

   Connecting the placenta to the embryo is the **umbilical cord**. The umbilical cord contains blood vessels that provide rapid transport of materials between the placenta and the embryo.

   Among the many examples of placental mammals are human beings, dogs, and whales.

b. Nonplacental mammals (marsupials) also carry on internal development, but do so without the aid of the placental connection, and therefore without direct maternal nourishment. Instead, marsupial eggs remain unattached to the mother's uterus and derive their nourishment from the small amount of yolk contained in the egg. Marsupial offspring are born in an immature state and complete their development attached to a mammary gland in an external maternal pouch.

Among the many examples of marsupials are kangaroos, opossums, and koalas.

QUESTION SET 4.3

1. Which equation best represents the change in chromosome number in the process of normal fertilization in animals?
   1. $n + 2n \rightarrow 3n$
   2. $n + n \rightarrow 2n$
   3. $n \rightarrow n$
   4. $2n + 2n \rightarrow 4n$

2. The fusion of monoploid gametes restores the diploid condition in organisms that reproduce
   1. asexually
   2. sexually
   3. vegetatively
   4. mitotically

3. In frogs, zygote formation normally occurs in the
   1. ovary
   2. oviduct
   3. external environment
   4. internal environment

4. In the early development of a zygote, the number of cells increases without an increase in mass by a process known as
   1. ovulation
   2. cleavage
   3. germination
   4. metamorphosis

5. The hollow-ball stage in the development of an invertebrate embryo is known as the
   1. blastula
   2. ectoderm
   3. gastrula
   4. endoderm

6. In a developing embryo, the process most closely associated with the differentiation of cells is
   1. gastrulation
   2. menstruation
   3. ovulation
   4. fertilization

7. From which embryonic layer do muscle and bone develop?
   1. epidermis
   2. mesoderm
   3. ectoderm
   4. endoderm

8. In most species of fish, a female produces large numbers of eggs during a reproductive cycle. This would indicate that reproduction in fish is most probably characterized by
   1. internal fertilization and internal embryonic development
   2. internal fertilization and external embryonic development
   3. external fertilization and internal embryonic development
   4. external fertilization and external embryonic development

9. Which type of fertilization and development is exhibited by birds and many reptiles?
   1. external fertilization and external development
   2. internal fertilization and internal development
   3. external fertilization and internal development
   4. internal fertilization and external development

10. Which structure is a source of food for embryos that develop externally?
    1. yolk
    2. placenta
    3. chorion
    4. amnion

11. Which embryonic membrane of a bird's egg contains metabolic wastes?
    1. allantois
    2. amnion
    3. placenta
    4. yolk sac

12 Embryos of both sea and land animals develop in a watery environment. The fluid for the developing land animal is found within the
    1. umbilical cord
    2. yolk sac
    3. amnion
    4. allantois

13. External fertilization and external development are characteristic of most
    1. fishes
    2. birds
    3. reptiles
    4. mammals

14. The embryos of marsupials, such as the kangaroo and opossum, complete their development externally. What is the source of nutrition for the last stages of a marsupial embryo's development?
    1. milk from maternal mammary glands
    2. nutrients diffused through the uterine wall
    3. concentrated food in the yolk stored in the egg
    4. food gathered from the environment and fed to the embryo

15. In mammals, the placenta is essential to the embryo for
    1. nutrition, reproduction, and growth
    2. nutrition, respiration, and excretion
    3. locomotion, respiration, and excretion
    4. nutrition, reproduction, and excretion

16. In which reproductive pattern does an egg cell of a vertebrate have the best chance of developing into an embryo and completing this stage of development?
    1. external fertilization and external development
    2. external fertilization and internal development
    3. internal fertilization and external development
    4. internal fertilization and internal development

17. The uterine lining of most mammals is prepared to receive a developing embryo by hormones secreted by the
    1. ovaries
    2. oviducts
    3. vagina
    4. pancreas

18. Which organ of a cow is used to feed a new-born calf?
    1. a mammary gland
    2. the uterus
    3. a placenta
    4. the gonad

RD

B. Human Reproduction and Development

1. Gametogenesis

1.1. In Males. In **gametogenesis**, sperm cells, the male gametes, are produced by meiosis in the male gonad, known as the **testis**. Two testes are located in an external pouch, the **scrotum**, which extends from the wall of the lower abdomen. Temperatures in the scrotum are generally 1 to 2 degrees (Celsius) cooler than normal body temperature. This reduced temperature is known to be optimum for sperm production.

When first formed, the sperm is inactive and is stored in the testis until activated and released. Once released, the sperm travels along a series of tubes to the exterior of the body. Along its route, a number of glands add various fluids to the sperm to activate it and increase its volume.

The **urethra** is a tube that carries the activated sperm along the last portion of its journey in the body. The urethra extends from the urinary bladder to the exterior of the body through the **penis**. In human beings, the penis is the structure that permits internal fertilization, through direct implantation of sperm into the female reproductive tract. The male reproductive system is diagramed in Figure 4.14.

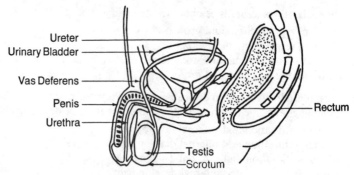

Figure 4.14   Human Reproductive System: Male

In addition to their sperm-producing role, the testes also act as endocrine glands, producing the male hormone **testosterone**, which controls the expression of male **secondary sex characteristics**. These sex characteristics include growth of hair on the face and body, deepening of the voice pitch, sperm production, and certain skeletal and muscular changes.

1.2. In Females. Ova (egg cells), the female gametes, are produced by meiosis in the female gonad, known as the **ovary**. Two ovaries are located in the lower abdomen.

At birth, each female child has already produced more egg cells than she will ever release in her lifetime. After it has been formed, the egg cell is stored in the **follicle** of the ovary until it matures and is released. Once released, the egg cell travels along a short tube, the **oviduct**, to the **uterus**. In human beings, internal fertilization normally occurs in the oviduct, also known as the **Fallopian tube**, in the upper end of the female reproductive tract.

In human beings the uterus acts as a development chamber and is a muscular organ bounded on the lower end by the **cervix**. The lower end of the female reproductive tract consists of the **vagina**, which functions to receive sperm during intercourse and later as the birth canal. The female reproductive system is diagramed in Figure 4.15.

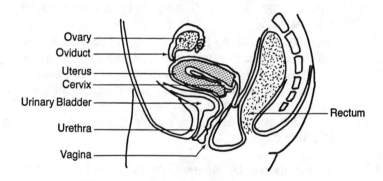

Figure 4.15 Human Reproductive System: Female

In addition to their egg-producing role, the ovaries also act as endocrine glands, producing the female hormones **estrogen** and **progesterone**, which control the expression of female **secondary sex characteristics**. These sex characteristics include growth of body hair, enlargement of the **mammary glands** (breasts), broadening of the pelvis, and certain other skeletal and muscular changes.

The **menstrual cycle**, whch begins at puberty as one of the secondary sex characteristics, is a hormone-controlled process in the human female that is responsible for the monthly release of mature eggs. The average duration of the menstrual cycle is 28 days, although the actual length varies from woman to woman.

The four stages of the menstrual cycle are as follows:

- **Follicle stage.** As the cycle begins, a single egg matures within the ovarian follicle under the influence of **follicle stimulating hormone,** or **FSH,** produced by the pituitary gland. The ovary begins to produce estrogen, which affects the lining of the uterus by stimulating its thickening and vascularization. This portion of the cycle requires approximately 14 days to complete.

- **Ovulation.** This portion of the cycle involves the release of the mature egg from the follicle into the oviduct. On average, this event occurs about day 14 of the cycle.

- **Corpus luteum stage.** After ovulation the cells that made up the ovarian follicle begin to change under the influence of **lutinizing hormone,** or **LH,** which is produced by the pituitary gland. The resulting structure, known as the **corpus luteum,** secretes a hormone, progesterone, which helps to ready the uterine lining for the possible inplantation of a fertilized egg. This portion of the cycle continues for about 8 to 10 days.

- **Menstruation.** If no fertilized egg is received in the uterus within a few days after ovulation, the uterine lining begins to break down. The disintegrated tissue and blood are expelled from the body via the vaginal canal in the process known as menstruation.

  This aspect of the cycle normally occurs over a period of a few days.

1.3. The **interaction of hormones** in the menstrual cycle provides an excellent illustration of the body's feedback mechanisms. The concept of "feedback" simply refers to the stimulation of one gland by the secretions of another gland. Initially, the hypothalamus, a small glandular structure at the base of the brain, secretes hormones that stimulate the pituitary gland (also near the brain) to release FSH. FSH stimulates the ovary to produce estrogen. Estrogen, in turn, makes its way back to the pituitary to stop the release of FSH and start the release of LH. LH stimulates the ovary (corpus luteum) to secrete progesterone, while inhibiting the release of estrogen. Progesterone, in turn, further depresses the production of FSH and LH.

This situation, in which high concentrations of certain hormones depress the production of other hormones, is an example of **negative feedback** control.

2. Fertilization

In human beings, **fertilization** normally occurs in the upper end of the oviduct within 24 hours of ovulation. If fertilization does not occur within this time period, the egg will usually deteriorate and be shed with the menstrual flow.

If fertilization is successful, the zygote undergoes the first of several cleavage divisions it will experience over the next 6 to 10 days as it continues down the oviduct to the uterus. Upon reaching the uterus, the embryo implants itself (see Figure 4.16) into the blood-rich lining of the uterus and begins to establish the placental connection.

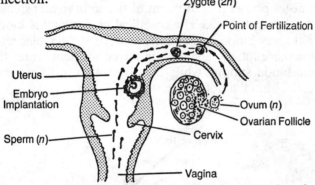

Figure 4.16   Human Reproduction: Fertilization and Implantation

A multiple birth (e.g., twins, triplets) may result if two eggs are released simultaneously and are fertilized separately (**fraternal twins**). It may also occur if a zygote, during cleavage, separates into two identical masses, the cells of which have identical genetic traits (**identical twins**).

Recent research into the process of fertilization has led to the development of techniques that allow fertilization to be accomplished outside the body in a laboratory. The fertilized eggs are then reimplanted into the mother's uterus and allowed to develop normally. Such fertilization, known as *in vitro* fertilization, may enable previously childless couples to have children.

3. Development

3.1. Prenatal development (see Figure 4.17) includes all the processes of embryonic development that occur before birth. Prenatal development begins as soon as the egg is fertilized and the resulting zygote begins to undergo cleavage. Within 10 days after fertilization, the ball of cells resulting from cleavage, now known as an embryo, implants itself in the lining of the uterus. Once embedded in the uterine lining, the embryo undergoes a process resembling gastrulation. At the same time, the cells of the embryo begin to differentiate into the specialized tissues and organs of the adult organism.

The tissues of the embryo and the mother grow together to form the **placenta**, the connection that allows nutrients and oxygen to pass from mother to embryo during the development process. Wastes such as carbon dioxide and urea pass by diffusion from the embryo's blood to that of the mother for excretion through the mother's excretory processes. The **amnion** is a membrane that surrounds the developing embryo and contains a fluid (**amniotic fluid**) that bathes it. Amniotic fluid promotes proper development of the embryo and shields it from mechanical shock. The **umbilical cord** contains blood vessels that carry blood between the embryo and the placenta. Within the placenta, oxygen and dissolved nutrients enter the embryonic blood supply, while dissolved wastes diffuse into the maternal blood supply.

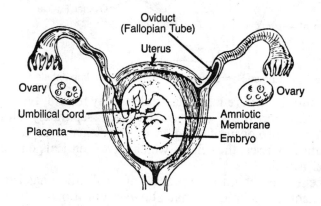

Figure 4.17 Placental Development

3.2. Birth occurs after a period of development, known as the **gestation** period, of approximately 9 months. Strong contractions of the uterus force the baby through the cervix and vagina to the outside of the body, where the baby begins to breathe, eat, and excrete wastes on his/her own.

3.3. Postnatal development refers to the continued growth and development of the baby after birth. During this process, different parts of the body grow at different rates at different times. Even after active growth has ceased in young adulthood, the tissues of the body continue to change.

The **aging** process involves a complex series of changes that continue throughout the life of the individual. Aging occurs differently in different people and is thought to be dependent on the interaction of factors such as environment and heredity.

Death is the final stage of the human life cycle. Although there is considerable debate over its exact definition, many physicians agree that irreversible cessation of the brain's activities signals the time of death.

QUESTION SET 4.4

1. In human beings, immediately after ovulation the egg normally enters the
   RD 1. follicle sac   3. oviduct
       2. cervix   4. uterus

2. Which group of terms indicates the correct sequence of stages in a menstrual cycle?
   RD 1. follicle stage, ovulation, corpus luteum stage, menstruation
       2. menstruation, corpus luteum stage, follicle stage, ovulation
       3. ovulation, menstruation, follicle stage, corpus luteum stage
       4. follicle stage, menstruation, ovulation, corpus luteum stage

3. The human male's testes are located in an outpocketing of the body wall known as
   RD the scrotum. An advantage of this adaptation is that
       1. the testes are better protected in the scrotum than in the body cavity
       2. sperm production requires contact with atmospheric air
       3. a temperature lower than body temperature is best for sperm production and storage
       4. the sperm cells can enter the urethra directly from the testes

4. In human males, the maximum number of functional sperm cells that is normally
   RD produced from each primary sex cell is
       1. 1   3. 3
       2. 2   4. 4

5. The liquid that contains male gametes and secretions from glands of the human
   RD male reproductive system is known as
       1. sperm   3. progesterone
       2. testosterone   4. semen

6. In which part of the human reproductive system does the development of an embryo
   RD normally occur?
       1. scrotum   3. ovary
       2. uterus   4. testes

Base your answers to questions 7 through 9 on the diagram below and on your knowledge of biology.

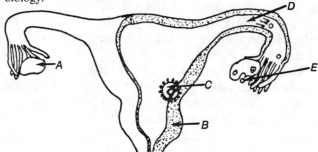

7. Which letter indicates the normal site of most prenatal development?

RD
1. *A*	**3.** *C*
2. *E*	**4.** *D*

8. What event is occurring at *E*?

RD
1. sperm production	**3.** ovulation
2. menstruation	**4.** gastrulation

9. Which letter indicates the structure that produces the hormones estrogen and RD progesterone?

1. *A*	**3.** *C*
2. *B*	**4.** *D*

---

10. A complete human gestation period normally occurs between the times of

RD
1. blastula formation and implantation
2. fertilization and birth
3. ovulation and cleavage
4. birth and puberty

11. The technique of uniting sperm cell with an egg cell in a test tube is an example of

RD
1. *in vitro* fertilization	**3.** gametogenesis
2. internal fertilization	**4.** artificial ovulation

C. Reproduction and Development in Flowering Plants

Flowering plants carry on a sexual reproductive process similar in many respects to that of animals. The plant structure specialized for sexual reproduction is the flower.

1. Flower Structure

Flower structure (see Figure 4.18) is specially adapted for carrying out the process of sexual reproduction. Many flowers contain both male and female reproductive organs. Others may contain only the male structures, while still others may house only the female reproductive structures. Flowers may or may not have accessory structures, such as **petals** or **sepals**.

The male reproductive organ is the **stamen**. The upper portion of the stamen is the **anther**, which produces **pollen** grains (the male gamete). Pollen grains contain monoploid nuclei that have resulted from the process of meiosis. The lower portion of the stamen is the **filament**, which holds the anther in the proper position for pollen distribution.

The female reproductive organ is the **pistil**. The upper portion of the pistil is the **stigma**, which serves to catch pollen during pollination. The lower structure in the pistil is the **ovary**, containing monoploid egg nuclei housed in **ovules**. Connecting the stigma to the ovary is a middle portion of the pistil known as the **style**.

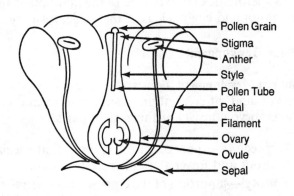

Pollen Grain
Stigma
Anther
Style
Pollen Tube
Petal
Filament
Ovary
Ovule
Sepal

Figure 4.18   Flower Structure

2.   Pollination

Pollination is the actual transfer of pollen grains from the anther to the stigma. If this transfer occurs within a single flower or between two flowers on the same plant, the process is referred to as **self-pollination**. If the transfer is accomplished between two flowers located on different plants, it is known as **cross-pollination**. Self-pollination tends to limit genetic variation, since the genetic makeups of the pollen grain nucleus and the egg nucleus are very similar. Cross-pollination, on the other hand, permits the mixing of traits of plants with different genetic makeups, increasing the likelihood of genetic variation in the offspring.

Agents of pollination include wind, water, insects, and animal carriers. These agents function to transport pollen from the anther of one flower to the stigma of another. The petals, fragrances, and nectars of flowers often serve as attractants to insect and animal carriers.

The tough, waterproof coat of the pollen grain is a special adaptation designed to help it survive the dry, often hostile conditions encountered in terrestrial environments. Without this coat, the cellular contents of the pollen grain would quickly dry up and die. Pollen grains have been known to survive for many years under such conditions and still be able to germinate.

After pollination, the pollen grain germinates. **Germination** involves the growth of the pollen grain and the formation of a **pollen tube**, which grows down into the tissues of the pistil toward the ovule. During the growth of the pollen tube, the monoploid nucleus divides into two sperm nuclei, one of which will fertilize the egg nucleus in the ovule when the pollen tube reaches it. The pollen tube may be thought of as an adaptation for "internal fertilization" in plants.

3. Fertilization

Fertilization occurs when the pollen tube reaches the ovule and releases its sperm nuclei into it. One of the monoploid sperm nuclei fuses with the monoploid egg nucleus in the ovule, forming a diploid zygote.

4. Plant Development

4.1. Embryo development consists of the processes of cell division, growth, and differentiation, resulting in formation of the embryo. The embryo consists of three main parts:

- **Hypocotyl**—a portion of the embryo that specializes to become the root and lower stem of the adult plant.
- **Epicotyl**—a portion of the embryo that specializes to become the upper stem, leaves, and other parts derived from the stem.
- **Cotyledon**—a portion of the embryo that specializes to become the storage area for nutrients necessary to support the growing plant after germination. The cotyledons of most plant species are temporary and quickly shrivel and drop from the plant after their food supply is depleted. Plants may contain either one or two cotyledons, depending on the type of plant.

4.2. Seed formation occurs as the embryo is developing. The coats of the ovules develop into the seed coat, a protective covering for the seed interior. The ovule itself develops into the seed, containing the parts listed above. The seeds are normally enclosed within a **fruit,** which develops from the tissues of the ovary after fertilization. The parts of the seed are shown in Figure 4.19.

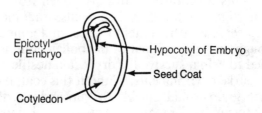

Figure 4.19 Seed Structure

4.3. **Seed dispersal** is accomplished in different ways, depending on the species of plant involved. Flowering plants enclose their seeds inside **fruits**, which aid in the dispersal process. Some fruits are eaten by animals, and seeds are dispersed after passing through the animal's digestive tract. Some seeds are specialized for transport by wind or water or in the fur or feathers of wildlife species to sites remote from the parent plant. Seed dispersal is important to the survival of the plant species, since it allows the plant to avoid overcrowding in the original site as well as to enlarge its range.

5. Germination

Germination involves cell division and growth, as well as the differentiation of cells into specialized tissues. This rapid cell growth results in the formation of new root and stem tissues, which grow from the seed and establish themselves in the environment. Germination occurs when a seed has been successfully dispersed into an environment favorable to growth. Conditions required for most seeds include sufficient moisture, proper temperature for efficient enzyme operation, and adequate levels of oxygen for the respiratory process.

6. Growth

Growth in plants is limited to particular areas of the plant. Specialized growth tissues are known as **meristems** and are located in two main regions of the plant:

6.1. **Apical meristems.** These growth regions are located in the stem root tips. Apical meristems are responsible for the annual growth in stem length, as well as the deep penetration of roots into the soil. The apical meristem is shown in Figure 4.20

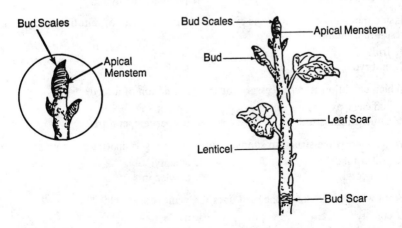

Figure 4.20   Stem (Bud) Structure

6.2. Lateral meristems. These growth areas are located just under the bark layers in woody plants and are responsible for the annual growth in stem and root diameter. The lateral meristem growth area is often called the **cambium** and is located between the xylem tissues and the phloem tissues.

All of the specialized parts of the plant are derived from the undifferentiated tissues of the plant meristems. Once cell division has occurred, the new cells elongate, mature, and take on the specialized tasks of the specialized tissues of which they are a part. Specialized tissues such as those previously described in the leaf, stem, root, and flower are derived in this manner.

QUESTION SET 4.5

---

1. In a flowering plant, the ovule develops within a part of the
   1. style
   2. anther
   3. pistil
   4. stigma

2. In flowering plants, sperm nuclei are formed in
   1. testes
   2. ovules
   3. pollen tubes
   4. polar bodies

3. Which structure produces pollen grains?
   1. anther
   2. stigma
   3. pollen tube
   4. ovule

4. In apple blossoms, the function of the stigma is to
   1. produce the pollen
   2. receive the pollen
   3. form sperm nuclei
   4. pollinate the ovule

5. Inside the ovule of a plant, the zygote undergoes a series of mitotic divisions that result in a multicellular
   1. fruit
   2. embryo
   3. spore
   4. seed

6. Which condition is *not* necessary for the germination of a seed?
   1. sufficient oxygen
   2. sufficient moisture
   3. proper soil
   4. proper temperature

7. Which embryonic structure supplies nutrients to a germinating bean plant?
   1. pollen tube
   2. hypocotyl
   3. epicotyl
   4. cotyledon

8. From which structure in the seed does the plant root develop?
   1. epicotyl
   2. cotyledon
   3. ovule
   4. hypocotyl

Base your answers to questions 9 through 11 on the diagram of the flower below and your knowledge of biology.

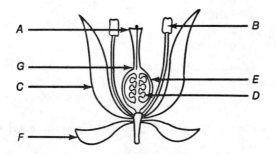

9. The process of meiosis occurs in structures
    1. *A* and *B*   **3.** *A* and *D*
    2. *B* and *C*   **4.** *B* and *D*

10. Which structure could develop into a seed?
    1. *A*   **3.** *E*
    2. *B*   **4.** *D*

11. A germinating pollen grain is represented by
    1. *E*   **3.** *G*
    2. *B*   **4.** *D*

12. In a bean seed, the part of the embryo that develops into the leaves and upper portion of the stem is known as the
    1. seed coat   **3.** hypocotyl
    2. epicotyl   **4.** cotyledon

13. The apical meristems of plants are located
    1. between the xylem and the phloem tissue of woody stems
    2. at the tips of roots and stems
    3. uniformly throughout the plant
    4. within root hairs of plants

UNIT 4 REVIEW

1. When a cell with 24 chromosomes divides by mitotic cell division, each of the resulting daughter cells will have a maximum chromosome number of
    1. 6   **3.** 24
    2. 12   **4.** 48

2. In human skin cells, the products of a normal mitotic cell division are
    1. 4 diploid cells   **3.** 2 monoploid cells
    2. 2 diploid cells   **4.** 4 monoploid cells

3. In an animal cell undergoing mitotic cell division, which of the following events does *not* normally occur?
   1. cell plate formation
   2. spindle fiber production
   3. centromere replication
   4. chromosomal disjunction

4. Each of the two daughter cells that result from the normal mitotic division of the original parent cell contains
   1. the same number of chromosomes, but has genes different from those of the parent cell
   2. the same number of chromosomes and has genes identical to those of the parent cell
   3. one-half the number of chromosomes, but has genes different from those of the parent cell
   4. one-half the number of chromosomes and has genes identical to those of the parent cell

5. The following list describes some of the events associated with normal cell division.

   A—Nuclear membrane formation around each set of newly formed chromosomes

   B—Separation of centromeres

   C—Replication of each chromosome

   D—Movement of single-stranded chromosomes to opposite ends of the spindle

   What is the normal sequence in which these events occur?
   1. $A \to B \to C \to D$
   2. $C \to B \to D \to A$
   3. $C \to D \to B \to A$
   4. $D \to C \to A \to B$

6. The diagram at the right represents a cell that will undergo mitosis. Which diagrams below best illustrate the nuclei of the daughter cells that result from a normal mitotic cell division of the parent cell shown?

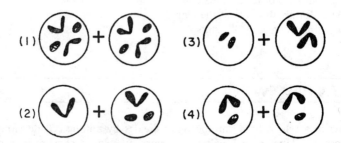

7. Double-stranded chromosomes are produced as a direct result of the
   1. synapsis of homologous chromosomes
   2. formation of spindle fibers
   3. replication of chromosomes
   4. formation of cell plates

8. The diagram below represents reproduction in a paramecium. Which specific process is represented by the diagram?

    1. sporulation   **3.** budding

    2. binary fission   **4.** regeneration

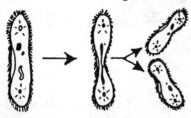

9. One apple tree can bear such varieties as Delicious, McIntosh, and Cortland apples on its branches. These three varieties of apples grow on the same tree as a result of a type of vegetation propagation known as

    1. cross-pollination   **3.** grafting

    2. fertilization   **4.** binary fission

10. Two organisms that can reproduce by budding are

    1. yeast and hydra

    2. yeast and earthworm

    3. bread mold and grasshopper

    4. grasshopper and goldfish

11. Mitotic and meiotic cell division are similar in that both processes

    1. produce diploid gametes from monoploid cells

    2. produce monoploid gametes from diploid cells

    3. involve synapsis of homologous chromosomes

    4. involve replication of chromosomes

12. The chromosome number in an egg cell nucleus of a plant is 14. The normal chromosome number in a root epidermal cell of the same plant is

    1. 7   **3.** 21

    2. 14   **4.** 28

13. The diagrams below represent the sequence of events in a cell undergoing normal meiotic cell division.

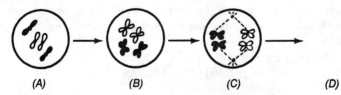

Which diagram most likely represents *D* of this sequence?

        3

14. During the normal meiotic division of a diploid cell, the change in chromosome number that occurs is represented as
   1. $4n \to n$
   2. $2n \to 4n$
   3. $2n \to n$
   4. $n \to \frac{1}{2}n$

15. In animals, polar bodies are formed as a result of
   1. meiotic cell division in females
   2. meiotic cell division in males
   3. mitotic cell division in females
   4. mitotic cell division in males

16. In a species of corn, the diploid number of chromosomes is 20. What is the number of chromosomes found in each of the normal egg cells produced by this species?
   1. 5
   2. 10
   3. 20
   4. 40

17. Cells with the monoploid number of chromosomes would normally be found in the
   1. stem of a dandelion
   2. liver cells of a horse
   3. skin cells of a human being
   4. anthers of a rose

18. Organisms that contain both functional male and female gonads are known as
   1. hybrids
   2. hermaphrodites
   3. phagocytes
   4. parasites

19. As a fertilized egg develops into an embryo, it undergoes a series of mitotic divisions known as
   1. parthenogenesis
   2. meiosis
   3. reduction division
   4. cleavage

20. Which structure is formed as a result of the process of cleavage?

   1. egg cell
   2. sperm cell
   3. zygote
   4. blastula

21. In nonplacental animals, most of the food for the embryo is found in the
   1. uterus
   4. umbilical cord
   3. amniotic sac
   4. yolk of the egg

22. The female of a certain species of fish releases several thousand eggs during reproduction. Based on this information, it would be reasonable to assume that this species
   1. provides a protective shell for the egg
   2. provides considerable parental care for its eggs
   3. has internal fertilization of the egg
   4. has a low survival rate for its eggs

23. In the diagrams below, $2n$ represents the diploid number of chromosomes in a cell of an organism, and n represents the monoploid number. Which diagram represents fertilization?

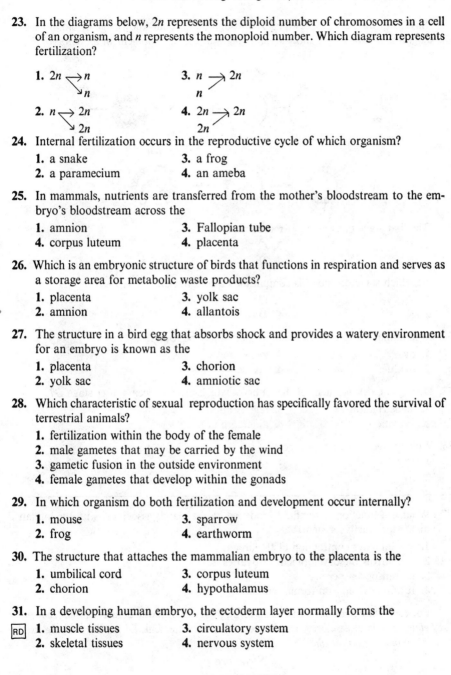

    1. $2n \searrow n$ **3.** $n \nearrow 2n$
            $\searrow n$ n

    2. $n \searrow 2n$ **4.** $2n \nearrow 2n$
             $\searrow 2n$ $2n$

24. Internal fertilization occurs in the reproductive cycle of which organism?
    1. a snake   **3.** a frog
    2. a paramecium   **4.** an ameba

25. In mammals, nutrients are transferred from the mother's bloodstream to the embryo's bloodstream across the
    1. amnion   **3.** Fallopian tube
    4. corpus luteum   **4.** placenta

26. Which is an embryonic structure of birds that functions in respiration and serves as a storage area for metabolic waste products?
    1. placenta   **3.** yolk sac
    2. amnion   **4.** allantois

27. The structure in a bird egg that absorbs shock and provides a watery environment for an embryo is known as the
    1. placenta   **3.** chorion
    2. yolk sac   **4.** amniotic sac

28. Which characteristic of sexual reproduction has specifically favored the survival of terrestrial animals?
    1. fertilization within the body of the female
    2. male gametes that may be carried by the wind
    3. gametic fusion in the outside environment
    4. female gametes that develop within the gonads

29. In which organism do both fertilization and development occur internally?
    1. mouse   **3.** sparrow
    2. frog   **4.** earthworm

30. The structure that attaches the mammalian embryo to the placenta is the
    1. umbilical cord   **3.** corpus luteum
    2. chorion   **4.** hypothalamus

31. In a developing human embryo, the ectoderm layer normally forms the
    RD **1.** muscle tissues   **3.** circulatory system
    2. skeletal tissues   **4.** nervous system

Base your answers to questions 31 through 35 on your knowledge of biology and the diagrams below, which show several stages in the development of an embryo.

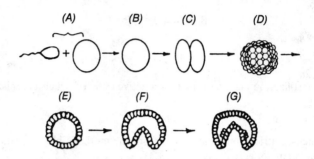

32. The first stage of cleavage is represented by
1. *E*   3. *C*
2. *B*   4. *D*

33. At which stage do the cells contain monoploid nuclei?
1. *A*   3. *C*
2. *F*   4. *D*

34. In mammals, where would stage *B* most likely be found?
[RD] 1. ovary   3. testis
2. oviduct   4. uterus

35. During the development of the embryo, which process occurs after stage *G*?
[RD] 1. menstruation   3. regeneration
2. ovulation   4. differentiation

36. Which stage represents a zygote?
1. *E*   3. *G*
2. *B*   4. *D*

---

37. What is a function of the fluid produced by the glands associated with the human
[RD] male reproductive system?
1. It stimulates contraction of the ureter.
2. It controls the development of ovaries.
3. It transports sperm.
4. It initiates nephron formation.

For each statement in questions 38 through 42 select the structure, *chosen from the list below*, that is most closely associated with that statement. (*A number may be used more than once or not at all.*)

Structure

(1)   Uterus
(2)   Ovary
(3)   Scrotum
(4)   Urethra
(5)   Oviduct

38. This organ produces monoploid cells.
RD

39. This structure provides an optimum temperature for sperm production and storage.
RD

40. In males, this structure serves both an excretory and a reproductive function.
RD

41. This structure serves as the normal site of fertilization.
RD

42. This organ produces the hormone estrogen.
RD

---

43. In human males, for every primary sex cell that undergoes meiosis, the number of
RD sperm cells produced is usually
   1. 1   **3.** 3
   2. 2   **4.** 4

44. In human beings, before fertilization occurs, a sperm travels from the
RD **1.** uterus to the vagina to the oviduct
   2. vagina to the oviduct to the uterus
   3. vagina to the uterus to the oviduct
   4. vagina to the ovary to the follicle

45. In human beings, gestation normally occurs while the developing organism is in the
RD **1.** uterus   **3.** vagina
   2. follicle   **4.** ovary

46. The cyclical nature of the menstrual cycle in the human female is maintained by the
RD hormonal feedback system between the ovaries and the
   1. pituitary   **3.** vagina
   2. placenta   **4.** oviducts

47. Flowers whose reproductive structures consist only of stamens are able to produce
   1. fruits with seeds   **3.** pollen
   2. fruits without seeds   **4.** ovules

48. The production of sperm nuclei in plants occurs in cells from the
   1. anther   **3.** pistil
   2. stigma   **4.** ovary

49. Heavy use of insecticides in springtime may lead to a decrease in apple production.
   This decreased apple production is most probably due to interference with the
   process of
   1. pollination   **3.** absorption
   2. cleavage   **4.** transpiration

50. Which are examples of male reproductive organs?
   1. stamens and pistils
   2. pistils and ovaries
   3. pistils and testes
   4. stamens and testes

51. Which reproductive structures are produced within the ovaries of plants?
    1. pollen grains
    2. sperm nuclei
    3. egg nuclei
    4. pollen tubes

52. Which structure of a seed provides the developing plant embryo with food?
    1. pollen grain
    2. epicotyl
    3. hypocotyl
    4. cotyledon

53. Two flowering holly plants grew in the same yard. Each year fruit developed on one plant, but never on the other. A correct explanation for this observation would be that the plant that never produced fruit
    1. had flowers with stamens only
    2. had flowers with pistils only
    3. produced cones
    4. produced spores

54. The factors necessary for maple seed germination are moisture, proper temperature, and
    1. oxygen
    2. soil
    3. chlorophyll
    4. darkness

55. In seed plants, an increase in the number of cells is restricted largely to specific regions known as
    1. stomates
    2. amnions
    3. endoderms
    4. meristems

HOW DO SPECIES TRANSMIT TRAITS FROM GENERATION TO GENERATION?

> **KEY IDEAS**
>
> This unit focuses on the methods by which living things maintain genetic continuity. The mechanisms by which genetic information is passed on to future generations, the patterns of genetic inheritance, the roles played by DNA and RNA in genetic inheritance, and the effect of genetic variation on species populations are emphasized.

KEY OBJECTIVES

Upon completion of this unit, you should be able to:

☐ Trace the history of genetic research from the work of Gregor Mendel to that of T. H. Morgan.

☐ Describe the mechanisms governing the transmission of traits in terms of the gene-chromosome theory.

☐ Describe the most common patterns of inheritance found in living things, and predict outcomes of genetic crosses in each pattern.

☐ List and describe the principal forms of mutation as they relate to the production of variations and genetic disorders.

☐ List some significant techniques of genetic research, and describe how they are used to aid in genetic counseling.

☐ Describe practical applications of theoretical genetics in the areas of plant and animal breeding.

☐ Relate how the environment may influence the expression of genetic traits in human beings and other organisms

☐ Describe a few genetically related disorders that affect human beings.

☐ Recognize the structure of the DNA molecule, and describe its role in the processes of replication and protein synthesis.

☐ Describe the factors of inheritance studied in population genetics and the role of the Hardy-Weinberg principle in predicting gene frequencies.

I. THE HISTORICAL DEVELOPMENT OF GENETIC THEORY

A. Mendelian Principles

Gregor Mendel, an Austrian who lived in the nineteenth century, accomplished fundamental studies in genetic theory that even today serve as the basis for modern concepts of genetics. Mendel's experimental organism was the **garden pea**, a species with a number of readily observable genetic traits, and one that could be easily propagated in large numbers in a garden "laboratory."

Mendel's experiments involved cross-pollinating different pea plants displaying specific, contrasting genetic traits. He wanted to see how these traits were inherited by the offspring plants of these crosses. Through his experiments, Mendel was able to establish mathematical ratios developed from comparing these offspring and making simple counts of the contrasting traits that they displayed. His ratios described the inheritance of genetic "factors" controlling these traits in terms of probability. It should be noted that Mendel's experimental work was accomplished before the discovery of genes and chromosomes.

B. Gene-chromosome Theory

The **gene-chromosome theory** of inheritance is a modern view of the way in which genetic inheritance is accomplished in living things. It is based in large measure on the work of Gregor Mendel in the nineteenth century, but adds new experimental evidence gained from studies of many different organisms. One of the most important experimental organisms in modern genetic studies is *Drosophila*, the common fruit fly. The fruit fly is ideal for use in such studies because of its small size, large number of offspring, and short generation period. These studies have enabled biologists to link the patterns of inheritance of genetic traits in the offspring of sexual reproduction to the movement of chromosomes during meiotic cell division.

Mendel's "factors," now known as **genes**, are thought to exist as discrete portions (loci) of chromosomes. It is believed that pairs of homologous chromosomes contain linear, matching arrangements of genes exerting parallel control over the same traits. Pairs of genes that exercise such parallel control over the same traits are known as **alleles** (see Figure 5.1). The control mechanism used by genes in genetic inheritance is now thought to be chemical in nature and is described in greater detail in Section VII of this chapter.

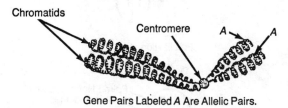

Gene Pairs Labeled *A* Are Allelic Pairs.

Figure 5.1   Double-Stranded Chromosome Structure

II.   SOME MAJOR GENETIC PATTERNS

A.   Dominance

The principle of **dominance** was first described by Mendel in his classic studies of inheritance patterns. It is known that, when genes of certain allelic pairs have contrasting effects on the same trait, only one may be expressed, while the other is "masked." The gene that is expressed in this case is known as the **dominant** allele; the masked gene is referred to as the **recessive** allele.

It is sometimes difficult to imagine what a gene is like and how it works. To make it easier to visualize the gene and to work out problems of genetic inheritance, scientists use symbols to represent genes. In working with genetics problems, certain conventions in the use of these symbols are followed to ensure that the results are readily understood. In dominance problems dominant alleles are usually symbolized by a capital italic letter (e.g., *A*), while the recessive allele for the same trait is normally assigned the lower case of the same italic letter (e.g., *a*). In the garden pea, the allele for "tall" plants is dominant over the allele for "short" plants. The letter commonly designated for the "tall" allele is *T*; the letter designation for the "short" allele is *t*.

In Unit 4, we learned that fertilization (see Figure 5.2) brings together homologous chromosomes separated in meiosis. Since the chromosome complement of the zygote includes two homologues of each type, two alleles for each trait are almost always inherited. When the two inherited alleles are alike (e.g., *TT* or *tt*), the combination is said to be **homozygous**. When the two members of the allelic pair are different (*Tt*), the combination is said to be **heterozygous** (or **hybrid**).

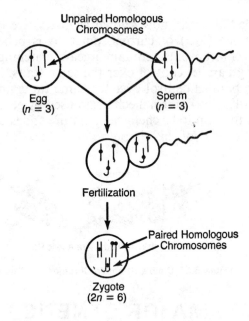

Figure 5.2   Fertilization

When the cross is accomplished between a homozygous dominant parent and a homozygous recessive parent, all the offspring will have the same **phenotype** (physical appearance) as the parent expressing the dominant characteristic in regard to that trait. This illustrates the principle of dominance.

However, all these offspring have a heterozygous **genotype** (gene combination), since they will have inherited the dominant gene from the "dominant" parent and the recessive gene from the "recessive" parent. The offspring generation is known as the "first filial" generation, or F_1.

When members of the F_1 generation are crossed, the new generation is known as the "second filial," or F_2.

The concepts of dominance and recessiveness may be illustrated by the following diagram:

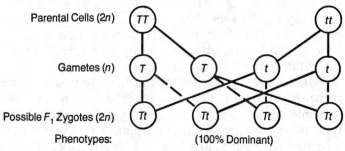

Key:
   T = Dominant Allele.
   t = Recessive Allele.

B. Segregation and Recombination

In Unit 4, we learned that during meiosis the chromosomes in each pair of homologous chromosomes separate from each other to form two monoploid sets of chromosomes. The exact pattern formed in this separation is entirely random and represents only one of many possible combinations. As homologous chromosomes separate, the genes in each allelic pair located on these homologous chromosomes also separate. The random separation of alleles during meiosis is known as **segregation**.

In the process of fertilization, two monoploid nuclei fuse to form a diploid nucleus. This process "recombines" pairs of homologous chromosomes and the pairs of alleles located on them that were separated during meiosis. **Recombination** is also entirely random, bringing together pairs of genes from two different organisms with different histories of genetic inheritance and forming new allelic combinations. The randomness of recombination is an important aspect of variation in sexually reproducing species.

The concepts of segregation and recombination may be illustrated by the following diagram:

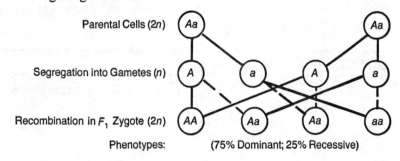

Key:
   A = Dominant Allele for "Axial Flowers"
   a = Recessive Allele for "Terminal Flowers"

This diagram illustrates the probabilities of obtaining different genotypes through recombination. It shows that, if large numbers of offspring are produced, the probability of the AA (homozygous dominant) genotype occurring is 25 percent, the probability of the aa (homozygous recessive) genotype occurring is 25 percent, and the probability of obtaining the Aa (heterozygous or hybrid) genotype is 50 percent in this particular cross. In general, in a **monohybrid cross**, the ratio of $AA : Aa : aa$ is $1 : 2 : 1$ in terms of the frequency of occurrence in the offspring of the cross.

The phenotypes of the AA and the Aa genotypes are indistinguishable for most traits controlled by simple dominance. To determine the genotype of an organism displaying the dominant phenotype, it may be necessary to perform a **test cross**, in which the organism in question is crossed with several homozygous recessive individuals. If the recessive phenotype appears in any of the offspring of such a cross, then the dominant phenotype individual cannot be homozygous recessive, but must be hybrid. The concept of the test cross may be illustrated by the following diagram:

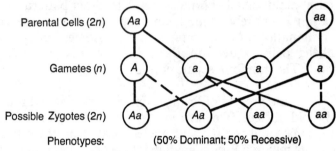

Phenotypes: (50% Dominant; 50% Recessive)

Key:
   A = Dominant Allele.
   a = Recessive Allele.

This diagram illustrates that, when an organism that displays a dominant phenotype (but is of uncertain genotype) is crossed with an organism known to be homozygous recessive, approximately 50 percent of the offspring will display the recessive phenotype if the "dominant" parent is not "pure" but is "hybrid." Thus the "dominant" parent may be "tested" for "purity" by means of the test cross.

QUESTION SET 5.1

1. One reason for Mendel's success with genetic studies of garden peas was that he
   1. used only hybrid pea plants
   2. used peas with large chromosomes
   3. studied large numbers of offspring
   4. discovered the sources of variations in peas

2. Each member of a pair of genes found in the same position on homologous chromosomes is known as
   1. an allele   3. a gamete
   2. a chromatid   4. an autosome

3. Curly hair in human beings, white fur in guinea pigs, and needlelike spines in cacti all partly describe each organism's
   1. alleles   3. chromosomes
   2. autosomes   4. phenotype

4. A student crossed wrinkled-seeded (*rr*) pea plants with round-seeded (*RR*) pea plants. Only round seeds were produced in the resulting plants. This illustrates the principle of
   1. independent assortment
   2. segregation
   3. dominance
   4. incomplete dominance

5. If one offspring is homozygous dominant and a second offspring is heterozygous for the same trait, then the two offspring will most likely
   1. have the same genotype
   2. have the same phenotype
   3. exhibit the recessive trait
   4. have different chromosome numbers

6. Sexually reproducing species show greater variation than asexually reproducing species because of
   1. lower rates of mutation
   2. the occurrence of polyploidy
   3. environmental changes
   4. the recombination of alleles

7. In cabbage butterflies, white color (W) is dominant and yellow color (w) is recessive. If a pure white cabbage butterfly mates with a yellow cabbage butterfly, all the resulting (F_1) butterflies are heterozygous white. Which cross represents the genotypes of the parent generation?
   1. $Ww + ww$
   2. $WW \times Ww$
   3. $WW \times ww$
   4. $Ww \times Ww$

8. A trait that is not visible in either parent appears in several of their offspring. Which genetic concept does this demonstrate?
   1. linked genes
   2. replication
   3. segregation
   4. sex determination

9. In guinea pigs, gene for black coat color is dominant over the gene for white coat color. In a cross between two hybrid black guinea pigs, what percentage of the offspring is likely to have the same coat color as the parents?
   1. 25%
   2. 50%
   3. 75%
   4. 100%

10. In sorghum plants, red stem (R) is dominant over green stem (r). If 1,000 seeds from a sorghum plant germinated to produce 760 red plants and 240 green plants, it would be most reasonable to assume that the parental gentoypes were
    1. $Rr \times Rr$
    2. $RR \times rr$
    3. $Rr \times RR$
    4. $Rr \times rr$

11. In pea plants, green pods are dominant over yellow pods. In which of the following groups would members have the same phenotype but different genotypes?
    1. pure green pods and homozygous green pods
    2. hybrid green pods and homozygous green pods
    3. yellow pods and hybrid green pods
    4. yellow pods and homozygous yellow pods

12. Which statement describes how two organisms may show the same trait, yet have different genotypes for that phenotype?
    1. One is homozygous dominant and the other heterozygous.
    2. Both are heterozygous for the dominant trait.
    3. One is homozygous dominant and the other homozygous recessive.
    4. Both are homozygous for the dominant trait.

13. In pea plants, the trait for tall stems is dominant over the trait for short stems. If two heterozygous tall plants are crossed, what percentage of the offspring would be expected to have the same *phenotype* as the parents?
   1. 25%   3. 75%
   2. 50%   4. 100%

14. Two pea plants, hybrid for a single trait, produce 60 pea plants. Approximately how many of these pea plants are expected to exhibit the recessive trait?
   1. 15   3. 30
   2. 45   4. 60

15. A cross between two tall garden pea plants produced 314 tall plants and 98 short plants. The genotypes of the tall parent plants were most likely
   1. *TT* and *tt*   3. *Tt* and *Tt*
   2. *TT* and *Tt*   4. *TT* and *TT*

16. Polydactyly is a human characteristic in which a person has six fingers per hand. The trait for polydactyly is dominant over the trait for five fingers. If a man who is heterozygous for this trait marries a woman with the normal number of fingers, what are the chances that their child would be polydactyl?
   1. 0%   3. 75%
   2. 50%   4. 100%

C. Intermediate Inheritance

Certain genetic traits do not show a clear pattern of dominance or recessiveness, but rather a "blending" of phenotypes of the parents. Such traits are classified as **intermediate inheritance** traits, since the phenotype of the offspring is often an "intermediate" blend of the traits of the homozygous parents.

One type of intermediate inheritance is **codominance**, which, as its name implies, results from the simultaneous expression of two dominant alleles. Since neither allele is recessive, neither can be easily masked, so both are expressed in the phenotype of the offspring. If the parents of such a codominant individual are homozygous and show contrasting phenotypes for the trait, then all their offspring will be heterozygous and display the "intermediate" phenotype. There are many well-documented examples of codominance, including "roan coat" (a mottled red and white coat) in cattle, pink petal color in four-o'clocks and sickle-cell anemia in human beings.

The concept of codominance may be illustrated by the following diagram:

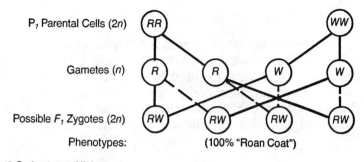

Phenotypes: (100% "Roan Coat")

Key:
  R = "Red Coat" Codominant Allele.
  W = "White Coat" Codominant Allele.
  RR = "Red Coat" Homozygous Genotype.
  WW = "White Coat" Homozygous Genotype.
  RW = "Roan Coat" Heterozygous Genotype.

When members of the F_1 generation of a codominant cross are mated, the resulting F_2 offspring show a redistribution of phenotypes as follows: approximately 25 percent will display the phenotype of one of the P_1 grandparents; approximately 50 percent will display the phenotype of the F_1 parents; approximately 25 percent will display the phenotype of the other P_1 grandparent (ratio of $1:2:1$).

The concept of F_2 offspring formation in codominance may be illustrated (for roan cattle) using the following diagram:

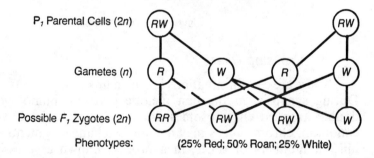

Phenotypes: (25% Red; 50% Roan; 25% White)

Key:
  R = "Red Coat" Codominant Allele.
  W = "White Coat" Codominant Allele.
  RR = "Red Coat" Homozygous Genotype.
  WW = "White Coat" Homozygous Genotype.
  RW = "Roan Coat" Heterozygous Genotype.

D. Independent Assortment

Genes located on different (nonhomologous) chromosomes are free to separate from each other in a random fashion during meiosis. Traits controlled by genes located on separate chromosomes are inherited independently of each other in the same random fashion. This mechanism, known as **independent assortment,** is a major source of genetic variation in living things.

E. Gene Linkage

Gene linkage (see Figure 5.3) provides a condition opposite to that of independent assortment. When nonallelic genes for different traits are located on the same homologous pair of chromosomes, they normally are inherited together, and so are referred to as being "linked." Gene linkage accounts for the fact that such traits as red hair and freckles are frequently found to be inherited together.

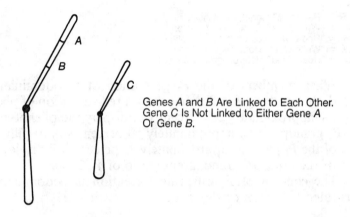

Genes *A* and *B* Are Linked to Each Other. Gene *C* Is Not Linked to Either Gene *A* Or Gene *B*.

Figure 5.3   Gene Linkage

F. Crossing Over

There are certain situations in which linked genes may be separated. During synapsis, in the first meiotic division, homologous pairs of double-stranded chromosomes are closely intertwined. While in this condition, chromosomes may actually exchange segments carrying significant numbers of genes, in a process known as **crossing-over** (see Figure 5.4). In crossing-over, linked genes may be separated to produce different genetic patterns, enhancing the likelihood of genetic variation.

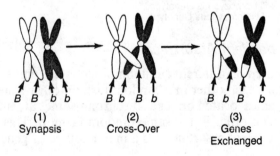

B B b b	*B b B b*	*B b B b*
(1)	(2)	(3)
Synapsis	Cross-Over	Genes Exchanged

Figure 5.4   Crossing Over

G. *Multiple Alleles*

There are some traits whose hereditary patterns do not seem to fit the rules of simple two-gene inheritance. Such a trait may be descibed by a model of gene action involving three or more alleles for the trait in question. A pattern of genetics of this type is known as **multiple alleles.** In utilizing such a pattern, it is important to remember that only two of the three (or more) alleles may be present in any one cell. This must be true since such cells would have been formed by the normal process of fertilization, which involves the fusion of two monoploid nuclei, each containing only one of each allele type.

The genetic trait most commonly used to illustrate the pattern of multiple alleles is the inheritance of the ABO blood group in humans. It is thought that three alleles are involved in controlling this trait: two codominant alleles (I^A and I^B), and a recessive allele (i). Allele I^A controls the production of a blood antigen known as A, allele I^B controls the production of a blood antigen known as B, and allele i, being recessive, does not function to produce any blood antigen. The six possible combinations of genes in this inheritance pattern are responsible for determining which of four different blood phenotypes (*A, B, AB,* or *O*) will be displayed in an individual human being. The following chart summarizes the production of these phenotypes:

Genotype	Antigens Produced	Blood Phenotype
I^A-I^A	A only	*A*
I^A-i^B	A only	*A*
I^B-I^B	B only	*B*
I^B-i	B only	*B*
I^A-I^B	Both A and B	*AB*
i-i	Neither A nor B	O

The concept of multiple alleles may be illustrated by the following diagram:

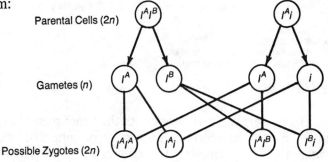

Parental Cells (2n)

Gametes (n)

Possible Zygotes (2n)

Phenotypes: (50% A; 25% AB; 25% B)

Key:
I^A = "A Antigen" Codominant Allele.
I^B = "B Antigen" Codominant Allele.
i = Recessive Allele.

H. Sex Determination

The diploid cells of most sexually reproducing species contain two different types of chromosomes. The chromosomes that contain genes controlling most "body" traits are known as **autosomes**. One pair is known to contain the genes controlling traits having to do with sex differences (differences in "maleness" and "femaleness"). These chromosomes are known as **sex chromosomes**.

In human beings, the diploid number of chromosomes is 46 (23 pairs). Of these, 22 pairs are autosome pairs. The remaining pair is a pair of sex chromosomes, known as the X and the Y chromosomes. Normally, each diploid human cell contains two of these chromosomes. The combination XX leads to the female condition, and the combination XY to the male condition. (The combination YY is not possible.)

The sex of a new organism is determined at the instant of fertilization. The human egg cell contains (in addition to 22 autosomes) a single X chromosome. The human sperm cell may contain (again, in addition to 22 autosomes) either an X or a Y chromosome. Roughly 50 percent of all human sperm cells carry the X sex chromosome; the remaining 50 percent carry the Y. When fertilization occurs, a diploid zygote results that contains 22 pairs of autosomes and either the XX or the XY combination of sex chromosomes, with a 50 percent probability of each occurrence.

The concept of sex determination in human beings may be illustrated by the following diagram:

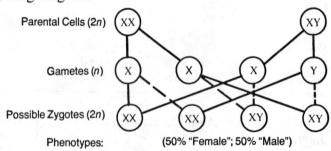

Phenotypes: (50% "Female"; 50% "Male")

Key:
X = X Sex Chromosome.
Y = Y Sex Chromosome.
XX = Female Homologous Chromosome Pair.
XY = Male Homologous Chromosome Pair.

I. Sex Linkage

Thomas Hunt **Morgan**, in his work with the experimental organism *Drosophila*, discovered that certain traits may be inherited more frequently by males than by females. His studies revealed that some of these traits are controlled by recessive genes located on the X sex chromosome only. No corresponding allele could be found on the Y sex chromosome. Such traits are known as **sex-linked** traits and display a unique inheritance pattern. The relative frequency of these sex-linked recessive traits in males is due to the fact that no "normal" dominant gene is present on the Y chromosome to mask the effects of the recessive gene. Therefore males who inherit the single recessive gene will display the trait, whereas the females, who inherit only a single gene, will not display the trait. Such female individuals are known as **carriers** of the trait.

In human beings, traits that are controlled by sex-linked genes include red-green color blindness and hemophilia. The following diagram illustrates the inheritance of sex-linked traits in human beings:

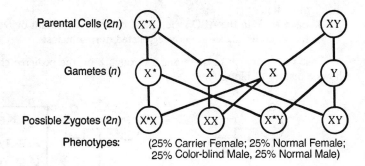

Phenotypes: (25% Carrier Female; 25% Normal Female; 25% Color-blind Male, 25% Normal Male)

Key:
    X = Normal X Sex Chromosome.
    X* = Carrier X Sex Chromosome.
    Y = Normal Y Sex Chromosome.
    XX = "Normal Female" Homologous Chromosome Pair.
    X*X = "Carrier Female" Homologous Chromosome Pair.

QUESTION SET 5.2

1. If the genes for two different traits assort independently, they probably are
    1. on the same chromosome
    2. on different chromosomes
    3. recessive
    4. dominant

2. A boy has brown hair and blue eyes, and his brother has brown hair and brown eyes. The fact that the boys have different combinations of traits is best explained by the concept known as
    1. multiple alleles
    2. incomplete dominance
    3. sex linkage
    4. independent assortment

3. If two roan cattle are crossed, what percent of the offspring are expected to show the parental phenotype for coat color?
    1. 25%   3. 75%
    2. 50%   4. 100%

4. When a mouse with black fur is crossed with a mouse with white fur, all F_1 generation offspring have gray fur. Which phenotypic results can be expected in the F_2 generation?
   1. 100% gray
   2. 25% black, 75% white
   3. 50% black, 50% white
   4. 25% black, 50% gray, 25% white

5. There are multiple alleles for the ABO blood group. Why are only two of these alleles normally present in any one individual?
   1. There are not enough nucleotides in a red blood cell to produce a third allele.
   2. Each parent contributes only one allele for the ABO blood group to the offspring.
   3. Each allele in the ABO group must be either dominant or recessive.
   4. Blood group alleles are not segregated during meiosis.

   Base your answers to questions 6 through 8 on the pedigree chart below, which shows a history of blood types.

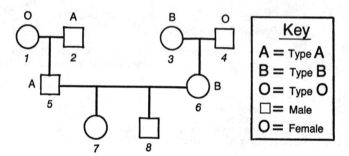

6. The genotype of the individual represented in the chart by *5* is
   1. $I^a I^a$ 3. *ii*
   2, $I^a i$ 4. $I^a I^b$

7. Which individuals represented by the chart *must* be homozygous for blood type?
   1. *1* and *2*   3. *3* and *4*
   2. *2* and *3*   4. *1* and *4*

8. The blood types of the individuals represented in the chart by *7* and *8* could be
   1. A or B only   3. A, B, or AB only
   2. AB only   4. A, B, AB, or O

9. Four children of the same parents each have a different blood type. Which can be assumed about the blood types of the parents?
   1. The mother is pure recessive, and the father is pure dominant.
   2. The mother is pure dominant, and the father is hybrid.
   3. One parent is type O, and the other is type A.
   4. One parent is type A, and the other is type B.

10. Three brothers have blood types A, B, and O. What are the chances that the parents of these three will produce a fourth child whose blood type is AB?
    1. 0%   3. 50%
    2. 25%   4. 100%

11. Which statement correctly describes the normal number and type of chromosomes present in human body cells of a particular sex?
    1. Males have 22 pairs of autosomes and 1 pair of sex chromosomes known as XX.
    2. Females have 23 pairs of autosomes.
    3. Males have 22 pairs of autosomes and 1 pair of sex chromosomes known as XY.
    4. Males have 23 pairs of autosomes.

12. The chance of a YY chromosome combination occurring in human beings as a result of normal meiotic division and normal gametic fusion is
    1. 0%   3. 50%
    2. 25%   4. 100%

13. Traits controlled by genes on the X chromosome are said to be
    1. sex-linked
    2. incompletely dominant
    3. homozygous
    4. mutagenic

14. When a color-blind woman marries a male with normal vision, all their daughters have normal vision and all their sons are color-blind. This is an example of which type of inheritance?
    1. multiple alleles
    2. codominance
    3. sex linkage
    4. autosomal dominance

15. A woman carrying the gene for hemophilia marries a man who is a hemophiliac. What percentage of their children can be expected to have hemophilia?
    1. 0%   3. 75%
    2. 50%   4. 100%

III. CHANGES IN GENETIC INFORMATION BY MUTATION

Any change in the genetic material is known as a **mutation**. Mutations occur at a predictable rate in all cells. When mutations occur in gamete-producing tissues, they may be passed on to the next generation in the normal reproductive process. Mutations occurring in body tissues other than gametic tissues may affect a small number of cells in the individual, but will not be passed on to future generations.

A. *Types of Mutations*

1. Chromosome Mutations

Chromosome mutations, in which there are changes in the number or in the structure of chromosomes, may involve thousands of genes and the traits that they control. They are usually very obvious because of the number of traits involved.

1.1. Chromosome Number. In Unit 4, we learned that the chromosomes in each homologous pair of chromosomes pull apart into two separate groups during the disjunction phase of meiosis. Under certain circumstances, the members of one or more pairs of homologous chromosomes may fail to separate normally. This failure of separation, known as **nondisjunction** (see Figure 5.5), is one of the principal causes of chromosome mutations that involve changes in chromosome number. The result of nondisjunction can be the formation of gametes that contain more (or less) than the monoploid (n) chromosome number. If such abnormal gametes participate in fertilization, the resulting zygote will receive more (or less) than the diploid ($2n$) chromosome number. Although this condition frequently results in the death of the abnormal zygote, in certain cases such zygotes develop into viable offspring with unusual genetic characteristics.

In human beings, the condition known as **Down's syndrome** may be caused by the nondisjunction of chromosome number 21 in the ovum. When this ovum is fertilized by a normal sperm cell, the resulting zygote contains 47 chromosomes arranged in 22 normal pairs and a triple chromosome 21 ($44 + 3 = 47$), instead of the normal 46. This condition occurs with far greater frequency in children of mothers over age 35 than in those of younger mothers.

In rare cases, nondisjunction of an entire homologous set of chromosomes occurs, leading to the production of diploid gametes. When such a gamete participates in fertilization with a normal monoploid gamete, a "triploid" ($3n$) zygote results. Such a condition, known as **polyploidy** ("many-number condition"), results in offspring with exaggerated characteristics. Polyploidy is quite common in certain plants and is responsible for the development of several varieties of human food crops. Many polyploid individuals display exaggerated physical characteristics, such as size, and reduced fertility. Polyploidy is rare in animals, where it is usually a lethal condition.

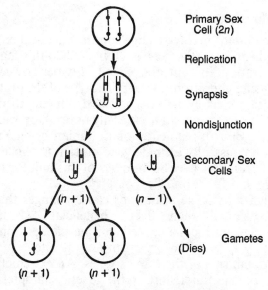

Nondisjunction of a single pair of homologous chromosomes in the first meiotic division results in abnormal gametes containing either too much ($n + 1$) or too little ($n - 1$) genetic information. Although many of these gametes die shortly after formation, some may survive to participate in fertilization. This will result in an abnormal zygote, which may develop into an abnormal organism.

Figure 5.5   Nondisjunction

1.2. Chromosome Structure. Under certain conditions, the structure of the chromosomes themselves may be altered by environmental conditions in the cell. The chromosome strand may break and reattach in a new configuration.

MG

Specific examples of such structural alterations include:

- **Translocation,** the transfer of a section of one chromosome to a nonhomologous chromosome.
- **Addition,** the gain by one chromosome of a portion of a homologous chromosome.
- **Deletion,** the loss of a broken portion of a chromosome.

2. Gene Mutations

Gene mutations involve changes in the chemical nature of the gene. The active chemical in the gene is DNA. When this material undergoes chemical alteration, its control over cell activities and cell characteristics changes, causing alterations in the phenotype of the organism. Although these changes are likely to be small and difficult to detect, they may occasionally be great enough to be easily noticed, or even to cause death. **Albinism** (lack of skin pigmentation in human beings) is an example of a human trait caused by a single gene mutation whose effects are quite obvious and dramatic. (See Section VII of this unit for further details on the structure and function of DNA.)

Although most gene mutations cause changes that are neutral or harmful, occasionally their effect is beneficial. A beneficial mutation, such as one that produces a needed enzyme, causes a phenotypic change that in some way gives an organism an advantage in its environment over other organisms of the same species. An extremely beneficial gene mutation can cause major shifts in the genetic characteristics of a species population, a phenomenon considered by many scientists to be a major mechanism of evolution (see Unit 6 of this book for a more detailed treatment of the science of evolution).

B Mutagenic Agents

Gene mutation can occur spontaneously by random chemical alteration of DNA during replication. In addition, a number of naturally occurring phenomena are known to cause or accelerate gene mutation. These phenomena include:

- Radiation, from sources such as cosmic rays, radon, X rays, ultraviolet radiation, and radioactive radiation.
- Chemicals, such as benzene and formaldehyde.

IV. PRACTICAL APPLICATIONS OF GENETIC THEORY

Geneticists have applied many of the theoretical concepts of the science of genetics to the practical areas of plant and animal breeding. Many of the most productive plant and animal breeds raised for human uses have been developed through careful **artificial selection** of breeding individuals, cross-breeding (**hybridization**) between related varieties, and **inbreeding** of perfected strains to maintain established traits. Such traits as reproductive potential, resistance to disease, adaptability to climate, and productivity, among others, are considered to be desirable traits for human uses. These and other traits are purposely bred into grains (corn, wheat), flowers (roses), fruits (apples, oranges), cattle, horses, and dogs and maintained by careful attention to breeding patterns.

V. INFLUENCE OF ENVIRONMENT ON HEREDITY

Genetic traits are determined largely through the precise information found in the cell's gene structure. Although this information provides a basis for each individual organism's characteristics, it is not the only force at work in shaping the actual phenotype. Another major force in shaping the final phenotype is the environment in which the gene has to operate. It is known that a variety of factors in the environment can actually alter the effects of a particular gene. Some examples of this altering effect are as follows:

- Effect of light on chlorophyll production. Although most plants have the genetic ability to produce chlorophyll, they will do this only in the presence of light. Without light, these plants produce only a light yellow pigment, and therefore appear pale and sickly until they are exposed to sunlight.

- Effect of temperature on hair color in the Himalayan rabbit. In their native Arctic environment, Himalayan hares have white body hair, with black hair on their extremities. However, when raised in warm climates, they are entirely white. In exploring this phenomenon, scientists shaved some hair off the hare's back (normally white) and strapped an ice pack on the bare skin. Under these experimental conditions, the hare's hair grew back black, indicating the role of skin temperature on the production of hair color in this species.

- Effect of environment on identical twins. As we learned in Unit 4, human identical twins result from the independent development of two cell masses of a single fertilization that have separated during cleavage. These two separate cell masses have identical genetic information, and therefore should be identical in all genetic respects. Through actual observation of identical twins raised apart, however, it is known that the social environment can have an effect on the expression of certain traits thought to have a genetic basis. Twin studies provide scientists with a unique opportunity to study how heredity and environment interact in human beings and other organisms.

- Effect of temperature on *curled-wing* in fruit flies (*Drosophila*). *Drosophila* homozygous for "curled wing" were raised at different controlled temperatures. Researchers found that the curled wing phenotype was expressed among offspring raised in "warm environment" cultures, but not among those raised in "cool environment" cultures. In this case, temperature was the environmental condition affecting expression of a genetic trait.

VI. UNDERSTANDING HUMAN HEREDITY

It is probable that human beings will never serve as the subjects for extensive genetic experimentation, since both ethical and practical considerations prevent this. What has been learned about human genetics has had to come, therefore, from the treatments of genetic disorders of various sorts and from continued genetic studies of other organisms. Even though the majority of genetic studies have been conducted on less complex organisms, such as pea plants and fruit flies, the frequency and probability ratios developed from these studies have been found to hold true for humans, as well. It is possible to apply all of the principles of genetics previously discussed to the study of human genetics.

MG

A. Techniques for Detection of Genetic Defects

Often, human disorders of a genetic nature have been identified through observation by physicians and other medical professionals during routine physical examination. Knowledge of the inheritance patterns followed by these disorders allows the physician to predict with reasonable accuracy the probability of the same disorder occurring in other members of the same family, including those of future generations. Frank discussion of these predictions with the members of the family enables them to make informed decisions concerning family planning. Such discussions are often referred to as **genetic counseling**.

Other laboratory techniques used by physicians to aid them in their observations include:

- **Genetic screening**. Genetic screening involves the chemical analysis of body fluids, such as blood and urine, for the presence of chemicals associated with the existence of certain allelic combinations.
- **Amniocentesis**. Amniocentesis involves the removal of a sample of the amniotic fluid surrounding a developing embryo and analysis of cells of the embryo that are found floating within it. This technique is useful for determining the chromosome complement of the embyro's cells, which may reveal chromosome abnormalities such as those associated with Down's syndrome.
- **Karyotyping**. Once cells have been removed from an organism by amniocentesis or another technique, they may be examined to determine their chromosome content by

karyotyping. Karyotyping involves the preparation of an enlarged photograph (**karyotype**; see Figure 5.6), showing the paired homologous chromosomes of a cell. If chromosome abnormalities, such as extra or missing chromosomes, are present, they may be detected by this means.

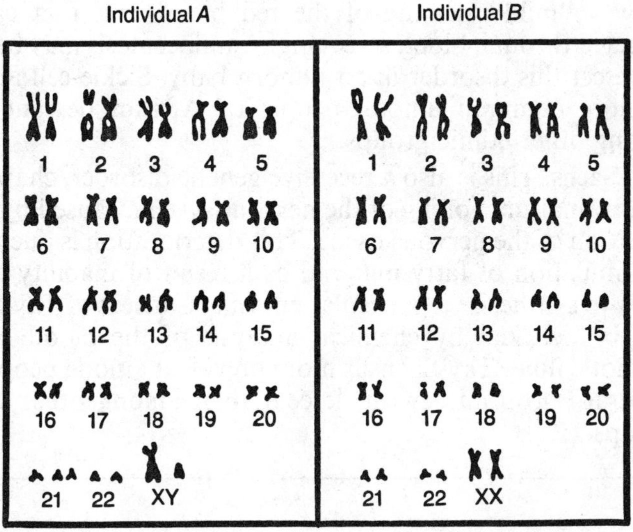

Karyotype of two human individuals, showing for *B* a normal female chromosome complement. Karyotype *A* shows an extra chromosome 21, which will lead to Down's syndrome in this male individual.

Figure 5.6   Karyotype

B. Genetically Related Disorders

A few of the identified human genetic disorders are as follows:

1. **Phenylketonuria (PKU)**. PKU is a genetic disorder that normally results in mental retardation. It is known to be caused by the homozygous pairing of two recessive mutant alleles. The recessive allele in question lacks the ability to synthesize a particular enzyme necessary for metabolism of the amino acid phenylalanine. As a result, phenylalanine builds up in the cells and is thought to be responsible for the mental retardation associated with PKU. If the disorder is detected early enough, through the use of genetic screening, it may be partially corrected through medication.

2.   **Sickle-cell anemia.** This genetic disorder results from the homozygous condition of a certain pair of mutant genes. These
genes produce an abnormal form of hemoglobin and red
blood cells of a "crescent" or "sickle" shape. These sickled red
blood cells do not carry oxygen as efficiently as do normal red
blood cells and may clog blood vessels because of their abnormal shape. Even individuals heterozygous for this disorder
show a "mild" sickling of the red blood cells that can be
detected through blood screening. Amniocentesis may be used
to detect this disorder in an unborn baby. Sickle-cell anemia
is more common among people of African descent than
among other ethnic groups.

3.   **Tay-Sachs.** This is also a recessive genetic disorder, characterized by malfunctioning of the nervous system, caused by deterioration of the nervous tissue. This deterioration is due to the
accumulation of fatty material as a result of inability of the
body to synthesize a particular enzyme. Carriers of Tay-Sachs
can be detected by chemical analysis of the blood or the
amniotic fluid. Tay-Sachs is more prevalent among peoples of
Central-European Jewish descent than among other ethnic
groups.

QUESTION SET 5.3

1.   Which terms best describe most mutations?
   1. dominant and disadvantageous to the organism
   2. recessive and disadvantageous to the organism
   3. recessive and advantageous to the organism
   4. dominant and advantageous to the organism

2.   Cosmic rays, X rays, ultraviolet rays, and radiation from radioactive substances
may function as
   1. pollinating agents   3. plant auxins
   2. mutagenic agents   4. animal pigments

3.   Which usually occurs in the first meiotic division of a primary sex cell?
   1. fertilization   3. crossing-over
   2. polyploidy   4. differentiation

4.   Which is a genetic disorder in which abnormal hemoglobin leads to fragile red
blood cells and obstructed blood vessels?
   1. phenylketonuria   3. leukemia
   2. sickle-cell anemia   4. Down's syndrome

5. Occasionally during meiosis, the members of a single homologous chromosome pair may fail to separate. A human gamete produced by such a nondisjunction will have a chromosome number of

1. 23   **3.** 25
2. 24   **4.** 26

6. The process by which homologous chromosomes exchange segments of DNA is

1. segregation
2. crossing-over
3. fertilization
4. independent assortment

7. The graph below shows the relationship between the number of cases of children with Down's syndrome per 1,000 births and maternal age.

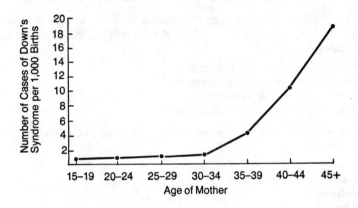

According to the graph, the incidence of Down's syndrome

1. generally decreases as maternal age increases
2. is about nine times greater at age 45 than at age 30
3. stabilizes at 2 per 1,000 births after age 35
4. is greater at age 15 than at age 35

8. The transfer of a section of one chromosome to a nonhomogolous chromosome is
MG known as

1. synapsis   **3.** translocation
2. deletion   **4.** disjunction

9. In human beings, the formation of a zygote containing three sex chromosomes (XXY) is most likely the result of

1. fission   **3.** nondisjunction
2. independent assortment   **4.** crossing-over

10. Mental retardation resulting from phenylketonuria (PKU) is caused by
MG **1.** lack of an enzyme necessary for normal metabolism
2. insufficient production of adrenalin
3. overproduction of insulin
4. bacterial infection of brain tissue

11. Race horses show many variations from the wild horse ancestors from which they were derived. It is most likely that these variations between race horses and their ancestors are due to

    1. use and disuse of organs
    2. artificial selection
    3. gene cloning
    4. chromosomal nondisjunction

12. Some studies of identical human twins show that their IQs, heights, and talents *may* be different. The best explanation for these differences is that

    1. the environment interacts with genes in the development and expression of inherited traits
    2. heredity and environment have no influence on the expression of phenotypes
    3. the genotype of twins depends on the interaction of diet and hormone control
    4. people are considered identical if at least half their genes are the same

13. A certain species of plant produces blue flowers when the soil pH is above 7.0. However, when the soil pH is below 7.0, the flowers are pink. Which statement best explains this color change?

    1. Mutagenic agents can alter genotypes.
    2. The environment influences gene action.
    3. Polyploidy produces 2n gametes.
    4. Chromosomal mutations produce color effects.

14. Which technique can be used to examine the chromosomes of a fetus for possible genetic defects?

    1. pedigree analysis
    2. analysis of fetal urine
    3. karyotyping
    4. cleavage

15. Which statement best describes amniocentesis?

    1. Blood cells of an adult are checked for fragility.
    2. Saliva of a child is analyzed for amino acids.
    3. Urine of a newborn is analyzed for the amino acid phenylalanine.
    4. Fluid surrounding an embryo is removed for cellular analysis.

VII. MODERN GENETIC CONCEPTS

A. DNA's Function as the Hereditary Material

As scientists have continued their study of genetics, they have learned more and more of the details of the mechanisms of genetics. In one of the most significant discoveries in genetic science, **deoxyribonucleic acid (DNA)** was revealed to be the chemically active agent of the gene. As is now known, it is DNA that replicates itself when chromosomes replicate in the early stages of cell division. It is DNA that is passed from generation to generation in the process of reproduction and that acts as Mendel's genetic "factors." It is DNA that interacts with the cell's chemical factory and produces the observable effects of the phenotype when genes are inherited by a cell or an organism. It is DNA that regulates the production of enzymes in the cell and thereby enables the cell to perform the complex cellular chemical reactions necessary to sustain its life.

1. DNA Structure

To understand how DNA operates, it is necessary to understand its chemical makeup, that is, **DNA structure**. DNA is a polymer made up of a repeating chemical unit known as the **nucleotide**. Thousands of these units are known to comprise a single DNA molecule, making DNA one of the largest of all organic compounds. DNA exists in hundreds of thousands of different forms, depending on the precise arrangement of nucleotides in the molecule. Its variability is the key to genetic variation in living things. Complementary nucleotides are shown in Figure 5.7.

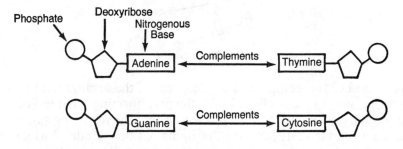

Figure 5.7   Complementary Nucleotides (DNA)

1.1. DNA nucleotides themselves are quite complex, being composed of three separate subunits:

- **Phosphate group**—a chemical group made up of phosphorus and oxygen.
- **Deoxyribose**—a five-carbon sugar made up of carbon, hydrogen, and oxygen.
- **Nitrogenous base**—a chemical unit composed of carbon, oxygen, hydrogen, and nitrogen. Bases found in DNA are **adenine**, **thymine**, **cytosine**, and **guanine**.

1.2. The **Watson-Crick model**, developed by **James Watson** and **Francis Crick**, is an attempt to describe the physical and chemical structure of DNA in a way that would explain its known characteristics, including its ability to replicate. Watson and Crick's model was developed using the best experimental evidence available at the time and involves the following points:

- Nucleotide units are joined end to end (see Figure 5.8), forming a long chain of alternating deoxyribose and phosphate units with nitrogenous bases sticking out to one side. The specific arrangement of nitrogenous bases on this chain makes up the "genetic code," which is discussed in greater detail in Section 3.3 of this unit.

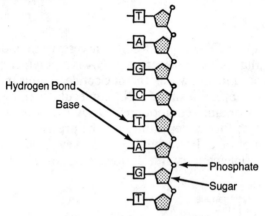

Figure 5.8   DNA: Single Strand

- A second chain, complementary in terms of the arrangement of nitrogenous bases, is aligned with the first, forming a ladderlike molecule (see Figure 5.9). In this formation, the repeating sugar and phosphate units form the "uprights" of the "ladder," while the pairs of nitrogenous bases form the "rungs" connecting the two strands.

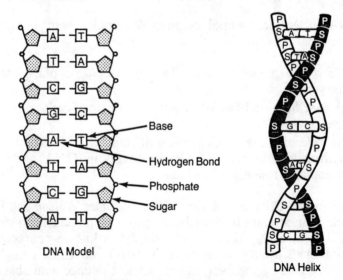

Figure 5.9   DNA: Double Strand

- The nitrogenous bases paired to form the "ladder rungs" are always adenine to thymine (A–T) and cytosine to guanine (C–G), because of the complementarity of their chemical structures. The bases are held together by weak **hydrogen bonds**.
- The DNA "ladder" is twisted to form a double-stranded **helix** shape.

MG

1.3. The structure of DNA is thought to be directly related to its major functions in the cell, which include:

- Replication of the genetic code, in order to maintain genetic continuity from generation to generation.
- Control of the production of cellular enzymes, in order to control the chemical activities of the cell and thereby the phenotypic characteristics of the organism.

2. DNA Replication

Replication, the exact self-duplication of the genetic material, has already been discussed in terms of the doubling of chromosomes in the early stages of both mitosis and meiosis. It is now known that chromosome replication is actually a function of the replication of the DNA making up the chromosome strand. **DNA replication** is thought to occur as follows:

- The two strands of the DNA molecule separate by "unzipping" between pairs of nitrogenous bases (see Figure 5.10).

"Unzipping" of Hydrogen
Bonds between Base Pairs

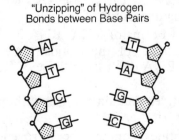

Figure 5.10   DNA Replication: Phase I

- Unbound nucleotides floating free in the cytoplasm are attracted to and incorporated into the unzipped portion of the DNA molecule. In the building process that follows, adenine nucleotides are attracted by thymine nucleotides, and cytosine nucleotides are attracted by guanine nucleotides. In this way, new complementary strands are produced that are exact duplicates of the original strands (see Figure 5.11).

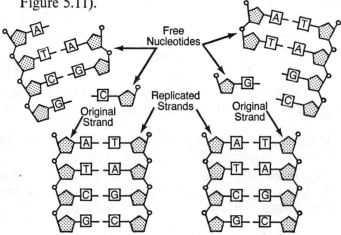

Figure 5.11   DNA Replication: Phase II

- When the "unzipping" process is complete and all bonding sites are filled with free nucleotides, two identical DNA molecules result; these are free to separate into two chromosome strands.

3. Gene Control of Cell Activities

3.1. Role of DNA and RNA. In gene control of cellular activities, DNA provides the information necessary to enable the cell to produce required enzymes. **Ribonucleic acid (RNA)** aids this process by carrying DNA's code to the protein-manufacturing apparatus of the cell.

3.2. Structure and Forms of RNA. RNA is a nucleic acid, like DNA, composed of nucleotide units. However, the molecules of RNA differ from those of DNA in the following ways:

3.2.1. RNA nucleotides contain the five-carbon sugar **ribose** instead of deoxyribose.

3.2.2. RNA nucleotides contain the nitrogenous bases **uracil**, adenine, cytosine, and guanine; uracil is substituted for thymine.

3.2.3. RNA molecules are single-stranded, unlike DNA, which is double-stranded.

3.2.4. RNA exists in three functionally different forms:

- **Messenger RNA**, which "reads" and carries the genetic code from DNA to the ribosome.
- **Transfer RNA**, which transports amino acid molecules to the ribosome.
- **Ribosomal RNA**, which makes up the ribosome.

3.3. Genetic Code. It is thought that the sequence of nitrogenous bases comprising a string of DNA provides a type of chemical "code" that is "understood" by the chemical mechanisms of the cell. The DNA code is used by these mechanisms to manufacture specific enzymes and other proteins through the process of protein synthesis (see below).

A DNA strand provides a **template** (pattern) for the formation of messenger RNA (m-RNA). The DNA code is transcribed ("read") by m-RNA as the latter is synthesized as a strand complementary to the DNA strand. The messenger RNA is pieced together, nucleotide by nucleotide, as unbound ("free") RNA nucleotides are attracted to complementary nucleotides on the DNA.

Each group of three nitrogenous bases, known as a **triplet codon**, provides the information necessary to code for a single, specific amino acid. The particular sequence of triplet codons on DNA (and transcribed to m-RNA) enables amino acids to be linked together in a specific sequence during protein synthesis.

3.4. Protein Synthesis. The actual process of protein synthesis begins when the m-RNA moves out of the nucleus where it was formed and attaches to a ribosome. The ribosome acts as the site of protein synthesis in the cell. The codon sequence of the m-RNA provides a pattern upon which a new polypeptide (protein) strand will be built.

Transfer RNA (t-RNA) molecules attach to specific free amino acids in the cytoplasm. The t-RNA molecules then transport these amino acids to the ribosome for addition to a growing chain of amino acids. Each codon on the m-RNA provides the specific information necessary for the placement of each such amino acid in the chain. The result of this process is the synthesis of a particular, specific polypeptide chain. Because protein molecules are manufactured through this process, the process is known as "protein synthesis" (see Figure 5.12).

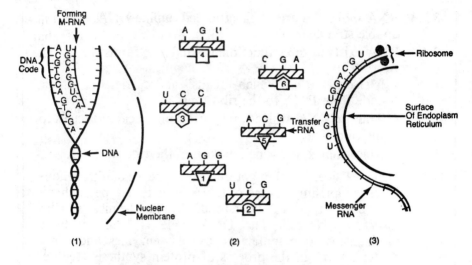

DNA provides the "genetic code" to produce new proteins in the cell (1). Messenger RNA (M-RNA) "reads" this code by forming a strand complementary to the DNA strand (1). Once formed, the M-RNA travels to the cytoplasm (2), where it associates with a ribosome (3). Transfer RNA molecules (*t*-RNA), carrying specific amino acid molecules, attach to the M-RNA at specific locations, forming a chain of amino acids know as a "polypeptide."

Figure 5.12   Protein Synthesis

3.5. One Gene-One Polypeptide Hypothesis. It is thought that each gene in the cell's nucleus contains the coded information required to synthesize a single polypeptide chain. A particular gene is believed to operate throughout the life of the cell to produce its specific polypeptide, and only that polypeptide. This concept is known as the **one gene-one polypeptide hypothesis.**

It is further thought that a gene will operate in the cell only when its polypeptide is required by the activities of the cell, and that at other times it is "switched off." The modern concept of the gene defines it as "that sequence of nucleotides and codons in a molecule of DNA necessary to code for a complete polypeptide chain."

3.6. Individual Traits as a Function of Their DNA. We have learned that DNA's principal role in the cell is to code for the production of specific polypeptides. Each enzyme or structural protein, composed of polypeptide chains, that is synthesized by a cell has a specific function to perform. These "functions" comprise the "characteristics" or "traits" of the cell. It follows logically, therefore, that the individual "traits" that a cell displays are a function of the particular combinations of DNA found in its cells. The particular combination of these traits gives organisms the individuality that enables us to recognize them as individual organisms. The following diagram will help to illustrate this concept:

DNA → gene → polypeptide → enzyme →

reaction → reaction product → "trait"

3.7. **Gene Mutations.** Gene mutations, discussed previously, may now be defined more precisely as being any changes in the base sequence of a molecule of DNA. When the base sequence of DNA is altered, the amino acid sequence of the polypeptide for which it codes will likewise be altered. Such an alteration may affect the operation of the resulting enzyme, preventing it from properly catalyzing its reaction, and thus preventing a "trait" from being produced in the cell.

Alterations of DNA may include the addition, deletion, or substitution of bases in the DNA strand.

B. Genetic Research

Genetic research assumes many forms, depending on the questions being investigated by the researcher. Areas of particular interest in genetic research include:

1. **Cloning.** This term refers to the production of a group of genetically identical "offspring" from the cells of a single "parent" organism that would normally reproduce by sexual means. Cloning has been attempted with varying success with a number of different organisms, both plant and animal. Although the process remains experimental in animals, it has proved to be quite successful in plants and is used extensively in the production of certain commercial crops. Its main advantage is that organisms with desirable combinations of traits, which would otherwise be changed in sexual reproduction, can be reproduced rapidly, with no alteration of their phenotype combinations. Each of the genetically identical "offspring" is known as a "clone."

2. **Genetic engineering.** This term refers to the series of techniques used to transfer genes from one organism to another (see Figure 5.13). It involves removing a small piece of DNA from a cell and adding it to the gene structure of another cell. The "new" DNA that results is known as **recombinant DNA**. The recombinant DNA will continue to produce its polypeptide product in the new cell, thus transferring to that cell a genetic ability it lacked before.

This technique has been used successfully to transfer to bacterial cells the ability to produce certain human biological products—**insulin, interferon** and **human growth hormone**, among others. These "genetically engineered" bacteria are then cultured in great numbers, and their new product is drawn off and purified for use in treating human disorders such as diabetes, autoimmune diseases, and hormonal deficiencies. This technique may be extended in the future to produce other "genetically engineered" organisms with desirable traits from human uses.

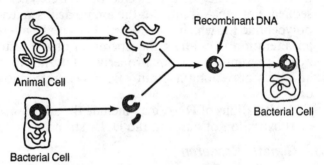

Genetic Engineering involves removing a desirable gene from a cell, adding it to the gene structure of a bacterial cell, and replacing the recombinant DNA into the bacterial cell. The bacteria then have the capability to produce the chemical produced by the original animal or plant cell.

Figure 5.13   Genetic Engineering

QUESTION SET 5.4

1. The diagram at the right represents a chromosome. Letters *A* and *B* indicate structures known as the

   1. chromatid and centromere
   2. spindle and cell plate
   3. centriole and centrosome
   4. stamen and pistil

2. The genetic material in living organisms is composed of organic molecules known as

   1. starches
   2. lipids
   3. nucleic acids
   4. fatty acids

3. When bonded together chemically, deoxyribose, phosphate and an adenine molecule make up

   1. a DNA nucleotide
   2. an RNA nucleotide
   3. a DNA molecule
   4. an RNA molecule

4. In a DNA molecule, hydrogen bonds are present between molecules of
   1. phosphate and adenine
   2. deoxyribose and cytosine
   3. phosphate and deoxyribose
   4. adenine and thymine

5. A nucleotide is composed of which substances?
   1. phosphate group, sugar, nitrogenous base
   2. phosphate group, starch, nitrogenous base
   3. sugar, nitrogenous base, enzyme
   4. phosphate group, sugar, enzyme

6. The diagram at the right represents the
   building block of a large molecule
   known as a

   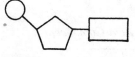

   1. protein   3. carbohydrate
   2. fatty acid   4. nucleic acid

7. Which nitrogenous bases tend to pair with each other in a double-stranded
   molecule of DNA?
   1. adenine–uracil   3. cytosine–thymine
   2. thymine–adenine   4. guanine–adenine

8. Which DNA strand below represents the
   base sequence complementary to the
   portion of a DNA strand represented
   in the diagram at the right?

   (1) G   (2) T   (3) A   (4) C
        C   A   T   A
        A   G   G   T
        T   C   C   G
        C   A   T   A
        G   T   A   C

9. Watson and Crick described the DNA molecule as a
   1. straight chain   3. double helix
   2. single strand   4. branching chain

10. The replication of a double-stranded DNA molecule begins when the strands
    MG separate at the
    1. phosphate bonds   3. deoxyribose molecules
    2. ribose molecules   4. hydrogen bonds

11. Which cellular process involves DNA replication?
    MG 1. mitosis   3. pinocytosis
    2. cyclosis   4. protein synthesis

12. DNA and RNA molecules are similar in that both contain
    MG 1. nucleotides   3. deoxyribose sugars
    2. a double helix   4. thymine

13. In regard to structure, in what way do DNA molecules differ from RNA molecules?
MG **1.** RNA contains the base uracil; DNA does not contain uracil.
   2. DNA is composed of two chains of nucleotides; RNA is composed of three chains of nucleotides.
   3. RNA is helical; DNA is branched.
   4. DNA is composed of four different bases; RNA is composed of three different bases.

Base your answers to questions 14 through 17 on the list of nucleic acid components below.

Nucleic Acid Component

(1)   Ribose
(2)   Deoxyribose
(3)   Adenine
(4)   Uracil
(5)   Phosphate
(6)   Thymine

14. Which components are found in both RNA and DNA molecules?
MG **1.** 1 and 2   **3.** 3 and 5
   2. 1 and 6   **4.** 3 and 6

15. Which components may be present in RNA molecules only?
MG **1.** 1 and 3   **3.** 3 and 6
   2. 1 and 4   **4.** 4 and 6

16. In RNA molecules, the genetic code is made up of specific sequences of
MG components. Examples of such components are
   1. 1 and 2   **3.** 3 and 4
   2. 2 and 3   **4.** 4 and 5

17. In DNA, which pair of components may be held together by relatively weak
MG hydrogen bonds?
   1. 1 and 5   **3.** 3 and 5
   2. 2 and 6   **4.** 3 and 6

18. A high concentration of an enzyme that breaks down RNA molecules is introduced
MG into a cell. Which cellular activity will probably be affected first?
   1. metabolism of fats   **3.** hydrolysis of ATP
   2. synthesis of proteins   **4.** oxidation of glucose

19. The specificity of genetic material is the result of the
MG **1.** type of sugar present in DNA
   2. type of phosphate found in a cell
   3. order of the nitrogen bases in DNA
   4. order of the amino acids in a protein

20. In a cell, the transfer of genetic information from DNA to RNA occurs in the
MG **1.** cell membrane   **3.** nucleus
     2. endoplasmic reticulum   **4.** nucleolus

21. The DNA code for a particular amino acid contains a sequence of how many
MG nucleotides?
     1. 5   **3.** 3
     2. 6   **4.** 4

22. The "one-gene, one-polypeptide" hypothesis deals most directly with the relation-
MG ship of genes to the synthesis of
     1. enzymes   **3.** lipids
     2. polysaccharides   **4.** carbohydrates

23. A change that affects the base sequence in an organism's DNA by the addition,
MG deletion, or substitution of a single base is known as
     1. DNA replication
     2. gene mutation
     3. chromosomal mutation
     4. independent assortment

For *each* phrase in questions 24 through 27, select the type of nucleic acid molecules, *chosen from the list below*, that is best described by that phrase. (*A number may be used more than once or not at all.*)

Types of Nucleic Acid Molecules

(1) DNA molecules only
(2) RNA molecules only
(3) Both DNA and RNA molecules
(4) Neither DNA nor RNA molecules

24. Transfer amino acids from the cytoplasm to the ribosomes
MG

25. Composed of polypeptides
MG

26. Involved in the synthesis of proteins within a cell
MG

27. Contain the base sequences AUG, CGA, UCG
MG

---

28. The function of transfer RNA molecules is to
MG **1.** transport amino acids to messenger RNA
     2. transport amino acids to DNA in the nucleus
     3. synthesize more transfer RNA molecules
     4. provide a template for the synthesis of messenger RNA

29. When complex plants are produced by cloning, which process is most directly
[MG] involved?

    1. mitotic cell division
    2. meiotic cell division
    3. gametogenesis
    4. budding

30. The formation of recombinant DNA results from the

[MG] **1.** addition of messenger RNA molecules to an organism
    2. transfer of genes from one organism to another
    3. substitution of a ribose sugar for a deoxyribose sugar
    4. production of a polyploid condition by a mutagenic agent

31. Which is a technique of genetic research in which genetic information is transferred
[MG] from cells of one organism to cells of another organism?

    1. genetic engineering   **3.** amniocentesis
    2. population genetics   **4.** chromatography

MG

C. Population Genetics

The genetic characteristics of the individual help to describe the individual. Similarly, the sum total of the genetic characteristics of all individuals in a species population helps scientists to describe the population. The study of the genetic characteristics of a sexually reproducing species population and of the factors that can affect the frequencies of genes in that population is known as **population genetics**. Important aspects of population genetics include:

1. **Population.** The species population, rather than the individual or the "family," is the basic unit studied in population genetics. A species population is defined as "all the members of a species in a given geographical location at a given time." Examples of populations include all the white-tail deer living in a mountain valley and all the dandelions inhabiting a vacant lot.

2. **Gene pool.** Among mature members of a species, the genes found in their reproductive cells represent those that may be passed on to the next generation. The sum total of all the heritable genes for the traits in a given population is known as the "gene pool."

3. **Gene frequency.** The population may be described in terms of its gene pool. By applying a simple statistical formula to experimental data, it is possible to predict the frequencies of alleles for many observable traits in sexually reproducing species. Gene frequency is defined as the percentage of each allele for a particular trait in a population. For example, a human population may be described in terms of the frequency of genes controlling the ability to taste the chemical PTC. If 60 percent of the genes controlling this trait in this population are recessive, then 40 percent of those genes must be dominant. It is even possible to predict the proportion of this population that is homozygous and that is heterozygous.

4. **The Hardy-Weinberg principle. G. H. Hardy** and **W. Weinberg**, mathematicians, studied populations of sexually reproducing species in an attempt to describe their gene pools mathematically. Their findings, known as the "Hardy-Weinberg principle," were based on the manner in which genes for single, two-gene traits are thought to assort themselves during meiosis and fertilization. The Hardy-Weinberg principle states that the gene pool (gene frequencies) of a population should remain stable over many generations as long as certain conditions are met. In this principle, the "ideal population" would have the following characteristics:

- The **population is large**, and sexes are represented in equal numbers.
- The members of the population **mate randomly**.
- **No migration** into or out of the population occurs.
- **No mutation** of genes or chromosomes occurs.

In reality, the "ideal" conditions required by the Hardy-Weinberg principle are rarely met. Mutation, for example, is a phenomenon that occurs at all times at a predictable rate; random mating may be prevented by pair bonding or by geographic barriers; migration is a behavior common to many species; population size may vary markedly from place to place. For these reasons, the genetic stability "predicted" by Hardy and Weinberg cannot normally occur. The result is that gene pools, far from being stable, are constantly in a dynamic state of change. This, along with variation brought on by genetic mechanisms, is thought by many scientists to be the driving force of evolution (see Unit 6 for a more detailed discussion of the science of evolution).

QUESTION SET 5.5

1. From an evolutionary standpoint, the greatest advantage of sexual reproduction
is the
   1. variety of organisms produced
   2. appearance of similar traits generation after generation
   3. continuity within a species
   4. small number of offspring produced

2. All the inheritable alleles for a particular population of a species are known as
the population's
   1. gene frequency
   2. gene pool
   3. reproductive isolation
   4. geographic isolation

3. Which level of biological organization is studied in the Hardy-Weinberg
principle?
   1. population   **3.** ecosystem
   2. community   **4.** organism

4. The gene pool in a population of *Rana pipiens* in a pond remained constant for
many generations. The most probable reason for this stable gene pool is that
   1. the population was small, with nonrandom mating and many mutations
   2. random mating occurred in a small population with many mutations
   3. no mutations occurred in a large, migrating population
   4. no migration occurred in a large population with random mating

5. Which set of conditions would most likely cause a change in gene frequency in
a sexually reproducing population?
   1. mutations and small populations
   2. large populations and no migrations
   3. random matings and large populations
   4. no mutations and no migrations

6. The frequency of traits that presently offer high adaptive value to a population
may *decrease* markedly in future generations if
   1. conditions remain stable
   2. the environment changes
   3. all organisms with the trait survive
   4. mating remains random

Base your answers to questions 7 and 8 on the graph below and on your
knowledge of biology. The graph illustrates changes in the percentage of two
varieties of a certain species.

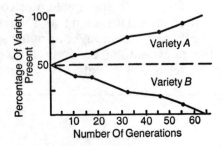

7. Which will contribute significantly to the future of the species gene pool?

MG
  1. variety *A* only
  2. variety *B* only
  3. both variety *A* and variety *B*
  4. neither variety *A* nor variety *B*

8. What is the most probable reason that the percentage of variety *A* is increasing
MG in the population of this species?
  1. There is no chance for variety *A* to mate with variety *B*.
  2. There is no genetic difference between variety *A* and variety *B*.
  3. Variety *A* has some adaptive advantage that variety *B* does not.
  4. Variety *A* is somehow less fit to survive than variety *B*.

UNIT 5 REVIEW

1. The principles of dominance, segregation, and independent assortment were first described by

  1. Darwin   **3.** Lamarck
  2. Watson and Crick   **4.** Mendel

2. The best way to determine the coat-color phenotype of a guinea pig is to

  1. X-ray the animal
  2. prepare a chromosome slide
  3. analyze a blood sample
  4. observe the organism

3. Only red tulips result from a cross between homozygous red and homozygous white tulips. This illustrates the principle of

  1. independent assortment
  2. dominance
  3. segregation
  4. incomplete dominance

4. In pea plants, the gene for tallness (T) is dominant over the gene for shortness (t). If 100% of the F_1 generation offspring are heterozygous tall, what were the most probable genotypes of the parent plants?

  1. *Tt* × *Tt*   **3.** *TT* × *Tt*
  2. *Tt* × *tt*   **4.** *TT* × *tt*

5. A pea plant that produces green pods is crossed with a pea plant that produces yellow pods. The resulting offspring has green pods. With respect to pod color, the genotype of the offspring is most likely

  1. heterozygous dominant
  2. pure recessive
  3. homozygous dominant
  4. homozygous recessive

6. If a trait that is *not* evident in the parents appears in their offspring, the parental genotypes are most likely

  1. pure recessive   **3.** homozygous
  2. monoploid   **4.** heterozygous

7. Which basic genetic concept states that chromosomes are distributed to gametes in a random fashion?

   1. dominance
   2. linkage
   3. segregation
   4. mutation

8. In a certain species of meadow mouse, dark coat color is dominant over cream coat color. If heterozygous dark-coated male mice are mated with cream-coated female mice, what will be the expected percentage of phenotypes in their offspring?

   1. 25% dark coated, 75% cream coated
   2. 50% dark coated, 50% cream coated
   3. 75% dark coated, 25% cream coated
   4. 100% dark coated

9. The appearance of a recessive trait in offspring of animals most probably indicates that

   1. both parents carried at least one recessive gene for that trait
   2. one parent was homozygous dominant and the other parent was homozygous recessive for that trait
   3. neither parent carried a recessive gene for that trait
   4. one parent was homozgyous dominant and the other parent was hybrid for that trait

10. Which structures code information for the inheritance of traits?

    1. nuclear membranes
    2. cell membranes
    3. vacuoles
    4. genes

11. Based on the gene-chromosome theory, the law of independent assortment assumes that certain genes are

    1. formed by chromosomal mutations
    2. located on the same chromosome
    3. formed in the cytoplasm
    4. located on separate chromosomes

12. White mice with fluffy (tufted) tails are mated with brown mice with hairless tails. In the F_2 generation, some of the white offspring have hairless tails, while some of the brown offspring have tufted tails. These results best demonstrate

    1. independent assortment
    2. sex linkage
    3. gene mutation
    4. intermediate inheritance

13. A cross of a red cow with a white bull produces all roan offspring. This type of inheritance is known as

    1. codominance
    2. mutation
    3. sex linkage
    4. multiple alleles

14. When two four-o'clock plants are crossed, 48 pink four-o'clocks and 52 white four-o'clocks are produced. The phenotypes of the parents were

    1. pink and white
    2. pink and red
    3. pink and pink
    4. red and white

15. A man who has blood type AB marries a woman who has blood type B. This couple would *not* normally have a child with which genotype?

1. $I^A i$

2. $I^B I^B$

3. $I^A I^B$

4. ii

Base your answers to questions 16 through 19 on the pedigree chart below and on your own knowledge of biology. The chart shows that Sally is a carrier for red-green color blindness.

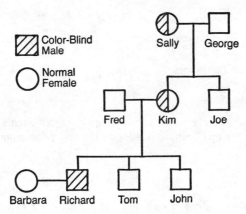

16. Which is most likely the chromosomal makeup of George's body cells?

1. 11 pairs of autosomes and one X chromosome

2. 11 pairs of autosomes and one Y chromosome

3. 22 pairs of autosomes and two X chromosomes

4. 22 pairs of autosomes, an X chromosome, and a Y chromosome

17. What is the probability that Barbara, who has no genes for color blindness, will have a color-blind daughter?

1. 0%

2. 25%

3. 50%

4. 100%

18. Which statement best describes Sally?

1. She has no genes for color blindness

2. She has one gene for color blindness, located on an X chromosome.

3. She has one gene for color blindness, located on a Y chromosome.

4. She has two genes for color blindness.

19. Richard is color-blind because he inherited the trait from his

1. father, Fred

2. grandfather, George

3. mother, Kim

4. uncle, Joe

---

20. A man heterozygous for blood type A marries a woman with blood type AB. The blood type of their offspring *cannot* be

1. A

2. B

3. O

4. AB

21. A child with blood type O has a mother with blood type A and a father with blood type B. The parental genotypes for blood types must be
    1. $I^A I^A$ and $I^B I^B$
    2. $I^A i$ and $I^B I^B$
    3. $I^A I^B$ and $I^B i$
    4. $I^A i$ and $I^B i$

22. The letters in the following crosses represent parental blood types. Which cross could produce offspring that represent all four blood types of the ABO blood group?
    1. $I^A I^A \times I^A I^B$
    2. $ii \times I^A i$
    3. $I^A I^B \times I^A I^B$
    4. $I^A i \times I^B i$

23. In human beings, sex is normally determined at fertilization by
    1. 1 pair of sex chromosomes
    2. 2 pairs of sex chromosomes
    3. 11 pairs of autosomes
    4. 22 pairs of autosomes

24. Based on the pattern of inheritance known as sex linkage, if a male is a hemophiliac, how many genes for this trait are present on the sex chromosomes in each of his diploid cells?
    1. 1
    2. 2
    3. 3
    4. 0

25. A color-blind woman marries a man who has normal color vision. What are their chances of having a color-blind daughter?
    1. 0%
    2. 25%
    3. 75%
    4. 100%

26. Which parental pair could produce a color-blind female?
    1. homozygous normal-vision mother and color-blind father
    2. color-blind mother and normal-vision father
    3. heterozygous normal-vision mother and normal-vision father
    4. heterozygous normal-vision mother and color-blind father

27. In which hereditary disease do the abnormal hemoglobin molecules differ from normal hemoglobin molecules by only a single amino acid?
    1. hemophilia
    2. albinism
    3. phenylketonuria
    4. sickle-cell anemia

28. A genetic disorder caused by homozygous combination of recessive mutant genes that may result in mental retardation is
    1. phenylketonuria
    2. sickle-cell anemia
    3. color blindness
    4. hemophilia

29. In the diagram below, A and B represent homologous chromosomes. What process is illustrated?

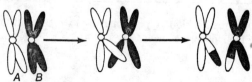

    1. crossing-over
    2. nondisjunction
    3. replication
    4. hybrid crossing

30. Down's syndrome is an inherited human defect that results in mental retardation. The most common cause of this defect is known to be
    1. gene linkage
    2. nondisjunction
    3. crossing-over
    4. sex linkage

31. In many human beings, exposing the skin to sunlight over prolonged periods of time results in the production of more pigment by the skin cells (tanning). This change in skin color provides evidence that
    1. ultraviolet light can cause mutations
    2. gene action can be influenced by the environment
    3. the inheritance of skin color is an acquired characteristic
    4. albinism is a recessive characteristic

32. Sickle-cell anemia results from the substitution of one base for another in a DNA molecule. This change is an example of
    1. crossing-over
    2. nondisjunction
    3. a polyploid condition
    4. a gene mutation

33. The members in a pair of chromosomes fail to separate during meiosis, producing a gamete with an extra chromosome. This process is known as
    1. crossing-over
    2. polyploidy
    3. nondisjunction
    4. recombination

34. Human disorders such as PKU and sickle-cell anemia, which are defects in the synthesis of individual proteins, are most likely the result of
    1. gene mutations
    2. nondisjunction
    3. crossing-over
    4. polyploidy

35. The development and expression of an inherited trait in an organism are influenced by
    1. the organism's genotype only
    2. the organism's environment only
    3. both the organism's genotype and environment
    4. neither the organism's genotype nor environment

36. The Himalayan rabbit has white fur over most of its body, but it has black fur on its tail, ears, and the tips of its legs and nose. Two rabbits that are homozygous for this hair pattern are mated. When their offspring are exposed to normal temperatures, they exhibit the normal Himalayan hair pattern. However, when the offspring are exposed to low temperatures (10°C), they have black hair covering their entire bodies. This illustrates
    1. that mutations can be caused by heat radiation
    2. that the traits of an organism are determined by its genes
    3. the importance of the environment in gene expression
    4. the law of incomplete dominance

37. Identical twins were separated at birth and brought together after 13 years. They varied in height by 2 inches and in weight by 20 pounds. The most probable explanation for these differences is that
    1. their environments affected the expression of their traits
    2. their cells did not divide by mitotic cell division
    3. they developed from two different zygotes
    4. they differed in their genotypes

38. Which combination of techniques can be used before birth to detect chromosomal
MG abnormalities?

    1. ultracentrifugation and chromatography
    2. screening and vaccination
    3. blood typing and vaccination
    4. amniocentesis and karyotyping

39. In a particular variety of corn, the kernels turn red when exposed to sunlight. In the
absence of sunlight, the kernels remain yellow. Based on this information, it can be
concluded that the color of these corn kernels is due to the

    1. effect of sunlight on transpiration
    2. law of incomplete dominance
    3. principle of sex linkage
    4. effect of environment on gene expression

40. In the diagram below, what is represented by the letter X?

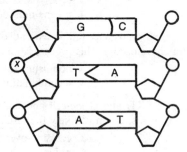

    1. ribose   **3.** phosphate
    2. deoxyribose   **4.** adenine

    Base your answers to questions 41 through 43 on the diagram below, which repre-
sents a segment of a DNA molecule, and on your knowledge of biology.

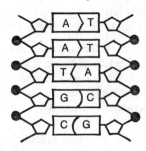

41. If the segment of DNA represented by the diagram was used as a template in the
MG synthesis of messenger RNA, which sequence represents the order of bases found in
the m-RNA molecule?

    1. U—U—A—C—G   **3.** A—A—T—C—G
    2. T—T—A—G—C   **4.** T—T—U—G—C

42. This DNA molecule acts as a template for RNA construction in the process of
MG **1.** gene replication   **3.** osmosis
    2. protein synthesis   **4.** synapsis

43. A change in the base sequence in this DNA molecule is known as
1. homeostatic control
2. gene segregation
3. disjunction
4. a gene mutation

44. Which series is arranged in correct order according to *decreasing* size of structures?
1. DNA, nucleus, chromosome, nucleotide, nitrogenous base
2. nucleotide, chromosome, nitrogenous base, nucleus, DNA
3. nucleus, chromosome, DNA, nucleotide, nitrogenous base
4. chromosome, nucleus, nitrogenous base, nucleotide, DNA

45. In human beings, a gene mutation results from a change in the
1. sequence of the nitrogenous bases in DNA
2. chromosome number in a sperm
3. chromosome number in an egg
4. sequence of the sugars and phosphates in DNA

For each phrase in questions 46 through 49, select the cell compound *chosen from the list below*, that is best described by that phrase. (*A number may be used more than once or not at all.*)
Types of Cell Compounds

(1) DNA
(2) Messenger RNA
(3) Transfer RNA
(4) ATP

46. A double-stranded molecule that contains instructions for the manufacture of cell protein

47. Provides energy for biochemical reactions

48. Replicates before mitosis

49. Carries protein building blocks to ribosomes

50. Which two bases are present in equal amounts in a double-stranded DNA molecule?
1. cytosine and thymine
2. adenine and thymine
3. adenine and uracil
4. cytosine and uracil

51. If one strand of a DNA molecule has the base sequence A—G—C—T—A, the complementary strand of DNA will have the base sequence
1. A—G—C—T—A
2. U—C—G—A—T
3. U—C—G—A—U
4. T—C—G—A—T

52. During replication, the strands of a double-stranded DNA molecule separate from each other when bonds are broken between their
1. nitrogenous bases
2. five-carbon sugars
3. phosphate groups
4. amino acids

53. Which molecules are composed of units known as nucleotides?
    MG 1. DNA molecules only
    2. messenger RNA molecules only
    3. transfer RNA molecules only
    4. both DNA and RNA molecules

54. Which is a major difference between messenger RNA molecules and transfer RNA molecules?
    MG
    1. Messenger RNA molecules contain ribose, and transfer RNA molecules contain deoxyribose.
    2. Messenger RNA molecules function in carrying coded information to the ribosomes, and transfer RNA molecules function in carrying amino acids to the ribosomes.
    3. Messenger RNA molecules contain thymine, and transfer RNA molecules contain uracil.
    4. Messenger RNA molecules function when they are double-stranded, and transfer RNA molecules function when they are single-stranded.

55. Which nitrogenous base is normally present in DNA but absent from RNA?
    MG 1. adenine   3. thymine
    2. cytosine   4. guanine

56. Specific proteins produced in a cell are directly related to the
    MG 1. number of mitochondria in the cell
    2. type of ribosomes in the cell
    3. sequence of sugars and phosphates in the cell
    4. sequence of nucleotides in the DNA of the cell

57. The sequence of nucleotides in DNA is responsible for the biochemical reactions in cells, since this sequence determines the arrangement of
    MG
    1. fatty acids in lipids
    2. amino acids in enzymes
    3. monosaccharides in starches
    4. organelles in cells

58. The genetic code for one amino acid molecule consists of
    MG 1. five sugar molecules
    2. two phosphates
    3. three nucleotides
    4. four hydrogen bonds

59. Substances that cause a chemical change in the DNA of a cell are known as
    MG 1. glycogens   3. chromatids
    2. mutagens   4. chromosomes

60. A mutation is any change that affects the nitrogenous base sequence in the structure of
    MG
    1. an ATP molecule   3. a ribosomal protein
    2. a nucleotide   4. a DNA molecule

61. Which compound carries amino acids from the cytoplasm to specific sites of protein synthesis?

MG

1. adenosine triphosphate
2. deoxyribose nucleic acid
3. messenger RNA
4. transfer RNA

62. By which process can a group of genetically identical plants be rapidly produced from the cells of a single plant?

MG

1. screening
2. chromosomal karyotyping
3. genetic engineering
4. cloning

63. Scientists have been able to synthesize insulin, interferon, and human growth hormone through work with

MG

1. gametogenesis
2. ribosomal induction
3. RNA hydrolysis
4. recombinant DNA

64. Which are two factors that change gene frequencies in a population?

MG

1. no mutations and large populations
2. no migration and no mutations
3. large populations and random mating
4. mutations and nonrandom mating

65. The term "gene pool" refers to

MG

1. some of the traits in a given population
2. some of the traits in a given species
3. the total of all the heritable genes for all the species in a given community
4. the total of all the heritable genes for all the traits in a given population

UNIT 6

HOW DO SPECIES CHANGE OVER TIME?

KEY IDEAS

Many theories, some scientific, some not, have been advanced to explain the diversity of living things on earth. The weight of scientific evidence indicates that organisms have been living and changing on earth for over 3 billion years. This unit examines such evidence and the scientific theories of evolution that have been developed in the past century by recognized scientists. It also focuses on the biological mechanisms studied in Units 4 and 5 as the principal forces at work in the evolutionary process.

KEY OBJECTIVES

Upon completion of this unit, you should be able to:

☐ Define evolution as a process by which living things change over long periods of time.

☐ Understand the interrelationships among many branches of science that have provided observations and other evidence supporting evolutionary theory.

☐ List and describe the principal evidences supporting modern scientific theories of evolution.

☐ Describe the significant aspects of organic evolution presented in the major scientific theories of evolution.

☐ Describe how modern science has theorized the events surrounding the early history of the earth and the early evolution of life.

I. ORGANIC EVOLUTION

Evolution is a process of change through time. **Organic evolution** refers specifically to the mechanisms thought to govern the changes in living species over geologic time. These changes may include variation within a species or the production of new species.

212

II. THE HISTORICAL DEVELOPMENT OF EVOLUTION THEORY

Since prehistoric times, human beings have no doubt asked many questions about their origins and the origins of other life forms. The earliest "theories" about the origin and evolution of life were shrouded in myth and superstition. In many cultures, nonscientific ideas about origin and evolution are still held. However, because these ideas cannot be directly supported by scientific evidence, they are beyond the scope of this course of study and will not be considered in this book. Instead, this unit focuses on the scientific theories of evolution developed in the past 200 years of scientific study.

The science of evolution, like life itself, has developed over time. Early scientists working in evolution science attempted to explain the diversity of life around them. Such diversity was observed in the differences in structure, function, and behavior among organisms throughout the world.

Early attempts at developing workable theories of evolution were thwarted by an incomplete body of scientific knowledge. Modern theories of evolution are built on earlier theories, but add to them many new ideas drawn from the growing quantity of supporting evidence available, much of it from the field of genetics. Theories of evolution assume that modern life forms have evolved from previously existing life forms.

A. Supporting Observations

As in most areas of science, the study of evolution is based on theoretical concepts and assumptions. This situation indicates that the science of evolution contains many ideas that remain to be proved beyond doubt. As in most areas of science, scientists studying evolution have developed their ideas concerning its mechanisms after studying the evidences available to them. These evidences have come from many different sources, including fossils from the geologic record and studies that show comparisons among different kinds of organisms in the areas of cytology (cell study), biochemistry, embryology, and anatomy.

1. **Geologic record.** There is no direct evidence of the exact date of the earth's formation. By studying the indirect evidence drawn from radioactive dating of the earth's rocks, however, geologists have estimated the age of the earth to be between 4.5 and 5.0 billion years. In this estimation, scientists have assumed that the earth is at least as old as the oldest rocks so far discovered.

   Found frequently within certain types of rocks are **fossils**, the preserved direct or indirect evidence of organisms that lived in the past. Fossils are most commonly discovered embedded in sedimentary rock, such as sandstone or limestone. However, the remains of organisms may also be found preserved in ice or permanently frozen soil or in naturally occurring tars or other chemical deposits. Knowing the age of the rock layers in which fossils are embedded enables scientists to determine with reasonable accuracy the age of those fossils. Fossils have been discovered that have been dated by scientific methods to be over 3 billion years old.

In undisturbed layers (**strata**) of sedimentary rock, the lowest layers were laid down first, the middle layers next, and the topmost layers last. It follows logically, then, that fossils found embedded in the lower strata are older than those in the upper strata. In fact, deeper layers of such rock are known to contain fossils of older, simpler life forms, whereas strata found near the surface contain younger and generally more complex forms (see Figure 6.1).

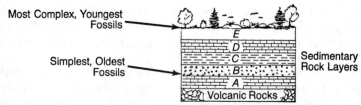

Figure 6.1   Geologic Record

Scientists have been able to identify a certain degree of continuity among the fossils in consecutive layers of fossil-bearing rock. It is possible to find fossils in upper strata that resemble those in lower strata, even though they are clearly of different species. This fact lends support to the theory that genetic links exist between modern life forms and ancient forms. It also suggests that genetic links exist among diverse modern life forms by virtue of their common links to ancestral species. It is thought that modern species having similar structures share these ancestral forms in common. Even diverse species displaying few obvious similarities, such as earthworms and mollusks, are thought to share distant common ancestors. This concept of **common ancestry**, in which two divergent forms can trace their inheritance to a single preexisting life form, is central to an understanding of the science of evolution.

2.   **Comparative cytology.** As we learned in Unit 1, the cell is the structure that all living things have in common. The organelles located within these cells function in much the same way in the cells of any organism. Despite the basic similarity of all cells, certain differences among cells of different species are known to exist. Organisms with very similar cell structure are usually considered to be more closely related than organisms whose cells show many differences.

3.   **Comparative biochemistry.** In Unit 5 we learned that each polypeptide in the cell is coded by a unique strand of DNA. We also

learned that the ability to produce such polypeptides may be passed from generation to generation through the processes of reproduction and genetic inheritance. Related organisms, therefore, having inherited their characteristics from common ancestors, may be expected to share many genes and their corresponding enzymes in common.

Biochemical analysis of enzymes and other proteins shows that a great deal of similarity exists in the biochemical makeup of organisms known to be related genetically. For example, the complex protein hemoglobin is found in the blood of many vertebrate species, whereas it is less common among invertebrates. Generally, the more closely related two organisms are, the more similar is their biochemical makeup. Likewise, organisms that are not as closely related share fewer biochemical similarities. However, even organisms that display little structural similarity may produce many enzymes in common, and thereby be shown to be closely related.

4.   **Comparative anatomy.** The determination of similarities in anatomical (structural) features is perhaps the most common method of demonstrating biological relationships among organisms. As we learned in Unit 1, this method provides the basis for biological classification, in which organisms are placed in the same kingdom, phylum, genus, or species on the basis of their degrees of anatomical similarity.

Similar organisms can be shown to have limbs, internal organs, or other structures that are constructed similarly. Such structures, called **homologous structures**, are believed to have originated from common ancestral forms of the same organs. Such structures may or may not function in the same way in all related organisms. Examples of homologous structures include the forelimbs of vertebrate species such as the bird, bat, horse, whale, and human being (see Figure 6.2). Because of their similar bone structures, the forelimbs of all these species are thought to have developed from a common ancestral forelimb. However, the forelimbs in these species take on quite different functions, depending on the physiological structure of each organism.

5.   **Comparative embryology.** Studies of the reproductive process in many different organisms reveal patterns of similarity among closely related organisms. In Unit 4 we learned about some of these patterns (e.g., the existence of the amnion in the egg of a bird and the uterus of a mammal). In fact, the embryos of most vertebrates show a great deal of similarity in the early stages of development (see Figure 6.3). It becomes possible to differentiate among the embryos of some species only in the later stages of development. This is thought to be an indicator of common ancestry among these species.

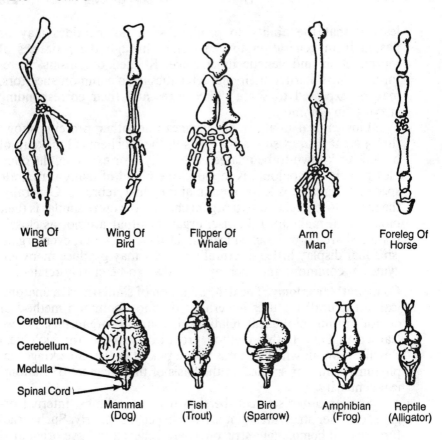

Wing Of Bat

Wing Of Bird

Flipper Of Whale

Arm Of Man

Foreleg Of Horse

Cerebrum
Cerebellum
Medulla
Spinal Cord

Mammal (Dog)

Fish (Trout)

Bird (Sparrow)

Amphibian (Frog)

Reptile (Alligator)

Figure 6.2   Homologous Forelimbs And A Comparison Of Vertebrate Brains

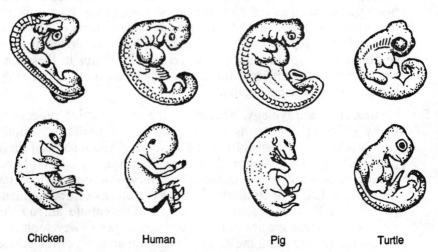

Chicken

Human

Pig

Turtle

Figure 6.3   Comparison Of Vertebrate Embryos

QUESTION SET 6.1

1. Organic evolution is best described as
   1. a process of change through time
   2. a process by which an organism becomes extinct
   3. the movement of large landmasses
   4. the spontaneous formation of all species

2. A geologist finds fossils in each of the undisturbed rock layers represented in the diagram below. The fossils are all structurally similar. Which is the most likely conclusion that the geologist would make?

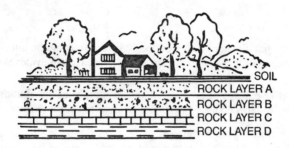

   1. All the fossils are of the same age.
   2. The relative ages of the fossils cannot be determined.
   3. The fossils in rock layer *D* are older than those in layer *A*.
   4. The fossils in rock layer *B* are older than those in layer *C*.

3. A scientist studying fossils in undisturbed layers of rock identified a species that, he concluded, had changed little over the years. Which observation probably would have led him to this conclusion?
   1. The simplest fossil organisms appeared only in the oldest rocks.
   2. The simplest fossil organisms appeared only in the newest rocks.
   3. The same kind of fossil organisms appeared in old and new rocks.
   4. No fossil organisms of any kind appeared in the newest rocks.

4. Biochemical analysis has shown that hemoglobin molecules found in monkeys are very similar to those found in human beings. Which concept is supported by this analysis?
   1. Homologous structures exist in all vertebrates.
   2. Embryonic development in human beings and monkeys is identical.
   3. Monkeys and human beings have a common ancestor.
   4. Invertebrates and vertebrates have a common ancestor.

5. Among many species, those most closely related to each other would probably
   1. live in the same geographic area
   2. contain similar enzymes and hormones
   3. have similar food requirements
   4. live during the same time period

6. From the information given in the chart below, which two organisms are most closely related?

Enzyme Type

	1	2	3	4
A	X		X	
B				X
C	X	X	X	X
D	X		X	X

Organism

X = Enzyme Present
In Organism

1. *A* and *B*   3. *C* and *D*
2. *B* and *C*   4. *D* and *B*

7. It is thought that all citrus fruit trees evolved from a common ancestor because of their common ability to synthesize citric acid. This type of evidence of evolution is known as

1. comparative embryology   3. geographical distribution
2. comparative biochemistry   4. anatomical similarity

8. The leg structures of many different vertebrates are quite similar in number and location of bones. Most scientists would probably explain this on the basis of

1. needs of the organism   3. chance occurrence
2. common ancestry   4. inheritance of acquired traits

9. Structures having a similar origin but adapted for different purposes, such as the flipper of a whale and the arm of a human being, are called

1. homozygous structures   3. homologous structures
2. identical structures   4. embryological structures

10. The observation that organisms have certain structural similarities most likely suggests that they

1. have similar genes
2. are more numerous than other types of organisms
3. are more advanced than other forms of life
4. are members of the same food chain

11. The modern classification system is based on structural similarities and

1. evolutionary relationships
2. habitat similarities
3. geographic distribution
4. Mendelian principles

12. Some starfish larvae resemble some primitive chordate larvae. This similarity may be used to suggest that primitive chordates

1. share a common ancestor with a starfish
2. evolved from modern-day starfish
3. evolved before starfish
4. belong to the same population as starfish

B. Theories of Evolution

Theories of evolution that have been proposed over time have attempted to explain the diversity of life forms on earth. Species diversity is actually based on the variety of adaptations found among these species. The term **adaptation** refers to any structural, functional, or behavioral characteristic of an organism that helps it to better survive in its environment. Most organisms are found to be extremely well adapted for survival in their particular environments. Attempts to describe the mechanisms by which these adaptations have come about have formed the basis of evolution theory.

1. Jean Lamarck

In the eighteenth century in France, **Jean Lamarck** proposed a theory of evolution that included two main ideas:

1.1. Use and Disuse. Lamarck theorized that adaptations developed in or disappeared from species as a function of the degree to which these adaptations were needed by members of the species. Lamarck further speculated that new traits not already present in a species would spontaneously appear on the basis of need. Lamarck believed that the more an organ was used (or needed), the larger and more efficient it would become. At the same time, an organ that was not used would diminish in size and eventually disappear.

1.2. Transmission of Acquired Traits. Traits thus developed as needed within the lifetime of an organism, according to Lamarck, could be passed on to future generations through the process of reproduction. This trait transmission was theorized to aid the species by adding new, favorable adaptations that would result in improved survival ability for the species.

Lamarck's theories were an important first step in the development of a scientific view of the evolutionary process. However, later experiments conducted by **August Weismann** and other scientists failed to support Lamarck's views. Weismann measured and surgically removed the tails of laboratory mice over several generations. He hypothesized that, if Lamarck's theory of transmission of acquired traits were correct, the offspring of each successive generation should show diminished tail length. Actual experimental data, however, showed no reduction of tail length in these mice. These experiments had the effect of disproving Lamarck's theory.

2. Charles Darwin

In the 19th century in England, a naturalist named **Charles Darwin** devised a theory of evolution based on **variation** and **natural selection**. This theory forms the basis of the modern theory of evolution. Included in this theory were five main ideas:

2.1. Overproduction. Darwin observed that naturally occurring species have a tendency to produce far more offspring than can possibly survive to become reproducing adults.

2.2. Competition. Darwin further observed that, despite the tendency to overproduce, the number of individuals in natural populations tends to remain relatively constant over many generations. This suggested to Darwin that, within each species, there is a struggle for survival that eliminates many individuals before they reach reproductive maturity. This struggle may be termed "intraspecies (within species) competition."

2.3. Variation. As a naturalist, Darwin was very familiar with the wide range of individual variation (difference of form) among individuals of a species. He observed and noted this variation despite the fact that he could not explain its genetic basis.

2.4. Survival of the Fittest. Darwin felt that variation and selection by natural forces were the keys to the evolutionary process. "Survival of the fittest" is a phrase commonly associated with this process. "Fitness" in this context may include an ability to resist disease, an ability to withstand some environmental condition, or an ability to produce more offspring than another variety. A particular set of adaptations that proves to be successful in promoting the survival of a species variety may give that variety a unique fitness over other varieties of the same species. Such fitness may be particularly advantageous in the face of harsh environmental conditions. In a sense, according to this view, the natural environment "selects" the varieties that will survive and those that will perish. This selection process is termed "natural selection" (see Figure 6.4).

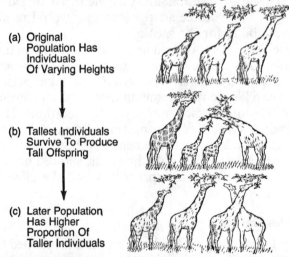

(a) Original Population Has Individuals Of Varying Heights

(b) Tallest Individuals Survive To Produce Tall Offspring

(c) Later Population Has Higher Proportion Of Taller Individuals

The animals (giraffes) that are less fit (have shorter necks) perish, leaving the better fit (those with longer necks) to survive and reproduce, passing on their favorable traits to the next generation.

Figure 6.4   Natural Selection

2.5. Reproduction. Since the surviving "fit" organisms will be the only members of the species left to reproduce, their favorable traits will be passed on to the next generation. By this means, the adaptive advantages of one generation may be passed on to the next, perpetuating them in the species. When enough new adaptations occur within an isolated population, a new species distinct from ancestral forms arises. The process by which new species arise is known as **speciation.**

The main weakness of Darwin's theory was its inability to explain the sources of the variations that he observed. It should be remembered that Darwin's work was performed before there was a good scientific understanding of the mechanisms of genetic inheritance.

QUESTION SET 6.2

1. Which is a concept of Lamarck's theory of evolution?
   1. Natural selection causes an organism to evolve.
   2. Mutations cause new variations.
   3. Acquired characteristics can be inherited.
   4. Genes control an organism's characteristics.

2. The pig has four toes on each foot. Two of the toes are very small and do not have a major function in walking. Lamarck probably would have explained the reduced size of the two small toes by his evolutionary theory of
   1. natural selection
   2. mutation
   3. use and disuse
   4. synapsis

3. An athlete explains that his muscles have become well developed through daily weight lifting. He believes that his offspring will inherit this trait of well-developed muscles. This belief would be most in agreement with the theory set forth by
   1. Darwin
   2. Lamarck
   3. Weismann
   4. Mendel

4. The wings of experimental fruit flies were clipped short each generation for 50 generations. The fifty-first generation emerged with normal-length wings. This observation would tend to disprove the theory of evolution based on
   1. inheritance of mutations
   2. inheritance of acquired characteristics
   3. natural selection
   4. survival of the fittest

5. Which concept is most closely associated with Darwin's original theory of evolution?
   1. Gene mutation is the basis for inherited variations.
   2. Gene sorting and recombination are the basis for variations.
   3. Survival of any species depends on the appearance of favorable mutations.
   4. The organisms best adapted to the environment are most likely to reproduce.

6. According to the theory of natural selection, genes responsible for new traits that are beneficial to the survival of a species in a particular environment will usually

1. decrease suddenly in frequency
2. decrease gradually in frequency
3. not change in frequency
4. increase in frequency

7. Of the 500 eggs produced by a certain female frog, only 10% developed into adult frogs. Which part of Darwin's theory does this best illustrate?

1. Favorable variations are not inherited.
2. There is a struggle for existence among organisms.
3. Mutations occur by chance.
4. Mating occurs in a random manner in a species.

8. Darwin's theory of evolution did *not* include the concept that

1. genetic variations are produced by mutations and sexual recombinations
2. organisms that survive are best adapted to their environment
3. population sizes remain constant because of a struggle for survival
4. favorable traits are passed from one generation to another

9. In a certain area, DDT-resistant mosquitoes now exist in greater numbers than 10 years ago. What is the most probable explanation for this increase in numbers?

1. Genetic differences permitted some mosquitoes to survive DDT use.
2. Mosquito eggs were most likely to have been fertilized when exposed to DDT.
3. DDT acted as a reproductive hormone for previous generations of mosquitoes.
4. DDT serves as a new source of nutrition.

10. The graph below represents the percent of variation for a given trait in four different populations of the same species. These populations are of equal size and inhabit similar environments.

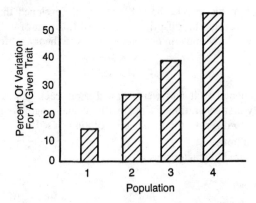

In which population is the greatest number of individuals most likely to survive significant environmental changes related to this trait?

1. 1	**3.** 3
2. 2	**4.** 4

11. The theory that birds and reptiles share common ancestry is supported by the evidence that they both

1. occupy similar niches
2. have similar environmental requirements
3. show structural similarities during their development
4. have evolved as separate groups at about the same time

3.   Modern Evolution Theory

The modern concept of evolution is based primarily on Darwin's theory of natural selection. In addition, it incorporates the information available from modern research into the mechanisms of genetic inheritance, both for individuals and for populations. (See Unit 5 for a more detailed discussion of the science of genetics.)

3.1. Production of Variations. We learned in Unit 5 that the major sources of variation in sexually reproducing species are the cellular processes of genetic mutation, meiosis, and fertilization. Mutation is important in providing the new genes that may lead to the production of new genetic traits. Although the majority of such new traits are harmful or neutral, a small percentage may provide significant adaptive advantages to a species. Within the reproductive process, meiosis and fertilization provide the mechanism by which new combinations of both old and new traits may be "tried out" as new varieties within a species.

3.2. Natural Selection. The modern theory of evolution, like Darwin's theory, recognizes the importance of natural selection in the evolutionary process. Each individual in a species population is engaged in a struggle for existence in its own environment. The individuals that survive (are "selected") are assumed to be those best adapted to survive under the particular set of environmental conditions in question. The individuals that perish are considered to be those less well adapted for survival. When the survivors later reproduce, they tend to pass on the genes associated with their adaptive advantages. In this way, nature provides selection pressures that limit or eliminate traits which do not promote individual survival. This is the primary role of natural selection in the process of evolution.

The frequencies of these favorable genes, then, increase in the gene pool relative to the frequencies of these genes controlling less favorable traits. This shift of gene frequencies in the gene pool of a species population is thought by scientists to constitute the mechanism of evolution. (See Unit 5 for a more detailed discussion of the science of population genetics.)

It is important to recognize that the adaptations in a population that are favorable under one set of environmental conditions may prove to be highly unfavorable under another set. At the same time, traits present in the gene pool of a population that have low or neutral survival value may increase markedly in value if the environment changes or if the population moves to a new environment. Here are two examples:

- In this century, the environments of many insect pests (e.g., houseflies, mosquitoes, roaches, and weevils) changed when new chemical insecticides, such as DDT, were introduced into these environments. Most of the insects that came into contact with the insecticide died, since they were not genetically resistant to it. A small number of the organisms were genetically resistant to the chemicals, however, and survived to reproduce offspring that were also genetically resistant. Today, such resistant strains present a problem to chemists attempting to develop other new insecticides to deal with insect infestations. In this case, the insecticide has acted as an agent of natural selection.

- A similar situation has occurred in the evolution of antibiotic-resistant strains of bacteria in some hospital environments. As in the insect example above, newly introduced antibiotics, such as penicillin, represented a change in the environment of disease-causing bacteria. Nonresistant bacteria died when they came into contact with the antibiotic, but a few resistant bacteria survived. The resistant bacteria gave rise to entire strains of the bacterial species that are resistant to the antibiotic. The selecting agent in this case is the antibiotic.

At any one time, the frequency of a particular gene may either be increasing, decreasing, or remaining constant, depending on its survival value relative to genes for contrasting traits in the gene pool.

3.3. Speciation. The processes of mutation, allelic recombination, and natural selection are constantly at work creating, testing, and selecting new adaptations in all species. When enough unique adaptations have been accumulated in a species population so that it becomes distinct from other populations of the same species, it may be classified as a new variety of the species. Under the right set of conditions, a species variety may become a new species. The process by which new species arise from parent species is known as **speciation**. Speciation is considered to have occurred when a species variety is no longer able to interbreed successfully with the parent species to produce fertile offspring. The process of speciation may be accelerated considerably if one of the following occurs:

3.3.1. Geographic isolation. As its name implies, **geographic isolation** (see Figure 6.5) refers to the separation of species populations by significant geographic barriers such as large bodies of water, mountains, deserts, canyons, or similar features. Geographic isolation aids speciation by segregating a small group from a main population. The small, isolated population is more likely than the large main population to experience shifts in gene pool frequencies for the following reasons:

- The gene pool frequencies of the isolated population may be different from those of the main population.

- Mutations that occur in the isolated population will be different from those occurring in the main population. A single mutation in the gene pool of a small population represents a significantly greater proportion of that gene pool than the same mutation would represent in the gene pool of a large population.

- The geographic separation of the two groups increases the likelihood that environmental (selection) pressures acting on them will be different.

Darwin's studies of species around the world provided many examples of the effects of geographic isolation. In one instance, Darwin documented more than a dozen species of finch (a small bird) that inhabited a remote island group (the Galapagos Islands) off the coast of Ecuador, South America. Each species displayed adaptations distinct from those of other finch species in the Galapagos and from the parent population on the mainland of South America. Darwin hypothesized that a small group had been geographically isolated from the parent population many years before his visit. He further hypothesized that this isolation had provided the opportunity for speciation by a mechanism similar to the one outlined above. Another example of speciation by geographic isolation is the existence of marsupial species in Australia that are thought to have arisen after the isolation of an ancestral marsupial species on the Australian continent thousands of years ago.

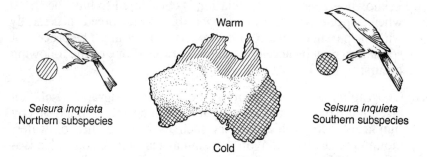

Warm

Seisura inquieta
Northern subspecies

Seisura inquieta
Southern subspecies

Cold

The Dry Central Region Of Australia Presents A Significant Barrier To The Migration Of Two Populations Of The Species *Seisura Inquieta,* Preventing Interbreeding. The Two Populations Already Show Significant Variation In Size. Over Time They Will Diverge Further And Become Separate Species.

Figure 6.5   Geographic Isolation.

3.3.2. Reproductive isolation. If two populations of a species are geographically isolated from each other for a sufficiently long period of time and they become significantly different in terms of their adaptations, it may happen that, even if the geographic barriers are removed, they will not be able to interbreed and produce fertile offspring. Under these conditions of **reproductive isolation** from each other, the two species are considered to have become distinct, separate species.

3.4. Time Frame for Evolution. Most scientists have come to a general agreement that the mechanisms controlling the evolutionary process are similar to those outlined above. However, considerable debate still exists concerning the time frame in which this mechanism operates. Two theories in regard to the time-frame question are as follows:

3.4.1. Gradualism. The theory of **gradualism** assumes that evolutionary change is slow, gradual, and continuous. A gradualistic view of evolution is supported by fossil records of species that display slight changes in each sedimentary layer, leading, however, to a significant divergence (difference) between specimens found in the bottom and the top layers.

3.4.2. Punctuated equilibrium. The theory of **punctuated equilibrium** assumes that species experience long geologic periods of stability (of a million years or more), in which little or no significant change takes place. This stability, then, is "punctuated" by brief periods (of a few thousand years) in which dramatic changes occur within species. During these brief periods of change, many species are thought to evolve very quickly from parent species. Such a view is supported by fossil evidence in which little change is noted between most sedimentary layers, but sudden "bursts" of change are evident in the fossils of a few sedimentary layers.

QUESTION SET 6.3

1. A possible conclusion based on the modern theory of evolution is that
   1. most species have changed
   2. all living things developed from fish
   3. most plants and animals can interbreed
   4. all dogs are more closely related to fish than to whales

2. The study of mutations is important to the modern theory of evolution because it helps to explain
   1. differentiation in embryonic development
   2. stability of gene pool frequencies
   3. the extinction of the dinosaurs
   4. the appearance of variations in organisms

3. Species that are most likely to adapt to a changing environment are those that reproduce by
   1. asexual means
   2. self-pollination
   3. cross-fertilization
   4. self-fertilization

4. According to modern biologists, hereditary variations are due to the
   1. use or lack of use of organs
   2. inheritance of characteristics acquired during development
   3. need for adapting to a changed environment
   4. genetic changes resulting from mutation and recombination

5. In areas of the American Southwest, certain insect species are quickly becoming resistant to continuous applications of chemical insecticides. The increase in the number of insecticide-resistant species is due to
   1. inheritance of acquired traits
   2. variability through asexual reproduction
   3. geographic isolation
   4. natural selection

6. Over a long period of time the organisms on an island changed so that they could no longer interbreed with the organisms on a neighboring island. This inability to interbreed is known as
   1. hybridization
   2. reproductive isolation
   3. artificial selection
   4. survival of the fittest

7. In an environment, barriers prevent an organism from entering other environments. This phenomenon illustrates the concept of
   1. punctuated equilibrium
   2. geographic isolation
   3. genetic variation
   4. natural selection

8. Populations of a species may develop traits different from each other if they are isolated geographically for sufficient lengths of time. The most likely explanation for these differences is that
   1. acquired traits cannot be inherited by offspring
   2. environmental conditions in the two areas are identical
   3. genetic recombination tends to be different in both populations
   4. mutations are likely to be the same in both populations

9. Which evidence for evolution is described as genetic change resulting from isolation?
   1. comparative anatomy
   2. comparative embryology
   3. fossil remains
   4. geographic distribution

10. The evolution of antibiotic-resistant strains of bacteria is an illustration of
    1. natural selection   3. random mating
    2. use and disuse   4. geographic isolation

11. Five species of an animal genus were found on an island 50 miles from the mainland. However, only two species of the same genus were found on the mainland. The most probable reason for the greater diversity on the island is
    1. random mating on the island
    2. more varied environment on the mainland
    3. increased mutation rates on the mainland
    4. genetic isolation on the island

III. HETEROTROPH HYPOTHESIS OF THE ORIGINS OF LIFE

Another question that has challenged scientist and nonscientist alike over time is the question of the origin of life. In Unit 1 we learned that the cell theory assumes that all cells arise from previously existing cells. But what gave rise to the "first cell"? Scientists have proposed the **heterotroph hypothesis** to help explain the origin of the first primitive life forms on the ancient earth. This scientific hypothesis assumes that the first primitive life forms were not able to manufacture their own food (i.e., were heterotrophic). The **heterotroph hypothesis** is consistent with much of the currently accepted scientific theory on the origins of the universe and with current understandings of the sciences of biology and biochemistry. However, like many hypotheses developed to explain phenomena that cannot be directly observed and measured, the heterotroph hypothesis is based on extensions of basic assumptions about the earth's origins.

A. Conditions on the Early Earth

Astronomical and geological evidence indicates that the earth formed from clouds of cosmic dust and gas over 5 billion years ago. After condensing to a semisolid form, the earth required hundreds of millions of years to cool to a point at which the chemical substances that comprised the earth became stable. During that period, the earth remained an extremely hot environment filled with inorganic and simple organic chemical substances such as water (H_2O), ammonia (NH_3), methane (CH_4), hydrogen gas (H_2), and various mineral salts. These substances were mixed together in the primitive atmosphere and oceans to form a "hot, thin soup," in which random chemical reactions could occur at a rapid rate. Gaseous oxygen and carbon dioxide are thought *not* to have been present in this early stage.

In addition to the abundance of raw materials present on the early earth, a variety of energy forms were available that are not as apparent on the earth today. Included among these energy forms were heat from volcanic activity, electrical energy from lightning discharges, X rays and ultraviolet radiation from unfiltered solar radiation, and radioactive radiation from newly formed mineral elements. These energy forms enabled many chemical reactions to occur that would not be possible outside a chemistry laboratory today.

B. Formation of Primitive Life Forms

The formation of the first primitive life forms is thought to have occurred in the following manner:

1. It is believed that random chemical reactions in the "hot, thin soup" led to the synthesis of simple biochemical molecules (e.g., simple sugars, lipids, and amino acids), which subsequently interacted to form longer "chain" molecules, such as polypeptides, polysaccharides, and hydrocarbon chains. The larger molecules, attracted to each other chemically, are thought to have come together to form **aggregates** (groupings or clusters) of such molecules. These aggregates are assumed to have constituted the first simple cell-like structures.

    This particular aspect of the heterotroph hypothesis was tested in the laboratory by scientist **Stanley Miller**, who set up a controlled environment that simulated the one described above (see Figure 6.6). After several days of continuous energy input, Miller's experimental flasks contained the precursors (beginning forms) of several simple organic substances, including amino acids, simple sugars, and nucleotides. In later experiments, **Sidney Fox** and other scientists demonstrated that Miller's precursors could be joined together into complex molecular arrangements and aggregated to form cell-like structures.

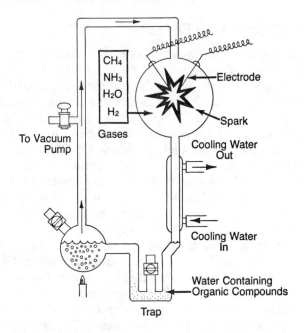

Figure 6.6 Stanley Miller's Experiment

2. It is thought that the molecular aggregates took on additional preformed organic molecules from the "hot, thin soup" of the seas and incorporated them into their structure. In a sense, such incorporation constituted a simple form of heterotrophic nutrition, since preformed "nutrients" were used to add to and maintain the structure of the aggregates.

3. Increasing structural complexity of the cellular aggregates, including the formation of complex proteins and nucleic acids, is thought to have led to the ability to reproduce new cellular aggregates. The ability to reproduce is considered to have represented the last critical step leading to the living condition, marking the difference between mere chemical aggregates and true living cells.

C. The Evolution of Autotrophic Nutrition

It is thought that the earliest living cells obtained their energy by means of a cellular process similar to fermentation (previously described in Unit 2), a natural by-product of which is carbon dioxide. This carbon dioxide, it is believed, built up in concentration in the earth's atmosphere as a result of widespread fermentative activity. Certain organisms, having spontaneously evolved an ability to use the newly introduced gaseous carbon dioxide to manufacture their own organic foods, became the earth's first food producers, or **pioneer autotrophs.**

D.　*The Evolution of Aerobic Respiration*

Extensive autotrophic nutritional activity, similar to the photosynthetic process discussed previously in Unit 2, added free molecular oxygen to the earth's atmosphere. It is likely that this new environmental condition proved toxic to many of earth's newly evolved species. A few species, having spontaneously evolved an ability to utilize this molecular oxygen in the respiratory process, became the earth's first aerobes.

E.　*Evolution and Species Diversity*

These earliest evolutionary events are at best incompletely understood. How they actually occurred and under what circumstances they took place can only be a matter of speculation supported by scientific evidence. What is known, however, is that modern species show wide divergence (difference) of form and function. Modern species may be autotrophic or heterotrophic; they may be aerobic or anaerobic; they may reproduce sexually or asexually; they may differ from other species in countless ways. It is thought that these varieties came about through the evolutionary process described above, filling the earth's available environments with species able to survive the various physical conditions they encountered. The "geological clock" is diagramed in Figure 6.7.

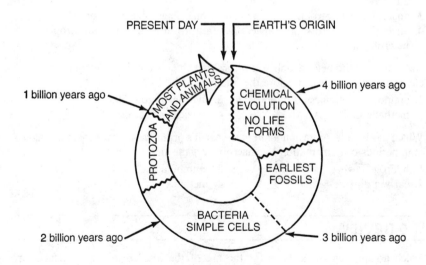

Figure 6.7　Geological Clock

The geological clock shows all of earth's history from its formation 5 billion years ago to the present. Human beings appeared at "11:59," "1 minute" before the present!

Figure 6.7　Geological Clock

QUESTION SET 6.4

1. The heterotroph hypothesis is an attempt to explain
    1. how the earth was originally formed
    2. why simple organisms usually evolve into complex organisms
    3. why evolution occurs very slowly
    4. how life originated on the earth

2. According to the heterotroph hypothesis, the earliest heterotrophs must have
    1. been able to synthesize organic molecules from inorganic compounds
    2. used oxygen from the atmosphere for respiration
    3. survived on existing organic molecules in the seas
    4. been unable to carry on anaerobic respiration

3. In an experiment by Stanley Miller, the chemicals methane, hydrogen, ammonia, and water vapor were subjected to a high-energy electrical sparking device at high temperatures. This experiment was an attempt to
    1. produce organic compounds
    2. produce elements
    3. duplicate aerobic respiration
    4. duplicate photosynthesis

4. According to the heterotroph hypothesis, the order in which organisms evolved was
    1. aerobic heterotroph → aerobic autotroph → anaerobic autotroph
    2. anaerobic autotroph → anaerobic heterotroph → aerobic and anaerobic autotroph
    3. anaerobic heterotroph → aerobic autotroph → anaerobic autotroph
    4. anaerobic heterotroph → anaerobic autotroph → aerobic autotroph and heterotroph

5. According to the heterotroph hypothesis, which substance was missing from the environment of the earth before the origin of life?
    1. ammonia molecules   3. hydrogen molecules
    2. methane molecules   4. oxygen molecules

6. Which gas became more abundant in the earth's primitive atmosphere as a result of long periods of fermentation by anaerobic organisms?
    1. hydrogen   3. ammonia
    2. carbon dioxide   4. methane

UNIT 6 REVIEW

1. Which assumption is the basis for the use of the fossil record as evidence for evolution?
    1. Fossils have been found to show a complete record of the evolution of all mammals.
    2. In undisturbed layers of the earth's crust, the oldest fossils are found in the lowest layers.
    3. All fossils can be found embedded in rocks.
    4. All fossils were formed at the same time.

2. The diagram below represents a section of undisturbed rock and the general location of fossils of several closely related species. According to currently accepted evolutionary theory, which is the most probable correct assumption to be made concerning species *A*, *B*, *C*, and *D*?

Species *C* And *D*
Species *C*
Species *A* And *B* And *C*
Species *A* And *B*
Species *A*

1. *A* is the ancestor of *B*, *C*, and *D*.
2. *B* was already extinct when *C* evolved.
3. *C* evolved more recently than *A*, *B*, and *D*.
4. *D* is the ancestor of *A*, *B*, and *C*.

3. The undisturbed upper layers of rocks usually contain fossils of organisms that are
1. more complex than those found in the lower layers
2. less complex than those found in the lower layers
3. identical to those found in the lower layers
4. different from any other fossils found in the same layer

4. Many related organisms are found to have the same enzymes and hormones. This suggests that
1. enzymes work only on specific substrates
2. enzymes act as catalysts in biochemical reactions
3. organisms living in the same environment require identical enzymes
4. these organisms may share a common ancestry

5. Digestive enzymes and hormones are found to be similar in many mammals. These findings are examples of
1. similar anatomical structures
2. similar homologous structures
3. embyrological similarities
4. biochemical similarities

6. Which is an example of evidence of evolution based on comparative biochemistry?
1. Sheep insulin can be substituted for human insulin.
2. The structure of a whale's flipper is similar to that of a human hand.
3. Human embryos have a tail-like structure at one stage in their development.
4. Both birds and bats have wings.

7. Which term describes appendages that may have different functions, but are similar in structure and are assumed to have the same evolutionary origin?
1. fossils   **3.** homologous
2. homozygous   **4.** mutations

8. The presence of gill-like slits in a human embryo is considered to be evidence for the
   1. theory that fish and mammals have a common ancestry
   2. theory that the first organisms on earth were heterotrophs
   3. close relationship between fish and mammalian reproductive patterns
   4. close relationship between human beings and annelids

9. Lamarck proposed that new organs evolved according to the
   1. needs of the organism
   2. process of natural selection
   3. role of mutation
   4. sorting out of genes

10. "The human earlobe, because it is not used, will probably disappear from the human population in the future."
    This statement reflects a theory proposed by
    1. Darwin   3. Mendel
    2. Watson   4. Lamarck

11. A supporter of the evolutionary theory set forth by Lamarck would probably theorize that the giraffe evolved a long neck because of
    1. need and inheritance of acquired traits
    2. mutations and genetic recombination
    3. variations and survival of the fittest
    4. overproduction and struggle for survival

12. "It is likely that ducks developed webbed feet because ducks needed webbed feet for efficient swimming."
    This attempt to explain the development of webbed feet in ducks most nearly matches the theory of evolution proposed by
    1. Jean Lamarck   3. Gregor Mendel
    2. Charles Darwin   4. Francis Crick

13. Which concept is part of the modern evolutionary theory, but *not* Darwin's original theory?
    1. Variations in traits are caused by mutation and recombination.
    2. Species tend to produce more offspring than can survive.
    3. Better adapted individuals survive to produce offspring.
    4. The environment is responsible for eliminating less fit individuals.

14. Certain insects resemble the twigs of trees on which they live. The most probable explanation for this resemblance is that
    1. the trees caused a mutation to occur
    2. no mutations have taken place
    3. natural selection has favored this trait
    4. the insects needed to camouflage themselves

15. Which factor has the greatest effect on the rate of evolution of animals?
    1. environmental changes
    2. use and disuse
    3. asexual reproduction
    4. vegetative propagation

16. In an attempt to explain the diversity of living things, Darwin's theory of natural selection
    1. proved evolution took place
    2. described how mutations produced variations
    3. showed that only the largest animals survive
    4. described how evolution could have occurred

17. One weakness in Darwin's theory of evolution was that he was *not* able to
    1. explain selection of favorable traits
    2. account for an increase in population
    3. explain the genetic basis for variation in populations
    4. understand competition among individuals of a species

18. Variations among offspring are most frequently produced by the combined effects of
    1. vegetative reproduction and chromosome mutation
    2. gene mutations and sexual reproduction
    3. binary fission and sexual reproduction
    4. gene mutations and parthenogenesis

19. Natural selection can best be defined as
    1. survival of the strongest organisms
    2. elimination of the smallest organisms by the largest organisms
    3. survival of the organisms genetically best adapted to the environment
    4. survival and reproduction of the organisms that occupy the largest area in an environment

20. Modern evolutionary theory has modified the theory of natural selection by
    1. considering survival of the fittest to be invalid
    2. showing that competition does not exist within a species
    3. including a genetic basis for change and variation
    4. accepting the theory of use and disuse

21. According to modern evolutionary theory, which factor *least* influences the pattern of evolution in a population?
    1. sexual reproduction
    2. environmental change
    3. geographic isolation
    4. use and disuse

22. Which reproductive process offers the greatest chance for variation within a species?
    1. fertilization   3. budding
    2. fission   4. spore formation

23. In areas of heavy use of the insecticide DDT, fly populations may show marked resistance to the DDT over a period of time. Someone who accepts the evolutionary theory of Lamarck would most likely explain this observation using the concept of
    1. natural selection
    2. inheritance of acquired characteristics
    3. overproduction of a species
    4. a change in the gene frequencies

24. Geographic isolation of organisms increases the likelihood of genetic differentiation. This genetic differentiation occurs because geographic isolation
  1. prevents interbreeding between populations
  2. prevents interbreeding within populations
  3. stimulates the production of different kinds of enzymes
  4. accelerates the production of new mutations

Base your answers to questions 25 and 26 on the graph and information below and on your knowledge of biology.

Scientists studying a moth population in a woods in New York State recorded the distribution of moth wing color as shown in the graph below. The woods contained trees whose bark color was predominantly brown.

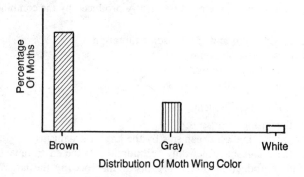

Distribution Of Moth Wing Color

25. A fungus infection affected nearly all trees in the woods so that the coloration of the tree bark was changed to a gray-white color. Which graph shows the most probable results that would occur in the distribution of wing coloration in this moth population after a long period of time?

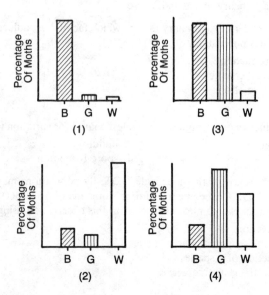

26. As a result of the fungus infection, the change in the moth wing color distribution would most probably occur by the
   1. inheritance of an acquired characteristic
   2. natural selection of favorable variations
   3. ingestion of pigmentation from fungus spores
   4. production of fungus-induced gene mutations

Base your answers to questions 27 and 28 on the paragraph below and on your knowledge of biology.

A fungicide was used to kill the mushrooms in a lawn. Some mushrooms were not affected by the fungicide. The resistant mushrooms reproduced.

27. The fungicide acted as a
   1. neurotransmitter
   2. saprophyte
   3. selecting agent
   4. biological control agent

28. The resistance of some of the mushrooms to the fungicide was
   1. caused by the existence of mutations
   2. transmitted to the mushrooms from the fungicide.
   3. transferred through the food web
   4. developed in response to the fungicide

29. The theory of continental drift hypothesizes that Africa and South America were once a single landmass, but have drifted apart over millions of years. The "Old World" monkeys of Africa, although similar, show several genetic differences from the "New World" monkeys of South America. Which factor is probably the most important for maintaining these differences?
   1. fossil records
   2. comparative anatomy
   3. use and disuse
   4. geographic isolation

30. According to the heterotroph hypothesis, oxygen was added to the primitive atmosphere as a result of
   1. fermentation
   2. lightning discharges
   3. photosynthesis
   4. anaerobic respiration

31. According to the heterotroph hypothesis, the earliest heterotrophs carried out what type of energy-releasing process?
   1. protein synthesis
   2. photosynthesis
   3. anaerobic respiration
   4. aerobic respiration

32. The heterotroph hypothesis states that heterotrophic forms appeared before autotrophic forms as the first living things. A major assumption for this hypothesis is that
   1. sufficient heat was not available for a food-making process
   2. heterotrophic organisms were able to use molecules from the sea as food
   3. lightning and radiational energy were limited to terrestrial areas
   4. moisture in liquid form was limited to aquatic areas

33. According to the heterotroph hypothesis, the first forms of life on earth probably obtained energy by anaerobic respiration. What material is thought to have been added to early earth's atmosphere by this process?
    1. methane
    2. carbon dioxide
    3. nitrogen
    4. hydrogen

Arm Of   Wing Of   Flipper Of
Human Being   Bird   Whale

34. Which type of evolutionary evidence is represented by these diagrams?
    1. homologous structures
    2. physiological likenesses
    3. biochemical similarities
    4. geographic distribution

HOW DO LIVING THINGS INTERACT WITH THEIR ENVIRONMENT?

KEY IDEAS	In preceding units, we explored the mechanisms by which individual organisms survive and pass their characteristics on to future generations. In this unit, we will discuss how species of diverse characteristics interact with each other and with their environments to contribute to the development and the maintenance of a stable set of environmental conditions. This unit also focuses on the place of human organisms in the environment, in regard to both how they affect and how they are affected by environmental conditions.

KEY OBJECTIVES

Upon completion of this unit, you should be able to:

☐ Explain the importance of interdependence of living things with each other and with their environments on the survival of all life on earth.

☐ List and describe the various levels of ecological organization devised by ecologists in their study of the environment.

☐ Describe the environmental factors responsible for the maintenance of the world environment, or ecosystem.

☐ Describe the changes that normally occur over time to the characteristics of ecological communities in response to environmental pressures.

☐ Describe how human beings, as part of the ecological community, affect and are affected by the balance of nature.

I. THE SCIENCE OF ECOLOGY

In preceding units we considered the conditions necessary to maintain the lives of individual organisms. It is important to realize that no living thing is independent of other living things. In fact, all living things need other organisms and a variety of nonliving environmental factors in order to survive. In this unit, we study the way that living things interact with each other and with their environments. The science that deals with the study of these interactions is **ecology**.

II. LEVELS OF BIOLOGICAL ORGANIZATION IN ECOLOGY

To learn more about ecological interaction, ecologists have subclassified the complexities of these interactions in terms of more easily understood levels of organization. Four of these levels are as follows:

A. Population

A **population** is defined as all the members of a given species inhabiting a given location at a particular time.

B. Community

In a defined area, all the populations of various types of organism that interact with each other are together considered to be an ecological **community**.

C. Ecosystem

The ecological community in interaction with the nonliving factors in the environment comprises the **ecosystem**. An ecosystem is a relatively self-sufficient and stable system.

D. Biosphere

The portion of the earth in which living things exist is known as the **biosphere**. The biosphere is made up of many complex and varied ecosystems.

III. THE BASIC UNIT OF ECOLOGICAL STUDY— THE ECOSYSTEM

When studying ecological interactions, the basic unit of study is the ecosystem. The ecosystem is the lowest level of ecological organization in which all environmental factors are represented and can interact freely.

A. Ecosystem Structure and Function

1. Components that Comprise the Ecosystem

These components include both living and nonliving factors.

1.1. Abiotic factors are the physical and chemical factors in the environment upon which life depends, but which themselves are nonliving. These factors often determine what type of plant and animal community can become established and thrive in a particular area.

Examples of abiotic factors include:

- Light intensity available for photosynthesis

- Temperature range

- Amount of available moisture

- Type of rock substratum under the soil

- Availability of minerals

- Availability of atmospheric gases

- Relative acidity (pH) of the system

Each environment is subjected to ranges of variation of these abiotic factors different from those of other environments. Because each species of living thing depends on a different "mix" of these factors, the relative amount of any factor can limit the species that can inhabit a particular environment. Factors that so limit the makeup of an ecological community are known as **limiting factors.**

Here are some examples of how limiting factors operate:

- The herb species inhabiting a forest differ markedly from those inhabiting an adjacent field because of the difference in light intensity in the two areas.

- Low temperature conditions common to extreme northern latitudes may prevent certain plant and animal species from living there, while favoring the survival of others.

- Certain fish species require an abundance of dissolved oxygen in their water environment. If oxygen concentrations drop, the species may die as a result of suffocation.

- Freshwater and saltwater environments play host to completely different species of fish, shellfish, and other aquatic species because of the difference in salinity between these two environments. Figure 7.1 shows a balanced aquatic ecosystem.

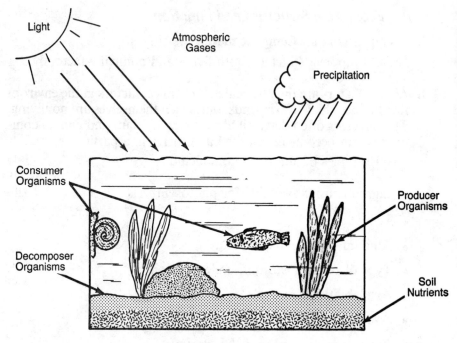

Light

Atmospheric
Gases

Precipitation

Consumer
Organisms

Producer
Organisms

Decomposer
Organisms

Soil
Nutrients

Figure 7.1 Balanced Ecosystem

1.2. Biotic factors include all the living components of the environment that affect the ecological community, either directly or indirectly, and help to limit the species that inhabit an area.

Examples of biotic factors include:

- The population levels of an individual species
- The particular set of requirements of a species
- The interactions that a species has with other species
- The wastes produced by the members of a species

1.2.1. Nutritional relationships between species involve the transfer of nutrient materials between one organism and another within the environment. Organisms may be classified as follows in terms of the type of nutritional relationships they have with other organisms:

  a. Autotrophs. As we learned in Unit 2, organisms capable of producing their own food are known as "autotrophs." Green plants and algae, the principal autotrophic organisms, manufacture organic molecules that serve as food for all the other organisms in the environment. This food is the main source of energy and structural components for all living things.

b. Heterotrophs. We also learned in Unit 2 that many species, particularly animals, protozoa, and fungi, are incapable of producing their own food. As such, they are dependent on other organisms for the food they consume. Heterotrophs may be further subclassified as follows:

- **Saprophytes** ("decay plants") include nongreen plants, fungi, and the decay bacteria that consume decaying organic matter and recycle its chemicals for use by other living things. Examples of saprophytes include mushrooms and bread mold.

- **Herbivores** ("plant consumers") are the animals that use only plant matter for food. This group contains many species of wild grazing animals, as well as many domesticated species. Examples of herbivores include rabbits and deer.

- **Carnivores** ("animal consumers") are animals that consume the bodies of other animals for food. Some carnivores kill their own prey (**predators**), while others consume the dead bodies of animals killed by predators or by other natural causes (**scavengers**). Carnivores range from the fiercest of jungle predators to the mildest of household pets. Examples of carnivores include wolves and mountain lions.

- **Omnivores** ("plant and animal consumers") are a small group of organisms that eat both plant and animal matter as regular parts of their diet. Human beings are classified in this group of consumers. Other examples of omnivores include bears and chimpanzees.

EC

1.2.2. Symbiotic relationships between species involve the ways that different types of organisms can live together in a close physical association. **Symbiosis** ("same life process") is the term used to describe such relationships. Types of symbiosis include the following:

- **Commensalism** is a form of symbiosis in which one organism is benefited, while its commensalistic partner is niether harmed nor helped. A symbolic representation of this relationship is " +, 0" which indicates the positive effect (" + ") coupled with the neutral effect ("0") at work in this type of symbiosis.

  Examples of commensalism include (1) the barnacle-whale relationship, in which the barnacle benefits from worldwide transport, while the whale is unharmed; and (2) the orchid-tree relationship, in which the orchid benefits from a stable growing environment on the tree, while the tree is unharmed.

- **Mutualism** is a form of symbiosis in which both organisms in the relationship benefit. The symbolic representation of this type of symbiosis is " +, +," indicating the mutually positive result of this association.

  Examples of mutualism include (1) the relationship between nitrogen-fixing bacteria and the roots of leguminous plants on which they live, in which the bacteria benefit by having a stable environment to reproduce and the legume benefits from the nitrates manufactured by the bacteria; and (2) the protozoa-termite relationship, in which the protozoa benefit from the nutrient-rich environment of the termite's intestine and the termite benefits from the wood-digesting action of the protozoa.

- **Parasitism** is a form of symbiosis in which one organism in the relationship benefits, while the other (the "**host**") is harmed. The symbolic representation of parasitism is " +, −," indicating the positive and negative impacts on the organisms in this relationship.

  Examples of parasitism include (1) athlete's foot fungus on human beings, in which the fungus derives nutrients from the human and the human host is harmed when the skin is opened to infection; (2) tapeworm infestation in a rabbit, in which the tapeworm gets nutrition from the digested food in the rabbit's intestine and the rabbit host is deprived of much of the food it eats and digests; and (3) heartworms in dogs, in which the worm parasite infects the heart muscle of the dog host and eventually kills the animal.

  The three types of symbiotic relationships are illustrated in Figure 7.2.

EXAMPLE	TYPE	SYMBOL

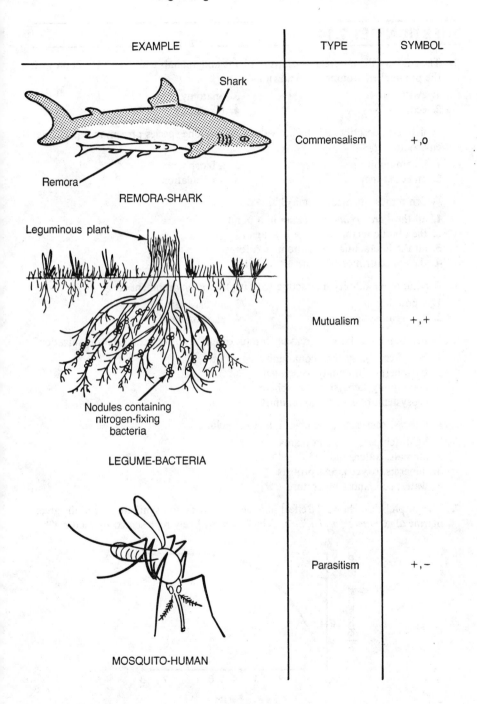

Shark / Remora / REMORA-SHARK	Commensalism	+,o
Leguminous plant / Nodules containing nitrogen-fixing bacteria / LEGUME-BACTERIA	Mutualism	+,+
MOSQUITO-HUMAN	Parasitism	+,−

Figure 7.2
Symbiotic Relationships

QUESTION SET 7.1

1. The study of the interrelationships of plants and animals and their interaction with the physical environment is known as
   1. evolution
   2. ecology
   3. anatomy
   4. taxonomy

2. Which term describes all the individuals of any one species present in a particular environment?
   1. a community
   2. an ecosystem
   3. a biosphere
   4. a population

3. Which represents a community?
   1. all the *Paramecium caudatum* in a pond
   2. the abiotic factors in Lake Michigan
   3. all the interacting populations in a forest
   4. the concentration of minerals in soil

4. A natural community interacting with its abiotic environment is
   1. a population
   2. an organ system
   3. an organism
   4. an ecosystem

5. Which sequence shows increasing complexity of levels of ecological organization?
   1. biosphere, ecosystem, community
   2. biosphere, community, ecosystem
   3. community, ecosystem, biosphere
   4. ecosystem, biosphere, community

6. In a forest ecosystem, the abiotic factors include
   1. light, temperature, and plants
   2. animals, water, and soil
   3. minerals, oxygen, and protists
   4. water, soil, and temperature

7. The graph below shows the effect of a factor on the photosynthetic rate of the green marine alga *Enteromorpha linza*. Which is most likely represented by factor *X*?

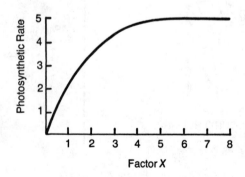

   1. light intensity
   2. water concentration
   3. competition level
   4. substratum type

8. An example of a biotic factor in a pond community is
    1. annual rainfall
    2. interspecies competition
    3. mineral concentration
    4. temperature change

9. Which is a biotic factor that affects the size of a population in a specific ecosystem?
    1. the average temperature of the ecosystem
    2. the number and kinds of soil minerals in the ecosystem
    3. the number and kinds of predators in the ecosystem
    4. the concentration of oxygen in the ecosystem

10. The trees in a forest aid in controlling floods chiefly because their
    1. branches store water in the form of sap
    2. leaves absorb moisture from the air
    3. root systems retain the soil substratum
    4. stems serve as reservoirs for food

11. Fungi and bacteria that depend on dead organic material for their existence are
    EC classified as
    1. decomposers   3. omnivores
    2. predators   4. herbivores

12. In the food chain below, what is the function of the rabbit?
    lettuce plant → rabbit → coyote
    1. parasite   3. consumer
    2. saprophyte   4. producer

13. Fly larvae consume the body of a dead rabbit. In this activity, they function as
    1. producers   3. herbivores
    2. scavengers   4. parasites

14. Which is true of most producer organisms?
    1. They are parasitic.
    2. They contain chlorophyll.
    3. They are eaten by carnivores.
    4. They liberate nitrogen.

15. At times hyenas feed on the remains of animals they themselves have not killed. At other times they kill other animals for food. On the basis of their feeding habits, hyenas are best described as
    1. herbivores and parasites   3. scavengers and parasites
    2. herbivores and predators   4. scavengers and predators

16. Growths of molds often appear on stale bread. This is an example of the relation-
    EC ship known as
    1. symbiosis   3. mutualism
    2. commensalism   4. saprophytism

17. Which term includes the other three?
    EC 1. symbiosis   3. parasitism
    2. mutualism   4. commensalism

18. Bacteria that live in the human intestine derive their nutrition from digested foods.
    [EC] From these nutrients digested by the human being, the bacteria synthesize vitamins usable by the human. This relationship demonstrates

1. commensalism	**3.** mutualism
2. saprophytism	**4.** parasitism

19. In which example of a nutritional relationship is an organism harmed?
    [EC]
    1. alga and fungus in a lichen
    2. nitrogen-fixing bacteria and clover
    3. remora and shark
    4. athlete's foot fungus and human being

20. The relationship between a leguminous plant and the nitrogen-fixing bacteria in its
    [EC] root nodules is an example of

1. succession	**3.** commensalism
2. competition	**4.** mutualism

2.   Energy Flow Relationships

These relationships, in which energy is brought into the environment and made available to all members of the ecological community, are essential to a self-sustaining ecosystem.

2.1. Energy. Energy is a necessary part of the life of each living thing. Each of the body's life functions requires energy to operate. The ultimate source of the energy used by living things is the energy of sunlight. Energy is transferred through the ecosystem by means of food chains and food webs involving nutritional relationships among living things.

2.2. Food Chains. Food chains begin when a green plant absorbs sunlight and converts it into chemical bond energy in the biochemical process of photosynthesis. An herbivore that consumes the plant represents the next "link" in the food chain by incorporating the stored energy of the plant into its own tissues. The next link in the food chain is a carnivore that consumes the body of the herbivore, thereby taking in the energy-containing molecules and releasing their energy for its own uses. Several more carnivorous animals may be involved as successive links of the food chain as animals consume other animals and in turn are consumed. The final link in any food chain is a saprophyte organism.

2.3. Food Webs. To illustrate more realistically the complex nature of nutritional relationships in a natural community, food webs are used. The food web concept recognizes that many plant species are present in any ecological community, all producing energy-rich organic compounds for consumption by many different species of herbivore. At the same time, multiple combinations of carnivorous and omnivorous species interact to consume the herbivorous species. Saprophytes of many different varieties are responsible for consuming the decaying bodies of plants and animals alike. In fact, different species may compete strongly for the same type of food available in the ecosystem. When the names of all the species types present in a community are written on a sheet of paper and lines are drawn between the species for which a nutritional relationship exists, a pattern resembling a web emerges.

The organisms in a food web may be categorized as follows:

2.3.1. Producers are the green plants and algae (the autotrophs) in the community responsible for trapping the sun's radiant energy and using it to manufacture organic compounds that are used for their own consumption and that of animals.

2.3.2. Consumers include herbivores, carnivores, and omnivores. Herbivores are known as **primary consumers** because they are the first consumers to tap the energy trapped by the producers. Carnivores are known as the **secondary consumers** because they do not tap plant energy directly, but obtain it through their consumption of primary-consumer organisms. Omnivores may be either primary consumers (when they eat plant matter) or secondary consumers (when they eat animal matter).

2.3.3. Decomposers include saprophytic fungi and bacteria responsible for breaking down the complex structure of the bodies of living things into simpler forms that can be used by other living things. In a sense, decomposers are responsible for operating a recycling system that reuses the chemical substances of life over and over. This action is essential to the continued functioning of the ecosystem.

A food chain and a food web are diagramed in Figure 7.3.

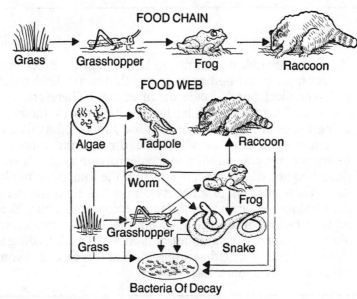

Figure 7.3   Food Chain And Food Web

2.4. Pyramid of Energy. The solar energy available to the producer organisms in a food chain is considerable. At each step of the food chain, however, some of this energy is lost. Some is used in the life processes, some is radiated as heat, some is lost in excretory waste. In any food chain, the producer level contains the greatest amount of energy. Primary consumers contain only about 10 percent of the energy found in the producers. In turn, secondary consumers contain only about 10 percent of the energy housed in the bodies of the primary consumers, and only about 1 percent of the energy of the producers. Since each feeding level contains less energy than the level below it, this phenomenon can be illustrated as a pyramid, known as a "pyramid of energy" (see Figure 7.4).

and will radiate as heat into the atmosphere. A constant resupply of energy from the sun is necessary to sustain the ecosystem.

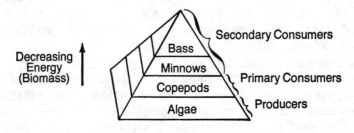

Figure 7.4   Energy (Biomass) Pyramid

EC

2.5. Biomass Pyramid. We learned in Unit 2 that the biochemicals that comprise the bodies of living things are held together with chemical bonds made up of energy. Therefore, as the amount of energy at each feeding level decreases, the mass of living material (**biomass**) that can be supported at that level generally decreases, as well. The diminishing amount of biomass is roughly parallel to the decreasing available energy described above. This fact means that the total mass of all the top-level consumers in an ecosystem is less than the mass of the producer organisms in the same ecosystem. Thus, like the representation of energy flow in the ecosystem described above, the decreasing amount of biomass at each feeding level can also be represented as a pyramid: in this case, a "biomass pyramid" (see Figure 7.4).

QUESTION SET 7.2

1. Which is the correct order of organisms in a food chain?
   1. carnivores → producers → herbivores
   2. producers → herbivores → carnivores
   3. herbivores → producers → carnivores
   4. producers → carnivores → herbivores

2. In the food chain below, which organisms are the primary consumers?
   weeds → grasshoppers → praying mantises → shrews → barn owls
   1. shrews   3. weeds
   2. praying mantises   4. herbivores

3. In a natural community, the greatest amount of energy is present in
   1. secondary consumers   3. producers
   2. carnivores   4. herbivores

4. Which is *not* essential in a self-sustaining ecosystem?
   1. a constant source of energy
   2. living systems capable of incorporating energy into organic compounds
   3. equal numbers of plants and animals
   4. a cycling of materials between organisms and their environment

5. Which statement best explains why a food web is a more realistic representation of nutritional patterns than a food chain?
   1. Energy is consumed in metabolic activities and gained at every feeding level.
   2. Decomposers return materials from the top-order predators to inorganic form.
   3. Energy always flows through the consumers to the producers.
   4. Practically all species are consumed by or feed on more than one species.

Base your answers to questions 6 through 8 on the information below and on your knowledge of biology.

An aquarium container is filled with water and colonies of aquatic plants and animals. Various protists are added, and the aquarium is then sealed and placed on a window ledge. After a period of time the aquarium appears to reach a state of balance.

6. The oxygen content of the tank is maintained by the
   1. autotrophs   3. fungi
   2. heterotrophs   4. carnivores

7. The energy needed to maintain this ecosystem originates from the
   1. fish   3. water
   2. green plants   4. sun

8. Which group of organisms in the aquarium contains the largest amount of energy?
   1. primary consumers   3. producers
   2. secondary consumers   4. herbivores

9. Which diagram best represents the usual relationships of biomass in a stable community?

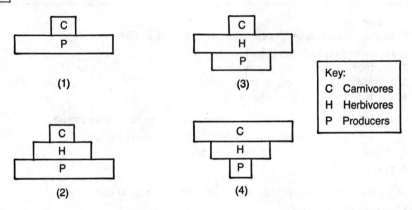

10. The pathway labeled IV represents
1. a food chain
2. a population
3. an ecosystem
4. an abiotic factor

Base your answers to questions 10 through 13 on the diagram below, which represents four possible pathways for the transfer of energy stored by green plants.

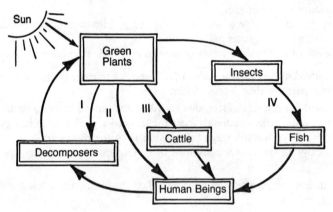

10. The pathway labeled IV represents
1. a food chain   3. an ecosystem
2. a population   4. an abiotic factor

11. Through which pathway would the sun's energy be most directly available to human beings?
1. I   3. III
2. II   4. IV

12. In this diagram, human beings are shown to be
1. herbivores only   3. omnivores
2. carnivores only   4. parasites

13. The cattle in the diagram represent
1. primary consumers   3. producers
2. secondary consumers   4. autotrophs

3. Material Cycles

Material Cycles function in nature to make chemical substances available to living things for their continued growth and reproduction. Material cycles consist of sequential chemical reactions that allow for such periodic recycling. Some examples of material cycles are as follows:

3.1. Carbon-Hydrogen-Oxygen Cycle (see Figure 7.5). In Unit 2, we studied the processes of respiration and photosynthesis, and discovered that they are quite similar in terms of the chemical substances involved. In photosynthesis, carbon dioxide and water combine with the aid of solar energy, producing glucose and oxygen gas. In respiration, glucose and oxygen combine to produce carbon dioxide and water, releasing cellular energy. One process produces waste materials that serve as raw materials for the other. In this way, environments that contain balanced communities of plants and animals should be self-sustaining in terms of the supply of the elements carbon, hydrogen, and oxygen.

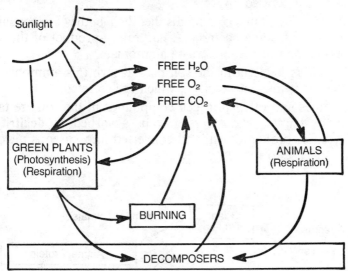

Figure 7.5
CARBON-HYDROGEN-OXYGEN CYCLE

3.2. Nitrogen Cycle (see Figure 7.6). Nitrogen is an elemental component of the class of compounds known as proteins, as we studied in Units 1 and 2. The nitrogen cycle makes nitrogen available to organisms for use in protein synthesis. As in the carbon cycle, living organisms are an important part of the nitrogen cycle. Decomposers and other soil bacteria are essential in converting the nitrogenous wastes of animals and plants into a form of nitrogen usable by plants as fertilizer. Animals of various types round out the nitrogen cycle as consumer organisms in a food web.

EC

The essential steps of the nitrogen cycle are as follows:

- Atmospheric nitrogen is absorbed by **nitrogen-fixing bacteria** as a gas and is converted into solid form as nitrate salts. The nitrogen-fixing bacteria are housed in small nodules on the roots of leguminous plants (a symbiotic relationship). These nitrate salts are released into the soil around the legume.
- Plants in the immediate vicinity take up these nitrates as dissolved minerals through their roots. The nitrates are then used by the plant in protein synthesis.
- The plants may be consumed by animal herbivores, and the plant proteins converted into animal protein within the cells of the animal. These proteins may in turn be consumed and converted by a series of carnivores and/or omnivores in a food web as described above.
- When these plants or animals die, their bodies are consumed by **decomposition bacteria**. A natural by-product of this process is the nitrogenous waste ammonia.
- **Nitrifying bacteria** in the soil may absorb this ammonia and convert it into nitrate salts, which are secreted.
- Nitrates and other nitrogen compounds in the soil are taken up by plants. The excess may be absorbed by **denitrifying bacteria**, broken down, and converted into gaseous nitrogen for release to the atmosphere.

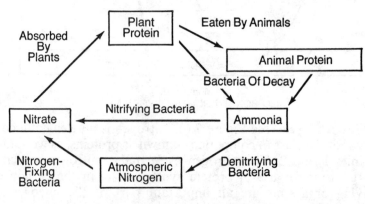

Figure 7.6   Nitrogen Cycle

3.3. Water Cycle (see Figure 7.7). As we learned in Unit 1, water is a vitally important material that must be available to all living things at all times. Many biotic and abiotic processes are involved with the cycling of water in the environment. Abiotic processes may include precipitation, evaporation, condensation, runoff, and percolation, which help to distribute water more or less evenly around the globe. Biotic processes are also involved in this distribution and may include photosynthesis, transpiration, respiration, and excretion. Each of these processes involves water as a reactant, product, or solvent and helps to cycle it through the environment to make it readily available to living organisms.

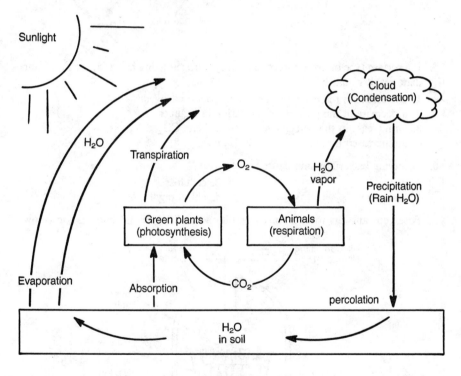

Figure 7.7
Water Cycle

QUESTION SET 7.3

1. The elements stored in living cells of organisms in a community will eventually be returned to the soil for use by other living organisms. The organisms that carry out this process are

    1. producers
    2. herbivores
    3. carnivores
    4. decomposers

2. In a balanced aquarium, the plants supply the fish and snails with
   1. molecular oxygen
   2. carbon dioxide
   3. hormones
   4. amino acids

3. Green plants generally absorb carbon in the form of
   1. carbon monoxide
   2. carbon dioxide
   3. bicarbonate ions
   4. amino acids

4. The diagram to the right illustrates some of the essential steps in
   EC
   1. the reproductive cycle
   2. the process of photosynthesis
   3. animal respiration
   4. the nitrogen cycle

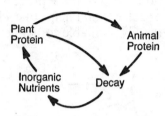

5. If a certain type of poison were to destroy nitrogen-fixing bacteria, the most imme-
   EC diate result would be
   1. a decrease in the percentage of atmospheric nitrogen
   2. a decrease in the nitrate concentration in legumes
   3. an increase in the percentage of atmospheric CO_2
   4. an increased number of healthier legumes

6. Nitrifying bacteria manufacture nitrates from
   EC
   1. water
   2. ammonia
   3. cellulose
   4. chlorophyll

Base your answers to questions 7 and 8 on the diagram and information below.

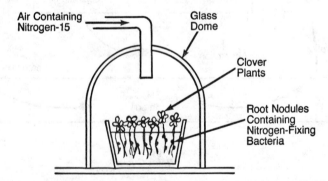

In the laboratory setup shown, clover plants containing nitrogen-fixing bacteria are cultured under a glass dome and receive air through an opening in the top. The air used contains nitrogen-15 in place of normal nitrogen. After 2 weeks, a grasshopper is allowed to feed on the clover plants.

7. Compounds containing nitrogen-15 pass from bacterial cells in root nodules to clover plant cells. These compounds are known as
   1. nitrates
   2. nitrogen molecules
   3. protein
   4. amino acids

8. The nitrogen-15 becomes part of the grasshopper when it is
   1. converted into uric acid by the muscle cells
   2. excreted into the intestine by Malpighian tubules
   3. used in the synthesis of polypeptides at a ribosome
   4. changed into normal atomic nitrogen in the nucleus

9. The processes of precipitation, evaporation, percolation, and runoff are part of the
   1. carbon cycle
   2. water cycle
   3. nitrogen cycle
   4. energy cycle

B. Ecosystem Formation

Ecosystems are not unchanging; they tend to undergo dynamic change with time as biotic and abiotic environmental factors alter. This dynamic change results in the establishment of an equilibrium state whose characteristics are determined by the particular set of conditions and limitations that affect the living community.

1. Succession

The term **succession** refers to a condition in which an established ecological community is gradually replaced by another, which in turn is replaced by others until a stable, self-perpetuating community is formed. This final, stable community is known as the **climax community**. The establishment of such successive community types is known as **ecological succession**.

EC

1.1. The beginning stages of an ecological succession frequently occur in barren, almost lifeless environments that may have been swept clean by glaciation, erosion, fire, or some other destructive event. The first living things to invade such an area and establish themselves in it are known as **pioneer organisms**. A typical pioneer organism that may be found populating bare rock is **lichen**. Lichen is a fungus-alga symbiotic association that can tolerate this harsh, dry soilless environment.

A pioneer organism such as lichen, although small, can significantly alter the environment it has invaded. The lichen produces a mild acid that acts on the rock surface and erodes it into grains of sand. At the same time, the material comprising the lichen adds organic matter to the sand grains, producing a crude form of soil. This "soil" gradually fills the rock crevices, providing favorable areas for seed germination and thereby paving the way for later stages of ecological succession.

1.2. The germinating seeds of grasses and other herbs give rise to organisms of the next succession stage. As this stage dominates over several generations, its members contribute organic matter to the soil and its root system protects the thin soil layer from erosion. Weathering and root pressure continue to fragment the parent rock substratum, adding still more substance to the growing soil layer. Eventually the soil layer is able to support the growth of shrubs and other woody plant species. When soil depth and quality are sufficient, varieties of trees and the animal species associated with them begin to invade and then dominate the area and continue to modify the environment.

In this way, each successive community of plants and animals modifies the environment to make it less favorable for its own offspring, but more favorable for the establishment of the next succession stage. In such a pattern, plant species (**flora**) tend to dominate, since they provide the basis of the food chain for the area. The animal species (**fauna**) that can live in the area are normally those that depend on the plant community for food. A succession stage is normally named for the dominant plant types in the environment, since these species exert the most influence on the environmental conditions of the area.

A typical series of succession stages in New York State (see Figure 7.8) might be as follows:

- Lichens
- Mosses
- Grasses
- Shrubs
- Coniferous woodlands
- Deciduous woodlands

1.3. Eventually in each succession, a community becomes established that is self-perpetuating and relatively stable. Such a community is known as a **climax community**. A climax community remains the dominant community for an indefinite period, maintaining a relatively stable set of environmental conditions for both plant and animal species.

A climax community will normally remain intact as long as conditions in the environment do not change appreciably. If, however, the environment changes drastically because of some catastrophic event, then the climax community may die out, making way for a new succession of communities. If the alteration is temporary, the same climax community may result; if the change is permanent, then a new group of climax organisms may be favored.

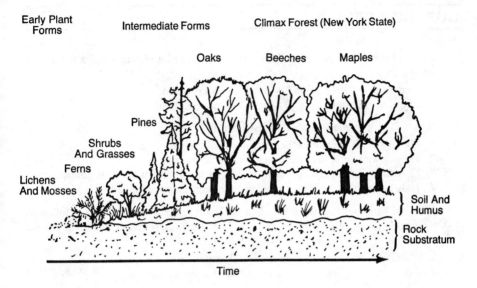

Figure 7.8   Forest Succession

Two typical climax communities found in New York State are as follows:

- Oak-hickory forest (common at lower elevations)
- Beech-maple-hemlock forest (common at higher elevations)

2. Competition

When different species living in the same environment (**habitat**) utilize the same limited resources, **competition** occurs. These resources may include requirements such as food, space, light, water, oxygen, and minerals. The more similar the requirements of the different species, the more intense the competition is likely to be. Such competition might be called "interspecies competition," since it involves members of different species (this should not be confused with the "intraspecies competition" discussed in Unit 6).

EC

An ecological **niche** is defined as the role that an organism plays in the environment. Interspecies competition usually limits the community to one species for each environmental niche. If two species attempt to inhabit the same niche, one is usually eliminated.

QUESTION SET 7.4

1. The replacement of one community by another until a climax stage is reached is known as
   1. predation
   2. adaptive radiation
   3. equilibrium
   4. ecological succession

2. Which is a characteristic of pioneer organisms?
   [EC]
   1. They have no effect on other organisms in their environment.
   2. They cannot survive under conditions of direct light.
   3. They modify their environment, making it favorable for the next community.
   4. They are always the last organisms to inhabit an environment.

3. A slab of bare rock is covered with lichens. In time, mosses cover the rock, followed
   [EC] by grasses, and finally by small shrubs and tree saplings. In this example, the lichens represent
   1. a climax community
   2. a dominant species
   3. secondary consumers
   4. pioneer organisms

4. In an ecological succession leading to the establishment of a pond community,
   [EC] which of the following organisms would be among the first to establish themselves?
   1. grasses
   2. algae
   3. minnows
   4. deciduous trees

5. After a major forest fire, an area that was once wooded is converted to barren soil.
   [EC] Which of the following schemes describes the most likely sequence of changes in vegetation in the area after the fire?
   1. shrubs → maples → pines → grasses
   2. maples → pines → grasses → shrubs
   3. pines → shrubs → maples → grasses
   4. grasses → shrubs → pines → maples

6. In New York State, a beech-maple forest in an example of
   [EC]
   1. a pioneer community
   2. a climax community
   3. an anaerobic environment
   4. an abiotic environment

7. Which is true of a climax community of a biotic succession?
   [EC]
   1. It persists until the environment or climate changes.
   2. It changes drastically from one year to the next.
   3. It is the first stage in the succession.
   4. It is the stage in which only plants are present.

8. The ecological niche of an organism refers to the
   [EC]
   1. biosphere in which the organism lives
   2. role the organism plays in the community
   3. position of the organism in a food web
   4. relation of the organism to human beings

9. In a freshwater pond community, a carp eats decaying material from around the
EC bases of underwater plants, while a snail scrapes algae from the leaves and stems of
   the same plants. They can survive at the same time because they occupy
   1. the same niche, but different habitats
   2. the same habitat, but different niches
   3. the same habitat and the same niche
   4. different habitats and niches

10. Cattails in freshwater swamps in New York State are being replaced by purple
    loosestrife plants. The two species have very similar environmental requirements.
    This observation best illustrates
    1. variation within a species
    2. competition between species
    3. isolation of species populations
    4. random recombination

11. In a community, the most severe competition develops among organisms that
    1. are active only during the night
    2. belong to two different genera
    3. depend on autotrophs for food
    4. occupy the same ecological niche

3. Biomes

Biomes are major groupings of similar ecosystems that cover substantial areas on the earth's surface. Each biome is made up of several ecosystems that can thrive under the climatic and other abiotic conditions of the area. Such biomes may be terrestrial (e.g., the Northeast Deciduous Forest Biome), aquatic (e.g., the ocean biome), or a mixture of these types.

EC

3.1. The characteristics of **terrestrial biomes** are determined by the major climate zones of the earth. These biomes are particularly sensitive to extremes in local climatic conditions. The plants and animals forming the ecological communities in a biome must be able to survive these conditions and to compete with other species attempting to inhabit the same niches.

The conditions that may vary in a terrestrial biome include temperature, duration and intensity of solar radiation, and available moisture. The availability of water is a major limiting factor for terrestrial biomes, since it is essential for the survival of all organisms.

3.1.1. Terrestrial biomes are commonly named for the dominant plant community (climax vegetation) of the predominant ecosystem in the region. In New York State the terrestrial biome native to most inland areas is known as the Northeast Deciduous Forest Biome, and is dominated by broadleaf hardwood trees and associated plant and animal species. The major terrestrial biomes, with their climatic characteristics, dominant flora, and typical index fauna, are presented in the chart below.

Major Terrestrial Biomes

Biome	Characteristics	Climax Flora	Climax Fauna
Tundra	Permanently frozen subsoil	Lichens, mosses, grasses	Caribou, snowy owl
Taiga	Long, severe winters; summers with thawing subsoil	Conifers	Moose, black bear
Temperate-deciduous forest	Moderate precipitation; cold winters, warm summers	Trees that shed leaves (deciduous trees)	Gray squirrel, fox, deer
Tropical forest	Heavy rainfall; constant warmth	Many species of broad-leaved plants	Snake, monkey, leopard
Grassland	Considerable variability in rainfall and temperature; strong prevailing winds	Grasses	Pronghorn antelope, prairie dog, bison
Desert	Sparse rainfall; extreme daily temperature fluctuations	Drought-resistant shrubs and succulent plants	Kangaroo, rat, lizard

3.1.2. Geographic features also affect the nature of the biomes in a particular area. The two major geographic factors affecting biome formation are **latitude** and **altitude**. It has been observed that increasing altitude has the same effect on the formation of biomes as does increasing latitude (in the Northern Hemisphere, latitude increases south to north). This phenomenon is illustrated in Figure 7.9.

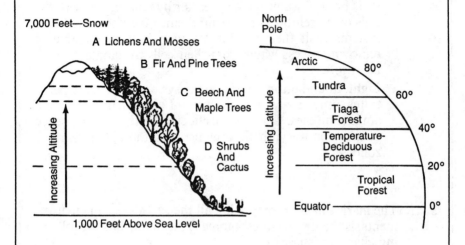

Figure 7.9   Effect Of Altitude And Latitude On Biome Formation

3.2. Aquatic biomes make up over 70 percent of the earth's surface and provide habitats for most of its species. Compared to terrestrial biomes, aquatic biomes are very stable environments. Because of the heat-holding ability of water, the temperature of a large aquatic ecosystem varies much less than does that of a land environment of comparable size. Since moisture conditions are virtually constant in aquatic environments, available moisture is not a limiting factor for most plant and animal species.

On the other hand, aquatic biomes may experience shortages of other materials or conditions that are only rarely absent in a terrestrial environment. For example:

- The availability of adequate gaseous oxygen and carbon dioxide may prove to be a limiting factor for some types of organisms. The gill structures of fish are examples of adaptations that permit their survival under conditions of low oxygen concentrations.
- The amount of suspended matter in water may affect the lives of plants by reducing light, and the lives of animals by clogging gills and other vital organs.

- Aquatic systems may have very different mineral contents, depending on the rock substratum underlying the system. Marine biomes, with their high mineral (salt) content, limit animal inhabitants to those able to regulate their bodies' water balance, either by adding water to the cells or by removing salt from them. Likewise, freshwater systems limit their inhabitants to organisms capable of removing excess water from their cells to maintain water balance.

- Light, abundant in terrestrial ecosystems, normally penetrates only a few meters beyond the surface of most aquatic biomes. This has the effect of limiting the presence of most photosynthetic organisms to the lighted surface zone.

3.2.1. The **marine biome**, comprising all the earth's ocean environments, is the single largest biome. This biome has the following characteristics:

- Provides temperature stability, due to the tremendous volume of water available to trap and hold solar heat, that is sufficient to modify the temperatures of nearby land environments.

- Provides mineral stability, due to the relatively constant salt content of seawater.

- Provides habitats for a tremendous number of organisms of many different kinds.

The oceans serve as a principal source of food for many different organisms, including human beings. The cold surface waters of the ocean near the coasts are home to countless microscopic organisms that serve as the basis for the oceanic food web. These waters are particularly well suited for such food production because they receive a constant flow of nutrients from rivers flowing from inland, as well as abundant light for photosynthesis. The deeper portions of the ocean are too dark to support algae and other photosynthetic organisms.

3.2.2. The **freshwater biome** consists of all the world's freshwater systems, including ponds, lakes, streams, rivers, and freshwater wetlands. Unlike the marine biome, the freshwater biome is not continuous, but is broken up by intervening land masses. Also unlike the marine biome, this biome shows considerable variation in conditions.

- The individual systems in the freshwater biome vary in size from the smallest ponds and streams to the largest lakes. Larger systems generally tend to be more stable than smaller systems.
- Temperature conditions may vary to a greater extent in freshwater systems than in marine systems, since smaller volumes of water are involved in the former.
- Small, warm bodies of water tend to lose dissolved oxygen concentration more readily than do waters that are large and cold. Therefore the content of this gas available for life can vary considerably from system to system.
- The content of suspended particles can vary widely from system to system. A high content of these particles can reduce light penetration considerably, adversely affecting the rate of photosynthetic food production.
- For stream and river environments, the velocity of the flow of water can be a critical factor for the survival of the organisms inhabiting them. Rapidly flowing streams promote the survival of certain communities of organisms, whereas slowly flowing rivers promote the establishment of a quite different set of plants and animals.
- The rate of change in freshwater systems is frequently more rapid than that in marine environments. As generations of aquatic plants die and decompose, their residue tends to accumulate on the bottom of a pond or lake, gradually filling it. When the system has become sufficiently shallow, terrestrial plants and animals may begin to invade, forming a "wetland" environment. As filling continues, the pond may actually begin to resemble a land environment more than a water environment. This change is a type of succession and, given enough time, may result in the formation of a terrestrial climax community.

QUESTION SET 7.5

1. Land biomes are characterized and named according to the
   EC 1. secondary consumers in the food webs
   2. primary consumers in the food webs
   3. climax vegetation in a region
   4. pioneer vegetation in a region

For *each* description in questions 2 through 6, select the biome, *chosen from the list below*, that is most closely associated with that description. (*A number may be used more than once or not at all.*)

Biome

---

(1) Grassland
(2) Tundra
(3) Temperate-deciduous forest
(4) Tropical rain forest
(5) Taiga
(6) Desert

2. This biome is found in the foothills of the Adirondack and Catskill Mountains of
EC New York State, and supports the growth of dominant vegetation including maples, oaks, and beeches.

3. The characteristic climax vegetation in this biome consists of coniferous trees com-
EC posed mainly of spruce and fir.

4. This biome receives less than 10 inches of rainfall per year. Extreme temperature
EC variations exist throughout the area over a 24-hour period. Water-conserving plants such as cacti, sagebrush, and mesquite are found.

5. This biome receives the least amount of solar energy. The ground is permanently
EC frozen (permafrost) throughout the year. During the summer season, plants quickly grow, reproduce, and form seeds during their short life cycle. Lichens and mosses grow abundantly on the surface of rocks.

6. A moderate, well-distributed supply of rain in this biome supports the growth of
EC broad-leaved trees, which shed their leaves as winter approaches.

---

7. In which biome are the factors of temperature, mineral content, and moisture most
EC constant?

1. ocean   3. rain forest
2. desert   4. tundra

8. In which of the following biomes does most of the photosynthesis taking place on
EC the earth occur?

1. deciduous forests   3. deserts
2. oceans   4. coniferous forests

Base your answers to questions 9 through 12 on the diagram below, which represents a comparison between latitudinal and altitudinal life zones (biomes) in the Northern Hemisphere, and on your knowledge of biology.

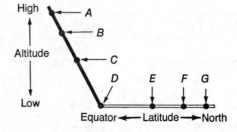

9. Which letter represents the biome that characteristically has permanently frozen
[EC] ground?
   1. *G*   3. *C*
   2. *F*   4. *D*

10. Which letter represents an area known as a taiga?
[EC] 1. *A*   3. *E*
    2. *B*   4. *D*

11. Which letters most probably represent similar biome areas?
[EC] 1. *A* and *G*   3. *C* and *F*
    2. *B* and *D*   4. *B* and *E*

12. Which letter most probably represents an area for a tropical rain forest?
[EC] 1. *E*   3. *F*
    2. *B*   4. *D*

IV. THE INTERACTIONS BETWEEN HUMAN BEINGS AND THE BIOSPHERE

Like other living organisms, human beings are dependent for survival on a balanced set of environmental conditions. Humans have been unique, however, in their ability to alter the very environment upon which they depend for survival. The extent to which humans can learn to preserve and restore their environment will determine their survival as a species, as well as the survival of most other species on the earth.

A. *The Past and Present*

The activities of a growing human population have had a tremendous impact on the natural environment over the past several centuries. Unlike most other species, human beings have systematically altered that environment for their own ends. In the past most of this alteration has had negative effects. In recent years, however, there has been an attempt to correct some of the abuses of the past.

1. Negative Aspects

Much of the negative impact of human activity on the natural environment has been due to the fact that human beings have lacked a good understanding of their role in nature. We have long understood our impact on other living things. What we have failed to understand is that we are, in turn, affected by, and entirely dependent on, other species for our continued survival.

The ecosystem has repeatedly demonstrated an ability to recover from minor natural and man-made disturbances and, given time, an ability to recover from major ones as well. The disruptions caused by human technology, however, have been of a fundamentally different kind from those due to natural forces. This technology has frequently resulted in the production of substances completely foreign to the natural environment. Their presence has introduced to the environment new selection pressures never before encountered by living things in their 3-billion-year history.

Every component of the ecosystem is connected to every other component in a dynamic equilibrium. A disruption of one component of the ecosystem ultimately has an effect on many other components because of these connections. A large number of shifts may occur before the equilibrium is reestablished.

1.1. Human population growth, unlike the increase in other species, has risen at a rapid rate over the past several centuries. As human technology has developed, many of the limiting factors that keep natural populations under control (e.g., disease, predation, hunger, exposure) have been progressively eliminated as checks on human population growth.

In many areas of the world, therefore, the human population has grown faster than the food supply, a condition that threatens to eliminate large portions of the population of these areas. In addition, the natural resources that support our technologies are being depleted at a rapid rate, imperiling our ability to maintain our advantage against nature's limiting factors. Apparently, the human species is rapidly approaching a point at which it will be unable to sustain continued growth. The trends of human population growth are illustrated in Figure 7.10.

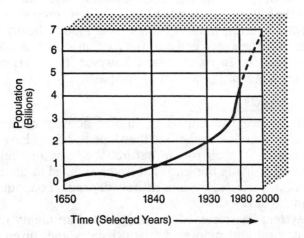

Figure 7.10 Human Population Growth

1.2. Human activities of various kinds have had a direct impact on the biotic portion of many ecosystems, leading to the endangerment or extinction of many species of plants and animals. Indirect impacts, which also affect the survival of species, include habitat destruction and environmental pollution. These activities include the following:

1.2.1. **Overhunting.** Overhunting of animals, uncontrolled by game laws, has resulted in the extinction of many species and has endangered still others. This activity still occurs in many parts of the world. Examples of extinct species include the dodo bird and the passenger pigeon. An example of an endangered species is the blue whale.

1.2.2. **Importation of organisms.** Organisms harmful to a particular environment have intentionally or unintentionally been introduced into that environment from other parts of the world where they are native. Because they often have no natural enemies, these newly introduced organisms have disrupted local communities of other organisms and, in some cases, totally eliminated native species. Examples of such imported organisms include the Japanese beetle (destructive of cultivated plants), gypsy moth (destructive of native hardwood trees), and a virus that causes dutch elm disease (destructive of the American elm tree).

1.2.3. **Exploitation of organisms.** Commercial trade in exotic plants and animals or their body parts has resulted in the endangerment or elimination of many species worldwide, as well as the disruption of their habitats. Examples of organisms exploited in this way include the African elephant and Pacific walrus (hunted for their ivory), Colombian parrots (traded as pets), and tropical hardwood trees (cut for the manufacture of plywood).

1.2.4. **Poor land-use management.** As the human population has grown, its need for living space has expanded. As cities and suburbs have developed, more and more open space and farmland have been used. This has resulted in the reduction of natural habitat important to native species, the outright elimination of many native species, the destabilization and erosion of soils, and the destruction of potential food-producing lands.

Poor agricultural practices have also had negative ecological consequences.

- **Overcropping,** that is, failure to allow soil to recover its nutrients and humus content between plantings, has caused the soil to lose its agricultural potential.
- **Overgrazing,** that is, the practice of allowing a large number of animals to graze on an area too small to support them, has resulted in destabilization of the soil.

- Failure to use **covercrops** to protect bare soil from erosion cycles has resulted in the loss of valuable topsoil, as well.

These and other improper agricultural methods have resulted in the loss of millions of cubic meters of topsoil from agricultural lands worldwide, land that will be needed in the future to provide food for a growing world population.

1.2.5. Technological oversight. When new technologies are developed, emphasis is often placed on their practical uses, without careful consideration of their potential impacts on the natural environment. Many such technological oversights have resulted in unplanned ecological consequences, such as the pollution of our air, water, and soil resources. As these resources have become progressively more polluted, our quality of life has become imperiled, our health threatened, and our future as a species made uncertain. Pollutants have also had a grave effect on other species around us, further threatening our existence.

- **Water pollution** has been worsened by the addition to our surface waters and groundwater of such pollutants as heat, sewage, chemical phosphates, heavy metals (e.g., mercury), polychlorinated biphenols (PCBs), and other chemical substances resulting from manufacturing and maintenance activities. These materials are destructive to both aquatic and terrestrial life forms, as well as to pipes, pumps, and other machines that use water.

- **Air pollution** has been worsened by the addition to our atmosphere of such pollutants as carbon monoxide, hydrocarbons, and particulate matter (from the burning of coal, oil, and gasoline). Nitrogen oxides and sulfur oxides, also produced from the burning of coal and petroleum products, have been found to combine photochemically with water in the atmosphere to form **acid precipitation** (see Figure 7.11). Acid precipitation is thought to be responsible for the destruction of populations of lake fish and forest trees in both North America and Europe.

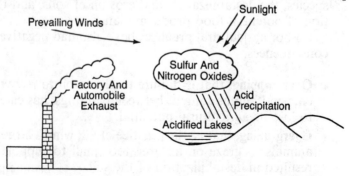

Figure 7.11 Acid Precipitation

Air pollutants and acid precipitation are also known to cause damage to machines and buildings, requiring consumers to pay millions of dollars in repair and replacement costs each year.

- **Biocide use** has introduced tremendous quantities of poisonous chemicals to the environment. Biocides (pesticides and herbicides) are used to control pests of various types that interfere with human activities. Once introduced into the environment, however, many of these biocides enter the food web and are passed from organism to organism, becoming more highly concentrated at each trophic level. In addition, their residues have increased the pollution of soil and water resources. The biocide DDT, widely used earlier in this century to control insect pests, is known to have disrupted the embryological processes of the bald eagle and other bird species.

- **Disposal problems** have resulted from the consumer-oriented societies of many technological nations. The products manufactured for consumption require the use of natural resources from all over the world, many of which are being depleted at an alarming rate. These products and their packaging materials are eventually discarded, adding to the solid-waste-disposal problem. When discarded, often in a landfill or in the ocean, all the materials and the energy required to manufacture and market them is lost. In addition, many of these products and the processes used to manufacture them leave chemical residues known to be toxic to living things, including human beings. Finding a safe method of disposing of such toxic residues is a problem modern society must solve. To an increasing degree, the safe disposal of waste from nuclear power plants is becoming a problem of global proportions.

2. Positive Aspects

We have begun to recognize the inherent dangers of continuing to abuse our environment. Increasing awareness of the environment and the role of human beings in it has led to attempts to change many of our current practices, as well as to correct some of the abuses of the past.

2.1. Population Control. Through education, many nations are seeking to encourage a responsible approach to family planning. Scientific and medical solutions to this problem continue to be developed.

2.2. Conservation of Resources. Erosion controls, such as reforestation and covercropping, are being used to reduce the loss of valuable agricultural soils. The importance of our water resources has been recognized, and water conservation measures have been implemented in many communities. Conservation of our energy resources, essential to our future survival as a nation, have proved successful in many instances. Recycling efforts have helped to slow the use of natural resources known to be in danger of depletion.

2.3. Pollution Controls. New technologies are being developed to control air, water and soil pollution at the source, rather than allowing it to escape into the environment. Many polluted areas are beginning to show signs of recovery since laws regulating the release of such pollutants were passed and are being enforced.

2.4. Species Preservation. The establishment of wildlife refuges and the enactment and enforcement of conservation laws have helped to reduce the pressure on some endangered species. By protecting their natural habitats, we allow these species to live and reproduce without interference, helping to ensure their continued survival. Game laws that limit or prohibit human destruction of these organisms also help to preserve these species for future generations.

Species formerly considered to be endangered, but currently responding to conservation efforts, include the American bison, the egret, the whooping crane, the bald eagle, and the peregrine falcon.

2.5. Biological Controls. Rather than controlling pests with chemical biocides, the use of biological controls is being encouraged. One method of biological control takes advantage of the natural enemies (parasites, predators, bacteria) of these pests as a means of eliminating them. Another method uses artificially produced sex hormones to attract and trap insect pests. These methods are less likely than biocide use to disrupt the environment, harm desirable species, or interfere with food webs.

2.6. Environmental Laws. Many federal, state, and local laws have been enacted that help to protect the environment and promote education concerning environmental issues.

B. The Future

As we have seen, the technological advances of human beings have often had a negative impact on the environment. Increasing awareness of the role of humans and all other organisms in the ecosystem, however, has begun to reverse this negative trend. Only through continued efforts to protect wild species, conserve resources, preserve natural habitat, control human population growth, and value all life forms as essential contributors to the maintenance of a healthy environment will we, as a global ecological community, survive and provide suitable living conditions for future generations.

QUESTION SET 7.6

1. The portion of the earth in which all ecosystems operate is known as the
  1. tropics
  2. community
  3. temperate zone
  4. biosphere

2. Compared to other organisms, human beings have had the greatest ecological impact on the biosphere because of their
   1. internal bony skeleton
   2. homeostatic regulation of metabolism
   3. adaptations for respiration
   4. ability to modify the environment

3. The rapid rise of the human population level over the past few hundred years has been due mainly to
   1. increasing levels of air and water pollution
   2. loss of topsoil from our farmable lands
   3. removal of natural checks on population growth
   4. increasing resistance level of insect species

4. Which of these human activities is quite often responsible for the other three human activities?
   1. increasing demand on limited food production
   2. rapid increase of loss of farmland due to soil erosion
   3. rapid increase in human population
   4. increasing levels of air pollution

5. The number of African elephants has been drastically reduced by poachers who kill the animals for the ivory in their tusks. This negative aspect of human involvement in the ecosystem can best be described as
   1. poor land-use management
   2. importation of organisms
   3. poor agricultural practices
   4. exploitation of wildlife

6. Gypsy moth infestations of rural areas of New York State may pose a potentially serious threat to many forested areas. Which would probably be the most ecologically sound method of gypsy moth control?
   1. widespread application of DDT
   2. introduction of a biological control
   3. removal of the organism's forest habitat
   4. contamination of its food sources

7. A poor land-use practice that usually leads to the loss of soil nutrients is
   1. reforestation   3. overcropping
   2. recycling   4. sewage control

8. Which is *not* generally a cause of increasing water pollution?
   1. increased growth of the human population
   2. increased use of pesticides
   3. increased industrialization
   4. increased use of biological controls

9. Which pollutant is produced by the burning of coal and oil and can result in the production of acid rain?
   1. phosphate   3. lead
   2. sulfur dioxide   4. hydrogen chloride

10. Before it was banned, the insecticide DDT was used to combat an organism called the red mite. An unexpected result of the use of DDT was that the population of the red mite increased rather than decreased, while the population of insect predators of the red mite declined. What is the most probable explanation of this phenomenon?

    1. Part of the red mite population was resistant to DDT and its predators were not.
    2. DDT is highly general in the kinds of insects it affects, killing both beneficial and harmful species.
    3. The red mite population could use DDT as a nutrient, while the predators could not.
    4. DDT triggered a mutation in the red mite population, making it immune to the effects of the chemical.

Base your answers to questions 11 through 13 on the graphs below, which show data on some environmental factors acting in a large New York lake.

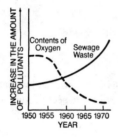 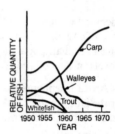

11. Which relationship can be correctly inferred from the data presented?
    1. As sewage waste increases, oxygen content decreases.
    2. As sewage waste increases, oxygen content increases.
    3. As oxygen content decreases, carp population decreases.
    4. As oxygen content decreases, trout population increases.

12. The greatest change in the lake's whitefish population occurred in the years between
    1. 1950 and 1955   3. 1960 and 1965
    2. 1955 and 1960   4. 1965 and 1970

13. Which of the fish species appears able to withstand the greatest degree of oxygen depletion?
    1. trout   3. walleye
    2. carp   4. whitefish

14. *Bacillus popilliae* is a bacterium that causes "milky disease" in the Japanese beetle. Using *Bacillus popilliae* to decrease a Japanese beetle population is an example of the
    1. abiotic control of insect pests
    2. use of biological control of insect pests
    3. use of artificial insecticides
    4. destruction of the abiotic environment

15. Erosion resulting from loss of topsoil due to poor farming techniques may be prevented by
    1. overgrazing pasturelands   3. overcropping farm fields
    2. removing trees, shrubs, and herbs   4. covercropping plowed fields

UNIT 7 REVIEW

1. Knowledge of ecology would be used most directly in studying the
   1. production of hormones and neurotransmitters in two related organisms
   2. current decline of bighorn sheep in the Rocky Mountains
   3. structure of subcellular organelles
   4. biochemical nature of genetic transmission

2. The members of the species *Microtus pennsylvanicus* living in a certain location make up a
   1. community
   2. succession
   3. population
   4. phylum

3. All the plant and animal life present in 1 cubic foot of soil makes up a
   1. community
   2. population
   3. species
   4. biosphere

4. Which is an example of an abiotic factor in a pond environment?
   1. the water
   2. a frog
   3. grasshopper
   4. a snake

5. Which is an example of a biotic factor that would limit the size of a deer herd?
   1. populations of predators
   2. severe summer drought
   3. lack of oxygen at high altitudes
   4. heavy winter snowfalls

6. Which would be considered a biotic factor in a pond ecosystem?
   1. snails
   2. water
   3. oxygen
   4. sunlight

7. The graph below represents a predator-prey relationship. What is the most probable reason for the increasing predator population from day 5 to day 6?

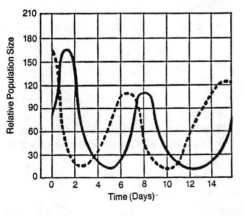

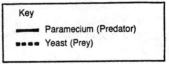

Key
— Paramecium (Predator)
■ ■ ■ ■ Yeast (Prey)

   1. an increasing food supply from day 5 to day 6
   2. a predator population equal in size to the prey population from day 5 to day 6
   3. the decreasing prey population from day 1 to day 2
   2. the extinction of the yeast on day 3 and day 10

8. Decomposition and decay of organic matter are accomplished by the action of
   1. green plants
   2. bacteria and fungi
   3. viruses and algae
   4. scavengers

9. Which term describes the bird and the cat in the following pattern of energy flow?

   sun → grass → grasshopper → bird → cat

   1. herbivores
   2. saprophytes
   3. predators
   4. omnivores

10. Which organisms are most directly and immediately necessary for the survival of secondary consumers?
    1. decomposers
    2. producers
    3. primary consumers
    4. scavengers

11. Forest trees can obtain nitrates through the action of decomposers on
    1. molecular nitrogen
    2. insecticides
    3. soil water
    4. dead plants

12. An organism that obtains its food at the expense of another living organism is
    EC known as a
    1. host
    2. saprophyte
    3. parasite
    4. scavenger

13. The wrasse, a small marine fish, periodically cleans harmful parasites from the
    EC mouth and body of the moray eel. The moray, in turn, protects the wrasse from larger predators and provides it with a constant supply of food. This is an example of the type of relationship known as
    1. mutualism
    2. parasitism
    3. commensalism
    4. saprophytism

14. A sudden increase in the number of producers in an ecosystem would first affect the population of
    1. carnivores
    2. herbivores
    3. saprophytes
    4. decomposers

15. Which food chain relationship illustrates the nutritional pattern of a primary consumer?
    1. seeds and fruits eaten by a mouse
    2. an earthworm eaten by a mole
    3. a mosquito eaten by a bat
    4. a mold growing on a dead frog

Base your answers to questions 16 and 17 on the key below and on your knowledge of biology. The symbols in the key represent possible effects on an organism of some nutritional relationships.

Key

(+)   Organism benefits.
(−)   Organism is harmed.
(0)   Organism neither benefits nor is harmed.

16. Which represents the effects of the relationship between nitrogen-fixing bacteria and
leguminous plants such as clover?
  1. nitrogen-fixing bacteria (−); clover (−)
  2. nitrogen-fixing bacteria (−); clover (+)
  3. nitrogen-fixing bacteria (0); clover (−)
  4. nitrogen-fixing bacteria (+); clover (+)

17. Which relationship would be illustrated by (+) for one organism and (0) for the
other organism?
  1. mutualism   **3.** commensalism
  2. parasitism   **4.** autotrophism

---

18. Food webs consist of many predator-prey relationships because many consumers
  1. are very large scavengers
  2. have anaerobic patterns of respiration
  3. have several alternative nutrient supplies
  4. are able to carry on photosynthesis

19. The rate of photosynthesis carried on by plants living in a body of water depends
chiefly on the
  1. amount of molecular oxygen in the water
  2. number of decomposers in the water
  3. amount of light that penetrates the water
  4. number of saprophytes in the water

20. In an ecosystem, the ultimate source of all energy is
  1. photosynthesis   **3.** fermentation
  2. oxygen   **4.** sunlight

Base your answers to questions 21 through 23 on the food chain represented below
and on your knowledge of biology.

rosebush → aphid → ladybird beetle → spider → toad → snake

21. Which organism in the food chain can transform light energy into chemical energy?
  1. spider   **3.** rosebush
  2. ladybird beetle   **4.** snake

22. At which stage in the food chain will the population with the *smallest* number of
animals probably be found?
  1. spider   **3.** ladybird beetle
  2. aphid   **4.** snake

23. Which organism in this food chain is herbivorous?
  1. rosebush   **3.** ladybird beetle
  2. aphid   **4.** toad

24. Nitrogen-fixing bacteria enrich the soil by producing nitrates beneficial to green
[EC] plants. The bacteria live in nodules located on the roots of legumes. These nodules
provide a favorable environment for the bacteria to grow and reproduce. The
relationship between these bacteria and the leguminous plant is an example of

1. parasitism   3. mutualism
2. commensalism   4. competition

25. In which form is nitrogen absorbed by the root hairs of an apple tree?
[EC] 1. free nitrogen   3. urea
2. ammonia   4. nitrates

Base your answers to questions 26 through 28 on the diagram below, which represents a food pyramid of organisms inhabiting a pond.

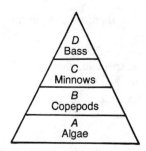

26. At which level of the food pyramids is the *smallest* percentage of total stored energy
found?

1. *A*   3. *C*
2. *B*   4. *D*

27. Which organisms in the food pyramid function as primary consumers?

1. bass   3. copepods
2. minnows   4. algae

28. What is the original source of energy for all organisms in this food pyramid?

1. water   3. the substratum
2. sunlight   4. carbon dioxide

---

29. Plant protein may be converted into simpler nitrogenous molecules by the process
[EC] of

1. synthesis   3. decomposition
2. osmosis   4. cyclosis

30. If a person traveled south from the Arctic Circle to the Equator, what would be the
[EC] most probable sequence of land biomes she would pass through?

1. temperate forest → taiga → tundra → tropical forest
2. taiga → tundra → temperate forest → tropical forest
3. tundra → tropical forest → taiga → temperate forest
4. tundra → taiga → temperate forest → tropical forest

Base your answers to questions 31 through 33 on the diagram below, which represents a cycle, and on your knowledge of biology.

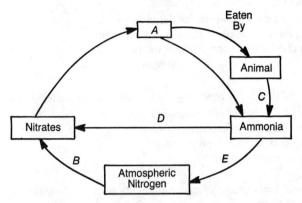

31. Nitrifying bacteria are represented by the letter
EC **1.** *A*   **3.** *C*
    2. *E*   **4.** *D*

32. The letter *B* most likely represents
EC **1.** bacteria of decay
    2. denitrifying bacteria
    3. a leguminous plant
    4. nitrogen-fixing bacteria

33. The cycle represented by the diagram is the
EC **1.** nitrogen cycle   **3.** water cycle
    2. carbon cycle   **4.** oxygen cycle

---

34. An earthworm lives and reproduces in the soil. It aerates the soil and adds organic material to it. Together these statements best describe an earthworm's
    1. habitat   **3.** niche
    2. nutrition   **4.** environment

35. In New York State, bluebirds and sparrows inhabit nearly the same ecological niche. In many areas, bluebirds are being replaced by sparrows as a result of
    1. symbiosis   **3.** mutualism
    2. competition   **4.** equilibrium

36. An aquarium in which the only living organisms are algae can be maintained for a longer period of time than an aquarium in which the only living organisms are protozoa. Which is the best explanation for this?
    1. The algae are able to carry on photosynthesis.
    2. The protozoa do not produce carbon dioxide.
    3. The protozoa produce carbohydrates.
    4. The algae are able to live in water.

37. What will most likely occur if two species of organisms living in the same environment have the same life requirements?
    1. Both species will compete for the same niche.
    2. One species will become parasitic on the other.
    3. Mutualism will develop between the two species.
    4. Both species will have the same reproductive rates.

38. Which condition is *not* necessary for an ecosystem to be self-sustaining?
    1. a greater number of consumers than producers
    2. the presence of decomposers
    3. the presence of autotrophic organisms
    4. a constant energy source

39. The natural replacement of one community with another until a climax stage is reached is known as
    1. ecological balance
    2. organic evolution
    3. dynamic equilibrium
    4. ecological succession

40. Which organisms can function as pioneer organisms on bare rock?
    1. scavengers
    2. parasites
    3. lichens
    4. shrubs

41. Which represents a natural climax community in New York State?
    1. a beech-maple forest
    2. an apple orchard
    3. a vegetable garden
    4. a cow pasture

42. The biome classification of a land region is determined by the region's
    1. animal life
    2. climax vegetation
    3. decomposers
    4. predators

43. A coniferous forest would be *least* likely to appear within the
    EC
    1. United States
    2. Arctic Circle
    3. Canadian Provinces
    4. U.S.S.R.

44. The major environmental factor limiting the presence of numerous autotrophs at
    EC great depths in the ocean is the
    1. type of substratum
    2. amount of light
    3. availability of minerals
    4. absence of biotic factors

45. Which is the most common sequence of major land biomes encountered from the
[EC] Equator to the polar region?
  1. tundra, taiga, temperate deciduous forest, tropical forest
  2. tropical forest, temperate deciduous forest, taiga, tundra
  3. temperate deciduous forest, tropical forest, taiga, tundra
  4. tropical forest, temperate deciduous forest, tundra, taiga

46. In which areas does the greatest amount of photosynthesis occur?
  1. mountains   **3.** deserts
  2. oceans   **4.** tundras

Base your answers to questions 47 through 51 on the chart below and on your
knowledge of biology.

A	Characteristics	Climax Flora	Climax Fauna
B	Long, severe winters	*D*	Moose, black bear
Tropical rain forest	Heavy rainfall	Many species of broadleaf plants	*E*
Desert	*C*	Succulent plants	Lizards

47. Which heading belongs in box *A*?
[EC] **1.** Land Biome   **3.** The Biosphere
   2. Aquatic Ecosystem   **4.** Succession Stage

48. Which name belongs in box *B*?
[EC] **1.** Tundra   **3.** Grassland
   2. Taiga   **4.** Temperate forest

49. Which characteristic belongs in box *C*?
[EC] **1.** Extreme daily temperature fluctuations
   2. Constant rainfall
   3. Seasonal animal migrations
   4. Strong prevailing winds

50. Which organisms belong in box *D*?
[EC] **1.** Maple trees   **3.** Lichens
   2. Cactus plants   **4.** Conifers

51. The climax fauna in box *E* would probably include
[EC] **1.** Lizards and caribou
   2. Red fox and whitetail deer
   3. Bison and antelope
   4. Monkeys and snakes

52. Contamination of the soil, atmosphere, and water by human beings is partially due to the use of

1. wildlife management
2. reforestation programs
3. chemical biocides
4. pollution controls

53. Why is the use of pesticides such as DDT being discouraged in many countries?
1. Crop production may increase because of their use.
2. They disrupt food webs.
3. All insects are immune to them.
4. They do not enter material cycles.

54. Human beings have been responsible for some of the negative changes that occur in nature because they
1. have controlled the use of chemical biocides
2. have passed laws to preserve the environment
3. are able to conserve scarce resources
4. are able to modify their physical environment

55. In which biome are the abiotic factors most constant?

EC 1. ocean
2. desert
3. deciduous forest
4. tundra

Base your answers to questions 56 through 58 on the graph below and your knowledge of biology. The graph represents the relationship between the capacity of the range (number of deer that could be supported by the range), the number of deer actually living on the range, and time.

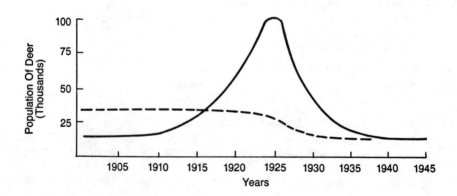

56. In what year was the number of deer living on the range equal to the capacity of the range?

1. 1905
2. 1915
3. 1920
4. 1930

57. What is the most likely reason why the capacity of the range to support deer decreased between 1920 and 1930?

1. The deer population became too large.
2. The number of predators increased between 1915 and 1925.
3. The deer population decreased in 1919.
4. An unusually cold winter occurred in 1918.

58. What might be one reason why the number of deer began to increase in 1910?

1. The deer's natural enemies were killed.
2. The capacity of the range increased.
3. The available vegetation of the area decreased.
4. The winter was longer than normal in 1905.

---

59. The most serious consequence of cutting down forests and overgrazing land is

1. the prevention of flooding
2. an increase in the chance of fire
3. an increase in the number of predators
4. the loss of topsoil cover.

60. Recent evidence indicates that lakes in large areas of New York State are being affected by acid rain. The major effect of acid rain in the lakes is

1. an increase in game-fish population levels
2. the stimulation of a rapid rate of evolution
3. the elimination of many species of aquatic life
4. an increase in agricultural productivity

61. Recent studies have found traces of the insecticide DDT accumulated in human fat tissue. A correct explanation for this accumulation is that

1. DDT is needed for proper metabolic functioning
2. DDT is passed along food chains
3. fat tissue absorbs DDT directly from the air
4. fat-tissue cells secrete DDT

62. Which illustrates the human population's increased understanding and concern for ecological interrelationships?

1. importing organisms in order to disrupt existing ecosystems
2. allowing the air to be polluted only by industries that promote technology
3. removing natural resources from the earth at a rate equal to or greater than the needs of an increasing population
4. developing animal game laws in order to limit the number of organisms that may be killed each year

Base your answers to questions 63 through 65 on the information below.

A scientist studied a river and forest area downstream from a large city with an increasing population. During the study, the scientist made many observations that could be classified as follows:

(1) Most likely a negative result of human activity.
(2) Most likely a positive result of human activity.
(3) Probably not being influenced by human activity to any extent.

For *each* observation in questions 63 through 65 select the phrase, *chosen from the list above*, which best fits that observation. (*A number may be used more than once or not at all.*)

63. Measurement of the levels of nitrates and phosphates in the river that flows through the forest showed the following results:

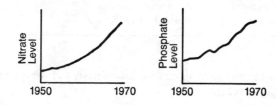

64. After 1970, the concentration of dissolved oxygen in the river increased, while the concentration of suspended particles decreased.

65. The population of hawks declined, while their food sources increased. High levels of insecticides were found in the reproductive tissues of the hawks.

Laboratory Skills and Current Events in Biology

The New York State Regents Examination in Biology tests the student's knowledge of both content (tested in Parts I and II) and practical skills (tested in Part III) in biology. In general, the skills required on Part III include basic laboratory technique, analysis of readings in science, written expression of biological concepts, and general awareness of current events in biology.

A. Laboratory Skills

The New York State Syllabus mentions 16 specific laboratory skills that should be mastered in preparation for the year-end Regents Examination. Students should be able to:

1. Formulate a question or define a problem, and develop a hypothesis to be tested in an investigation.
2. Given a laboratory problem, select suitable lab materials, safety equipment, and appropriate observation methods.
3. Distinguish between controls and variables in an experiment.
4. Identify parts of a light microscope and their functions, and focus in low and high power.
5. Determine the size of microscopic specimens in micrometers (microns).
6. Prepare wet mounts of plant and animal cells, and apply staining techniques using iodine and methylene blue.
7. Identify cell parts under the compound microscope, such as the nucleus, cytoplasm, chloroplast, and cell wall.
8. Use and interpret indicators, such as pH paper, Benedict's (Fehling's) solution, iodine (Lugol's) solution, and bromthymol blue.
9. Use and read measurement instruments, such as metric rulers, centigrade thermometers, and graduate cylinders.
10. Dissect plant and animal specimens for the purpose of exposing major structures for suitable examination. Suggestions of specimens include seeds, flowers, earthworms, and grasshoppers.
11. Demonstrate safety skills involved in heating materials in test tubes or beakers, using chemicals, and handling dissection instruments.
12. Collect, organize, and graph data.

13. Make inferences and predictions based on data collected and observed.
14. Formulate generalizations or conclusions based on the investigation.
15. Assess the limitations and assumptions of the experiment.
16. Determine the accuracy and repeatability of the experimental data and observations.

These laboratory skills may be subdivided into three broad areas:

1. The methods by which a laboratory experiment is designed.
2. The techniques used in conducting a laboratory experiment.
3. The skills related to interpreting the results of a laboratory experiment.

1. Methods Used in Designing a Laboratory Experiment (Skills 1–3)

Skill 1: Students should be able to express an experimental problem as a statement or as an experimental **question** to be answered. Such questions should be written so as to indicate a quantity to be measured in a laboratory experiment. Students also should be able to state the expected results of such an experiment in the form of a **hypothesis**.

Skill 2· When presented with a hypothetical experiment to perform, students should be able to select the group of tools that would be most appropriate to use for conducting that experiment. For example, an experiment involving comparative anatomy would probably be carried out most effectively with dissecting instruments; one involving the chemical nature of an unknown food would be performed most effectively using chemical indicators; one involving plant growth might use a variety of measuring instruments.

Skill 3: Given an outline of an experiment, students should be able to identify the **variables**, which are the (changing) quantities being measured in the experiment. Certain variables are manipulated by the investigator (independent variables), whereas others vary as a result of experimental factors (dependent variables). Students should also be able to identify the aspects of the experiment, known as **controls**, that are designed to exclude possible interference by unwanted variables.

2. Techniques Used in Conducting Laboratory Experiments (Skills 4–11)

Skill 4: Students should have a basic familiarity with the **compound light microscope**—its parts, the function of each part, its use as a tool for measuring small objects, the procedure for determining its magnification, and the appearance of objects within its visual field. Students should be prepared to describe each of these aspects in sentence form. (See Unit 1.)

Skill 5:   Students should be able to express the sizes of microscopic objects in metric units. An understanding of the **micrometer** (μm), a metric unit of linear measure, is required. Students should be able to convert measurements in micrometers to millimeters (or centimeters) and from larger units to micrometers [1 μm = 0.001 mm = 0.0001 cm = 0.000001 m]. (See Unit 1.)

Skill 6:   Students should be able to describe, in sentence form, how to prepare a wet-mount slide for examination under the compound light microscope. A basic familiarity with the application of biological stains and the stains **iodine** and **methylene blue** is also required. (See Unit 1.)

Skill 7:   Students should be able to recognize and label the major organelles of typical plant, animal, and protozoan cells as shown in photomicrographs (i.e., photographs taken through a microscope). Students should also be prepared to describe the major functions of these organelles in sentence form, as well as to judge their sizes in micrometers. (See Unit 1.)

Skill 8:   Students should be familiar with the use of indicators used to determine the chemical characteristics of food samples and solutions. Students should be prepared to describe, in sentence form, the use of each of the following indicators:

- **pH paper** is used to determine the relative acidity (pH) of a solution. A pH paper containing litmus will turn red in acid solution and blue in basic solution.
- **Bromthymol blue** turns yellow under acid conditions and remains blue under basic conditions. It may be used to detect the presence of carbon dioxide in solution, since carbon dioxide forms a weak acid when dissolved in water.
- **Benedict's (Fehling's) solution** is used to detect the presence of simple sugars in food samples or solutions. Benedict's solution is blue when first prepared, but when heated in the presence of simple sugar it turns color, ranging from yellow to brick red.
- **Iodine (Lugol's) solution**, normally light tan, turns blue-black when applied to a food sample or solution containing starch.

Skill 9:   Students should be able to determine quantities and dimensions, using a variety of metric measuring instruments. The dimensions of common objects should be determinable in millimeters and centimeters, using metric rulers and scales. Students should be able to read correctly the volume of liquid in a graduate cylinder by sighting the bottom of the meniscus. Students should be able to read the temperature indicated on a centigrade thermometer.

Skill 10:   Students should be able to recognize, identify, and label the major organs and organ systems of common dissection specimens, including earthworms, grasshoppers, seeds, and flowers. The organs identified in Unit 2 as being significant for these organisms are the ones with which students should be familiar.

Skill 11:   Students should be able to describe the proper methods of dealing with a variety of laboratory situations. This includes the correct means of handling chemicals, so as not to cause harm to oneself or others. Also included is the safe use of dissecting tools, such as scalpels and other sharp instruments. Students should also be able to describe the approved techniques for heating liquids and for handling hot objects. Students should be prepared to describe these methods in sentence form.

3.   Skills Used in Interpreting Experimental Results (Skills 12–16)

Skill 12:   Given unorganized raw data, students should be able to collect and organize these data in chart form according to increasing values of the independent variable. Students should also be familiar with the proper techniques to use in representing data in graph form. Knowledge of the correct methods for constructing both bar charts and line graphs is required. Students should be able to properly label and increment graph axes, appropriately title graphs, and correctly plot graphic data points.

Skill 13:   Students should be able to analyze the data that result from a laboratory experiment and to draw **inferences** (i.e., conclusions based on facts) that help to solve the experimental problem or answer the experimental question. Students should also be able to predict the outcome of experiments that broaden the range of the independent variable. In order to do this, students should understand how to interpret data organized in either chart or graph form. Students should be prepared to describe their inferences and predictions in sentence form.

Skill 14:   Given an experiment whose data have been properly organized and analyzed, students should be able to develop generalizations (i.e., broad **conclusions**) concerning the effect of the test variable on the experimental question. It should then be possible to project these generalizations onto situations outside the laboratory where similar variables are interacting. Students should be prepared to express such generalizations in sentence form.

Skill 15:   Students should recognize the limitations of their experimental methods in the context of professional science. This requires an understanding of the limits of accuracy of measuring equipment, the errors that may be introduced through incompletely developed laboratory skills, the ambiguities that result from inadequate experimental controls, and other limitations affecting experimental results.

Skill 16: Once a laboratory experiment is completed, students should be able to determine the accuracy of the experimental results through a review of the experimental methods. Where appropriate, the calculation of percent error may assist in the determination of experimental accuracy. The experimental methods should allow for repeatability of the experiment to assist in the verification of experimental accuracy.

B. Analysis of Readings in Science

Part III of the New York State Regents Examination may include items that require the analysis of short readings in science. These readings may deal with New York State Syllabus understandings or with concepts related to, but outside of, the syllabus. Students are expected to comprehend the meanings of technical terms that appear in the syllabus (those terms included in the glossary of this book). If other technical terms are used, they will be defined within the passage.

Students are expected to be able to read through the passage and answer questions based on it. Students should also be able to draw on their knowledge of syllabus understandings to answer some questions on these reading passages.

C. Written Expression of Biological Concepts

Students in the Regents Biology course are expected to be able to express themselves in complete sentences concerning biological principles. Although this requirement is not meant to be a test of the grammatical skills of the student, it is a test of the student's ability to express himself/herself clearly in scientific terms. Students should be prepared to answer Part III questions in sentence form.

D. Current Events in Biology

Students should maintain an awareness of current events in biology that have reached statewide, national, or international prominence. Students will not be tested directly on current events, but rather through use of reading comprehension and analysis of graphs and other data representations. Topics that may be selected for inclusion on Part III include environmental situations (e.g., acid precipitation, toxic waste disposal), advances in genetic research (e.g., genetic engineering), aspects of biomedical research (e.g., immunology, AIDS research), and others of a similar nature.

Following are typical Part III questions, grouped by the skills tested, that have appeared on actual Regents Examinations in recent years. Students can find additional practice questions in the Regents Examinations at the end of this book.

QUESTION SET

CATEGORY A—Laboratory Skills
Skill 1:

1.1. A student reported that a wilted stalk of celery became crisp when placed in a container of ice water. The student then suggested that water entered the stalk and made it crisp. This suggestion is considered to be
1. a control
2. a hypothesis
3. an observation
4. a variable

Skill 2:

Base your answers to questions 2.1. through 2.3. on the four sets of laboratory materials listed below and on your knowledge of biology.

Set *A*
Light source
Colored filters
Beaker
Test tubes
Test tube stand

Set *C*
Scalpel
Forceps
Scissors
Pan with wax bottom
Pins
Stereomicroscope
Goggles

Set *B*
Droppers
Benedict's solution
Iodine
Test tubes
Starch solution
Sugar solution
Test tube holder
Test tube rack
Heat source
Goggles

Set *D*
Compound light microscope
Glass slides
Water
Forceps

2.1. Which set should a student select in order to test for the presence of a carbohydrate in food?

2.2. Which set should a student select to determine the location of the aortic arches in the earthworm?

2.3. Which set should a student use to observe chloroplasts in elodea (a green waterplant)?

Skill 3:

3.1. Some scientists have concluded that stressful situations cause a decrease in the normal operation of the immune system in human beings. In a recent study, people who were under severe stress were examined to measure how well their immune systems were functioning. These people showed poorer immune system function during times of severe stress than when they were under less stress.

If the experimental group studied consisted of truck drivers who drove daily for 8 hours in very heavy traffic, a corresponding control group would most likely consist of truck drivers who drove

1. daily for 12 hours in very heavy traffic
2. every *third* day for 8 hours in very heavy traffic
3. every other day for 12 hours in very light traffic
4. daily for 8 hours in very light traffic

3.2. A student is studying the effect of temperature on the hydrolytic action of the enzyme gastric protease, which is contained in gastric fluid. An investigation is set up using 5 identical test tubes, each containing 40 milliliters of gastric fluid and 20 millimeters of glass tubing filled with cooked egg white. After 48 hours, the amount of egg white hydrolyzed in each tube is measured. Which is a variable in this investigation?

1. gastric fluid
2. length of glass tubing
3. temperature
4. time

Skill 4:

4.1. The diagram below represents a compound light microscope. Choose *one* of the numbered parts. *In a complete sentence*, name the part selected and describe its function.

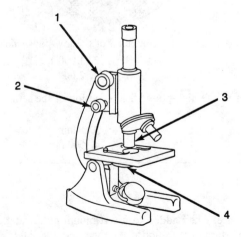

4.2. To view cells under the high power of a compound microscope, a student places a slide of the cells on the stage and moves the stage clips over to secure the slide. She then moves the high-power objective into place and focuses on the slide with the coarse adjustment. Two steps in this procedure are incorrect. For this procedure to be correct, she should have focused under

1. low power using the coarse and fine adjustments, and then under high power using only the fine adjustment
2. high power first, then under low power using only the fine adjustment
3. low power using the coarse and fine adjustments, and then under high power using the coarse and fine adjustments
4. low power using the fine adjustment, and then under high power using only the fine adjustment

Skill 5:

Base your answers to questions 5.1. and 5.2. on the information below and on your knowledge of biology.

A student was using a microscope with a 10× eyepiece and 10× and 40× objective lenses. He viewed the edge of a metric ruler under low power and observed the following field of vision.

5.1. What is the diameter, in micrometers, of the low-power field of vision?

1. 1	**3.** 1,000
2. 2	**4.** 2,000

5.2. The diameter of the high-power field of vision of the same microscope would be closest to

1. 0.05 mm	**3.** 5 mm
2. 0.5 mm	**4.** 500 mm

Skill 6:

6.1. Which substance, when added to a wet mount containing starch grains, would react with the starch grains and make them more visible?

1. litmus solution	**3.** distilled water
2. iodine solution	**4.** bromthymol blue

Base your answers to questions 6.2. and 6.3. on the diagrams below and on your knowledge of biology. The diagrams show wet-mount microscope slides of fresh potato tissue.

Slide *A*   Slide *B*

6.2. The formation of air bubbles on slide *A* could have been prevented by
  1. using a thicker piece of potato and less water
  2. using a longer piece of potato and a cover slip with holes in it
  3. holding the cover slip parallel to the slide and dropping it directly onto the potato
  4. bringing one edge of the cover slip into contact with the water and lowering the opposite edge slowly

6.3. A drop of stain is put in contact with the left edge of the cover slip on slide *B*, and a piece of absorbent paper is placed in contact with the right edge of the cover slip. What is the purpose of this procedure?
  1. It prevents the stain from getting on the ocular of the microscope.
  2. It prevents the water on the slide from penetrating the potato tissue.
  3. It allows the stain to penetrate the potato tissue without the removal of the cover slip.
  4. It helps increase the osmotic pressure of the solution.

Skill 7:

Base your answers to questions 7.1. and 7.2. on the drawing below, which shows a piece of tissue stained with iodine solution, as viewed with a microscope under high power.

7.1. The tissue represented in the drawing is most likely made up of
  1. onion epidermal cells
  2. ciliated protists
  3. cardiac muscle cells
  4. blue-green algae

7.2. The organelle labeled *B* in the drawing is most likely a
1. mitochondrion
2. centriole
3. lysosome
4. cell wall

7.3. Diagram *A* represents the appearance of a wet mount of plant tissue as seen through a compound light microscope. Diagram *B* represents the appearance of the same field of view after the fine adjustment knob is turned. What is the best conclusion to be made from these observations?

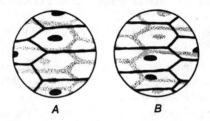

A   B

1. The tissue is composed of more than one layer of cells.
2. The tissue is composed of multinucleated cells.
3. The cells are undergoing mitotic cell division.
4. The cells are undergoing photosynthesis.

Skill 8:

8.1. A student was testing the composition of exhaled air by exhaling through a straw into a solution of bromthymol blue. The presence of carbon dioxide in the exhaled air would be indicated by
1. a color change in the solution
2. a change in atmospheric pressure
3. the formation of a precipitate in the solution
4. the release of bubbles from the solution

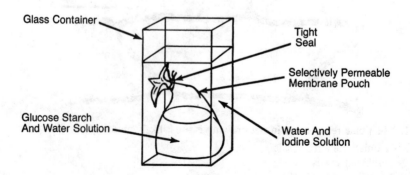

Glass Container

Tight Seal

Selectively Permeable Membrane Pouch

Glucose Starch And Water Solution

Water And Iodine Solution

8.2. A student tested a sample of the fluid in the glass container for glucose 30 minutes after the apparatus had been set up. Which indicator should be used for this test?

1. iodine solution
2. bromthymol blue
3. Benedict's solution
4. pH paper

Skill 9:

9.1. What is the total volume of water indicated in the graduated cylinder illustrated below?

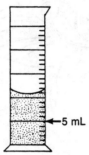

←5 mL

1. 10 mL   3. 12 mL
2. 11 mL   4. 13 mL

9.2. The diagram below represents a segment of a metric ruler and part of an earthworm. What is the length of the part of the earthworm shown? You must include the correct units in your answer.

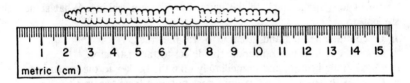

metric (cm)

9.3. Which group of measurement units is correctly arranged in order of increasing size?

1. micrometer, millimeter, centimeter, meter
2. millimeter, micrometer, centimeter, meter
3. meter, micrometer, centimeter, millimeter
4. micrometer, centimeter, millimeter, meter

Skill 10:

10.1. In earthworms and grasshoppers, which structure is ventral to the esophagus?

1. gizzard   3. intestine
2. brain   4. nerve cord

Base your answers to questions 10.2. and 10.3. on the illustration below of the flower of an amaryllis plant.

10.2. Name the circled part of the stamen.

10.3. Using a complete sentence, state a process carried out within the circled structure.

Skill 11:

11.1. A student performing an experiment noticed that the beaker of water she was heating had a slight crack in the glass, but was not leaking. What should the student do?
1. Discontinue heating and attempt to seal the crack.
2. Discontinue heating and report the defect to the instructor.
3. Discontinue heating and immediately take the beaker to the instructor.
4. Continue heating as long as fluid does not seep from the crack.

11.2. Which would be the proper laboratory procedure to follow if some laboratory chemical splashed into a student's eyes?
1. Send someone to find the school nurse.
2. Rinse the eyes with water and do not tell the teacher because he or she might become upset.
3. Rinse the eyes with water; then notify the teacher and ask further advice.
4. Assume that the chemical is not harmful and no action is required.

11.3. While a student is heating a liquid in a test tube, the mouth of the tube should always be
1. corked with a rubber stopper
2. pointed toward the student
3. allowed to cool
4. aimed away from everybody

Skill 12:

A student was investigating the relationship between different concentrations of substance X and the height of bean plants. He started with six groups, each of which contained the same number of bean plants with identical heights. Conditions were kept the same except that each group was watered with a different concentration of substance X for a period of 2 weeks. Then the concentration of substance X used in watering each group of plants and the average height for each group of plants were recorded by the student as follows:

Group A—6%, 32.3 cm	Group D—8%, 37.1 cm
Group B—0%, 28.7 cm	Group E—4%, 31.5 cm
Group C—2%, 29.4 cm	Group F—10%, 30.7 cm

For questions 12.1. and 12.2., organize the above data by filling in the Data Table below, following the directions given in the questions.

12.1. Label column III with an appropriate heading. [*Include the proper unit of measurement.*]

12.2. Complete all three columns in the Data Table so that the concentrations of substance X are increasing from the top to the bottom of the Data Table.

Data Table

I	II	III
Group	Concentration of Substance X (%)	

The data table shows the wolf and moose populations recorded at the end of June from 1970 to 1980 on an isolated island national park where no hunting by human beings is allowed. Before the arrival of wolves on the island (1965), the moose population had increased to over 300 members. Wolves have been observed many times on this island hunting cooperatively to kill moose.

Data Table

| Year | Number of Members | |
	Wolf Population	Moose Population
1970	10	90
1972	12	115
1974	20	145
1976	25	105
1978	18	95
1980	18	98

For questions 12.3. through 12.6., use the information in the data table to construct a line graph on the grid, following the directions given in the questions.

12.3. Mark an appropriate scale on the axis labeled "Number of Members of Each Population."

12.4. Mark an appropriate scale on the axis labeled "Year."

12.5. Plot the data for the wolf population on the graph. Surround each point with a small triangle and connect the points.

Example

12.6. Plot the data for the moose population on the graph. Surround each point with a small circle and connect the points.

Example

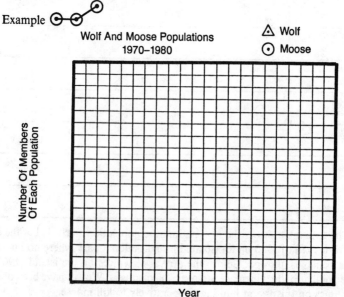

Wolf And Moose Populations
1970–1980

△ Wolf
⊙ Moose

Number Of Members Of Each Population

Year

Skill 13:

Base your answers to questions 13.1. through 13.3. on the following graph representing survival rates of fish species at various pH levels.

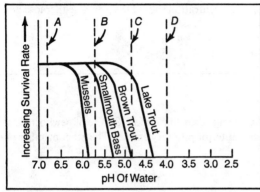

The Effect Of pH On Survival Rates
Of Selected Species In Certain Adirondack Lakes

Key:
A - pH Of A Certain Group of
     Adirondack Lakes, 1880

B - pH Of Rainfall, 1880

C - pH Of The Same Group Of
     Adirondack Lakes, 1980

D - pH Of Rainfall, 1980

— *National Geographic* (Adapted)

13.1. Which species can tolerate the highest level of acidity in its water environment?

  1. mussels   **3.** brown trout

  2. smallmouth bass   **4.** lake trout

13.2. In the years between 1880 and 1980, which species would most likely have been eliminated first because of the gradual acidification of Adirondack lakes?

  1. mussels   **3.** brown trout

  2. smallmouth bass   **4.** lake trout

13.3. What is the total change in the pH value of rainwater from 1880 to 1980?

  1. 1.3   **3.** 5.3

  2. 1.7   **4.** 9.7

Base you answers to questions 13.4. and 13.5. on the information and the graph given below, which shows the effect of sugar concentration on osmotic balance.

Four pieces of apple were cut so that all were the same mass and shape. The pieces were placed in four different concentrations of sugar water. After 24 hours, the pieces were removed and their masses determined. The graph below indicates the change in the mass of each piece.

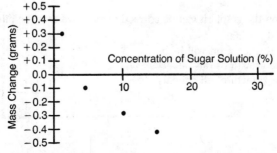

13.4. What was the change in mass of the apple piece in the 10% sugar solution?
1. a decrease of 0.45 gram
2. an increase of 0.30 gram
3. a decrease of 0.30 gram
4. an increase of 0.10 gram

13.5. At approximately what sugar concentration should the pieces neither lose nor gain weight?

1. 6%   3. 3%
2. 10%   4. 20%

Skill 14:

14.1. The graph below shows the average growth rate for 38 pairs of newborn rats. One of each pair was injected with anterior pituitary extract. The other member of each pair served as a control.

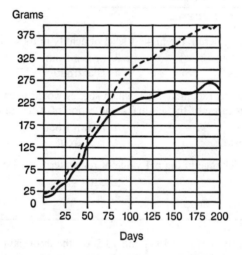

Grams

Days

———————— Average Growth Of 38 Untreated ?
(Control)

– – – – – Average Growth Of 38 Rats Injected With
Anterior Pituitary Extract (Experimental)

Based on the graph, it can be correctly concluded that the pituitary extract
1. is essential for life
2. determines when a rat will be born
3. affects the growth of rats
4. affects the growth of all animals

Based on the information in the graph below, which is a correct conclusion about plant hormones?

14.2. Plant hormones are chemical regulators that stimulate or inhibit growth depending on their concentration and the type of tissue in which they are found.

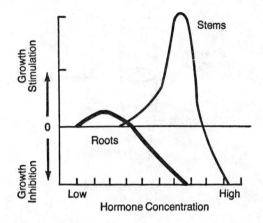

1. They stimulate maximum root growth and stem growth at the same concentration.
2. They stimulate maximum stem growth at low concentrations.
3. They most strongly inhibit root growth at low concentrations.
4. They stimulate maximum root and stem growth at different concentrations.

14.3. The data below are based on laboratory studies of male *Drosophila*, showing the inherited bar-eye phenotype.

Culture Temperature (°C) During Development	15	20	25	30
Number of Compound Eye Sections	270	161	121	74

Which is the best conclusion to be drawn from an analysis of these data?
1. The optimum temperature culturing *Drosophila* is 15°C
2. *Drosophila* cultured at 45°C will show a proportionate increase in the number of compound eye sections.
3. Temperature determines eye shape in *Drosophila*.
4. As temperature increases from 15°C to 30°C, the number of compound eye sections in male *Drosophila* with bar-eyes decreases.

14.4. The diagram below represents the result of spinning a suspension of broken cells in an ultracentrifuge. Which is a correct conclusion?

1. Ribosomes are more dense than mitochondria.
2. Nuclei are more dense than mitochondria.
3. Mitochondria and ribosomes are equal in density.
4. The cell consists of only solid components.

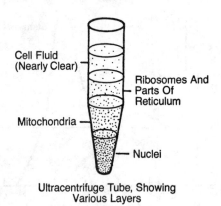

Cell Fluid
(Nearly Clear)

Ribosomes And
Parts Of
Reticulum

Mitochondria

Nuclei

Ultracentrifuge Tube, Showing
Various Layers

CATEGORY B—Readings in Science

Base your answers to questions 1 through 5 on the passage below.

Gene Splicing

Recent advances in cell technology and gene transplanting have allowed scientists to perform some interesting experiments. Some of these experiments have included splicing a human gene into the genetic material of bacteria. The altered bacteria express the added genetic material.

Bacteria reproduce rapidly under certain conditions. This means that bacteria with the gene for human insulin could multiply rapidly, resulting in a large bacterial population which could produce large quantities of human insulin.

The traditional source of insulin has been the pancreases of slaughtered animals. Continued use of this insulin can trigger allergic reactions in some humans. The new bacteria-produced insulin does not appear to produce these side effects.

The bacteria used for these experiments are *E. coli*, bacteria common to the digestive systems of many humans. Some scientists question these experiments and are concerned that the altered *E. coli* may accidentally get into water supplies.

For each statement below, write the number 1 if the statement is true according to the paragraph, the number 2 if the statement is false according to the paragraph, or the number 3 if not enough information is given in the paragraph.

1. Transplanting genetic material into bacteria is a simple task.
2. Under certain conditions bacteria reproduce at a rapid rate.
3. Continued use of insulin from other animals may cause harmful side effects in some people.
4. The bacteria used in these experiments are normally found only in the nerve tissue of humans.
5. Bacteria other than *E. coli* are unable to produce insulin.

CATEGORY C—Writing in Science

Base your answers to questions 1 through 3 on the reading passage below. Write your answers in complete sentences.

Time Frame for Speciation

Evolution is the process of change through time. Theories of evolution attempt to explain the diversification of species existing today. The essentials of Darwin's theory of natural selection serve as a basis for our present understanding of the evolution of species. Recently, some scientists have suggested two possible explanations for the time frame in which the evolution of species occurs.

Gradualism proposes that evolutionary change is continuous and slow, occurring over many millions of years. New species evolve through the accumulation of many small changes. Gradualism is supported in the fossil record by the presence of transitional forms in some evolutionary pathways.

Punctuated equilibrium is another possible explanation for the diversity of species. This theory proposes that species exist unchanged for long geological periods of stability, typically several million years. Then, during geologically brief periods of time, significant changes occur and new species may evolve. Some scientists use the apparent lack of transitional forms in the fossil record in many evolutionary pathways to support punctuated equilibrium.

1. Identify one major difference between gradualism and punctuated equilibrium.
2. According to the theory of gradualism, what may result from the accumulation of small variations?
3. What fossil evidence indicates that evolutionary change may have occurred within a time frame known as gradualism?

---

4. An organism contains many structures that enable it to survive in a particular environment. The human body has many such structures that have adaptive value. Several of these structures are listed below.

- sweat gland
- pancreas
- liver
- epiglottis
- capillary
- villus
- kidney
- platelet

Choose *five* of the structures listed above. For *each* one chosen, write the name of the structure and then, using a complete sentence, describe one of its adaptive values to the human body.

CATEGORY D—Current Events

Base your answers to questions 1 through 5 on the information below and on your knowledge of biology.

Acid rain is a serious environmental problem in large areas of Canada and the northeastern United States, including New York State. It is partly created as rain "washes out" sulfur and nitrogen pollutants from the air. Acid rain alters the fundamental chemistry of sensitive freshwater environments and results in the death of many freshwater species. The principal sources of this pollution have been identified as smokestack gases released by coal-burning facilities located mainly in the midwestern United States.

"Unpolluted" rain normally has a pH of 5.6. Acid rain, however, has been measured at pH values as low as 1.5, which is more than 10,000 times more acidic than normal. Commonly, acid rain has a pH range of 3 to 5, which changes the acidity level of the freshwater environment into which it falls. The effect of the acid rain depends on the environment's ability to neutralize it. Evidence is accumulating, however, that many environments are adversely affected by the acid rain. As a result, the living things within lakes and streams that cannot tolerate the increasing acidity gradually die off.

There are many environmental problems that result from acid rain. Most of these problems center around the food web upon which all living things, including humans, depend. If freshwater plants, animals, and protists are destroyed by the acid conditions, then terrestrial predators and scavengers dependent on these organisms for food are forced to migrate or starve. These changes in a food web can eventually affect the human level of food consumption.

1. The accompanying scale shows the pH values of four common household substances. Acid rain has a pH closest to that of which of these substances?

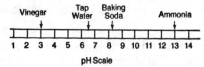

1. ammonia
2. tap water

3. baking soda
4. vinegar

2. What is most likely the source of acid rain in New York State?
   1. far western United States
   2. midwestern United States
   3. far eastern Canada
   4. far western Europe

3. Which food chain includes organisms that would most immediately be affected by acid rain?
   1. grass → rabbit → fox → decay bacteria
   2. algae → aquatic insect → trout → otter
   3. shrub → mouse → snake → hawk
   4. tree → caterpillar → bird → lynx

4. Acid rain is generally considered a negative aspect of human involvement with the ecosystem. As such, it would most correctly be classified as a type of
   1. biological control
   2. conservation of resources
   3. technological oversight
   4. land-use management

5. A strain of fish that could survive under conditions of increased acidity could best be obtained by
   1. binary fission
   2. vegetative propagation
   3. selective breeding
   4. budding

GLOSSARY OF PROMINENT SCIENTISTS

(with cross references to units and areas)

Crick, Francis   A 20th-century British scientist who, with **James Watson**, developed the first workable model of DNA structure and function.   (UNIT: 5; AREA: CORE)

Darwin, Charles   A 19th-century British naturalist whose theory of organic evolution by natural selection forms the basis for the modern scientific theory of evolution. (UNIT: 6; AREA: CORE)

Fox, Sidney   A 20th-century American scientist whose experiments showed that **Stanley Miller's** simple chemical precursors could be joined to form more complex biochemicals.   (UNIT: 6; AREA: CORE)

Hardy, G. H.   A 20th-century British mathematician who, with **W. Weinberg**, developed the Hardy-Weinberg principle of gene frequencies.   (UNIT: 5; AREA: MG)

Lamarck, Jean   An 18th-century French scientist who devised an early theory of organic evolution based on the concept of "use and disuse."   (UNIT: 6; AREA: CORE)

Linnaeus, Carl   An 18th-century Dutch scientist who developed the first scientific system of classification, based on similarity of structure.   (UNIT: 1; AREA: CORE)

Mendel, Gregor   A 19th-century Austrian monk and teacher who was the first to describe many of the fundamental concepts of genetic inheritance through his work with garden peas.   (UNIT: 5; AREA: CORE)

Miller, Stanley   A 20th-century American scientist whose experiments showed that the simple chemical precursors of life could be produced in the laboratory.   (UNIT: 6; AREA: CORE)

Morgan, Thomas Hunt   A 20th-century American geneticist whose pioneering work with *Drosophila* led to the discovery of several genetic principles, including sex linkage.   (UNIT: 5; AREA: CORE)

Watson, James   A 20th-century American scientist who, with **Francis Crick**, developed the first workable model of DNA structure and function.   (UNIT: 5; AREA: CORE)

Weinberg, W.   A 20th-century German physician who, with **G. H. Hardy**, developed the Hardy-Weinberg principle of gene frequencies.   (UNIT: 5; AREA: MG)

Weismann, August   A 19th-century German biologist who tested Lamarck's theory of use and disuse and found it to be unsupportable by scientific methods.   (UNIT: 6; AREA: CORE)

GLOSSARY OF IMPORTANT BIOLOGICAL TERMS

(with cross references to units and areas)

abiotic factor Any of several nonliving, physical conditions that affect the survival of an organism in its environment. (UNIT: 7; AREA: CORE)

absorption The process by which water and dissolved solids, liquids, and gases are taken in by the cell through the cell membrane. (UNIT: 2; AREA: CORE)

accessory organ In human beings, any organ that has a digestive function but is not part of the food tube. (See **liver; gallbladder; pancreas**.) (UNIT: 3; AREA: CORE)

acid A chemical that releases hydrogen ion (H^+) in solution with water. (UNIT: 1; AREA: CORE)

acid precipitation A phenomenon in which there is thought to be an interaction between atmospheric moisture and the oxides of sulfur and nitrogen that results in rainfall with low pH values. (UNIT: 7; AREA: CORE)

active immunity The immunity that develops when the body's immune system is stimulated by a disease organism or a vaccination. (UNIT: 3; AREA: HP)

active site The specific area of an enzyme molecule that links to the substrate molecule and catalyzes its metabolism. (UNIT: 1; AREA: CORE)

active transport A process by which materials are absorbed or released by cells against the concentration gradient (from low to high concentration) with the expenditure of cell energy. (UNIT: 2; AREA: CORE)

adaptation Any structural, biochemical, or behavioral characteristic of an organism that helps it to survive potentially harsh environmental conditions. (UNIT: 6; AREA: CORE)

addition A type of chromosome mutation in which a section of a chromosome is transferred to a homologous chromosome. (UNIT: 5; AREA: MG)

adenine A nitrogenous base found in DNA and RNA molecules. (UNIT: 5; AREA: CORE)

adenosine triphosphate (ATP) An organic compound that stores respiratory energy in the form of chemical-bond energy for transport from one part of the cell to another. (UNIT: 2; AREA: CORE)

adrenal cortex A portion of the adrenal gland that secretes steroid hormones which regulate various aspects of blood composition. (UNIT: 3; AREA: HP)

adrenal gland An endocrine gland that produces several hormones, including adrenalin. (See **adrenal cortex; adrenal medulla**.) (UNIT: 3; AREA: HP)

adrenal medulla A portion of the adrenal gland that secretes the hormone adrenalin, which regulates various aspects of the body's metabolic rate. (UNIT: 3; AREA: HP)

adrenalin A hormone of the adrenal medulla that regulates general metabolic rate, the rates of heartbeat and breathing, and the conversion of glycogen to glucose. (UNIT: 3; AREA: HP)

aerobic phase of respiration   The reactions of aerobic respiration in which two pyruvic acid molecules are converted to six molecules of water and six molecules of carbon dioxide.   (UNIT: 2; AREA: BC)

aerobic respiration   A type of respiration in which energy is released from organic molecules with the aid of oxygen.   (UNIT: 2; AREA: CORE)

aging   A stage of postnatal development that involves differentiation, maturation, and eventual deterioration of the body's tissues.   (UNIT: 4; AREA: RD)

air pollution   The addition, due to technological oversight, of some unwanted factor (e.g., chemical oxides, hydrocarbons, particulates) to our air resources.   (UNIT: 7; AREA: CORE)

albinism   A condition, controlled by a single mutant gene, in which the skin lacks the ability to produce skin pigments.   (UNIT: 5; AREA: CORE)

alcoholic fermentation   A type of anaerobic respiration in which glucose is converted to ethyl alcohol and carbon dioxide.   (UNIT: 2; AREA: CORE)

allantois   A membrane that serves as a reservoir for wastes and as a respiratory surface for the embryos of many animal species.   (UNIT: 4; AREA: RD)

allele   One of a pair of genes that exist at the same location on a pair of homologous chromosomes and exert parallel control over the same genetic trait.   (UNIT: 5; AREA: CORE)

allergy   A reaction of the body's immune system to the chemical composition of various substances.   (UNIT: 3; AREA: HP)

alveolus   One of many "air sacs" within the lung that function to absorb atmospheric gases and pass them on to the bloodstream.   (UNIT: 3; AREA: CORE)

amino acid   An organic compound that is the component unit of proteins.   (UNIT: 1; AREA: CORE)

amino group   A chemical group having the formula $-NH_2$ that is found as a part of all amino acid molecules.   (UNIT: 1; AREA: BC)

ammonia   A type of nitrogenous waste with high solubility and high toxicity.   (UNIT: 2; AREA: CORE)

amniocentesis   A technique for the detection of genetic disorders in human beings in which a small amount of amniotic fluid is removed and the chromosome content of its cells analyzed. (See **karyotyping**.)   (UNIT: 5; AREA: MG)

amnion   A membrane that surrounds the embryo in many animal species and contains a fluid to protect the developing embryo from mechanical shock.   (UNIT: 4; AREA: RD)

amniotic fluid   The fluid within the amnion membrane that bathes the developing embryo.   (UNIT: 4; AREA: RD)

amylase   An enzyme specific for the hydrolysis of starch.   (UNIT: 3; AREA: CORE)

anaerobic phase of respiration   The reactions of aerobic respiration in which glucose is converted to two pyruvic acid molecules.   (UNIT: 2; AREA: BC)

anaerobic respiration   A type of respiration in which energy is released from organic molecules without the aid of oxygen.   (UNIT: 2; AREA: CORE)

anal pore   The egestive organ of the paramecium.   (UNIT: 2; AREA: CORE)

anemia A disorder of the human transport system in which the ability of the blood to carry oxygen is impaired, usually because of reduced numbers of red blood cells. (UNIT: 3; AREA: HP)

angina pectoris A disorder of the human transport system in which chest pain signals potential damage to the heart muscle due to narrowing of the opening of the coronary artery. (UNIT: 3; AREA: HP)

Animal One of the five biological kingdoms; it includes multicellular organisms whose cells are not bounded by cell walls and which are incapable of photosynthesis (e.g., human being). (UNIT: 1; AREA: CORE)

Annelida A phylum of the Animal Kingdom whose members (annelids) include the segmented worms (e.g., earthworm). (UNIT: 1; AREA: CORE)

antenna A receptor organ found in many arthropods (e.g., grasshopper), which is specialized for detecting chemical stimuli. (UNIT: 2; AREA: CORE)

anther The portion of the stamen that produces pollen. (UNIT: 4; AREA: CORE)

antibody A chemical substance, produced in response to the presence of a specific antigen, which neutralizes that antigen in the immune response. (UNIT: 3; AREA: CORE)

antigen A chemical substance, usually a protein, that is recognized by the immune system as a foreign "invader" and is neutralized by a specific antibody. (UNIT: 3; AREA: CORE)

anus The organ of egestion of the digestive tract. (UNIT: 3; AREA: CORE)

aorta The principal artery carrying blood from the heart to the body tissues. (UNIT: 3; AREA: HP)

aortic arches A specialized part of the earthworm's transport system that serves as a pumping mechanism for the blood fluid. (UNIT: 2; AREA: CORE)

apical meristem A plant growth region located at the tip of the root or tip of the stem. (UNIT: 4; AREA: CORE)

appendicitis A disorder of the human digestive tract in which the appendix becomes inflamed as a result of bacterial infection. (UNIT: 3; AREA: HP)

aquatic biome An ecological biome composed of many different water environments. (UNIT: 7; AREA: EC)

artery A thick-walled blood vessel that carries blood away from the heart under pressure. (UNIT: 3; AREA: CORE)

arthritis A disorder of the human locomotor system in which skeletal joints become inflamed, swollen, and painful. (UNIT: 3; AREA: HP)

Arthropoda A phylum of the Animal Kingdom whose members (arthropods) have bodies with chitinous exoskeletons and jointed appendages (e.g., grasshopper). (UNIT: 1; AREA: CORE)

artificial selection A technique of plant/animal breeding in which individual organisms displaying desirable characteristics are chosen for breeding purposes. (UNIT: 5; AREA: CORE)

asexual reproduction A type of reproduction in which new organisms are formed from a single parent organism. (UNIT: 4; AREA: CORE)

asthma A disorder of the human respiratory system in which the respiratory tube becomes constricted by swelling brought on by some irritant. (UNIT: 3; AREA: HP)

atrium   In human beings, one of the two thin-walled upper chambers of the heart that receive blood.   (UNIT: 3; AREA: CORE)

autonomic nervous system   A subdivision of the peripheral nervous system consisting of nerves associated with automatic functions (e.g., heartbeat, breathing).   (UNIT: 3; AREA: HP)

autosome   One of several chromosomes present in the cell that carry genes controlling "body" traits not associated with primary and secondary sex characteristics.   (UNIT: 5; AREA: CORE)

autotroph   An organism capable of carrying on autotrophic nutrition.   (UNIT: 2; AREA: CORE)

autotrophic nutrition   A type of nutrition in which organisms manufacture their own organic foods from inorganic raw materials.   (UNIT: 2; AREA: CORE)

auxin   A biochemical substance produced by plants that regulates growth patterns.   (UNIT: 2; AREA: CORE)

axon   An elongated portion of a neuron that conducts nerve impulses, usually away from the cell body of the neuron.   (UNIT: 2; AREA: CORE)

base   A chemical that releases hydroxyl ion (OH^-) in solution with water.   (UNIT: 1; AREA: CORE)

bicarbonate ion   The chemical formed in the blood plasma when carbon dioxide is absorbed from body tissues.   (UNIT: 3; AREA: CORE)

bile   In human beings, a secretion of the liver that is stored in the gallbladder and that emulsifies fats.   (UNIT: 3; AREA: CORE)

binary fission   A type of cell division in which mitosis is followed by equal cytoplasmic division.   (UNIT: 4; AREA: CORE)

binomial nomenclature   A system of naming, used in biological classification, that consists of the genus and species names (e.g., *Homo sapiens*).   (UNIT: 1; AREA: CORE)

biocide use   The use of pesticides that eliminate one undesirable organism but that have, due to technological oversight, unanticipated effects on beneficial species as well.   (UNIT: 7; AREA: CORE)

biological controls   The use of natural enemies of various agricultural pests for pest control, thereby eliminating the need for biocide use—a positive aspect of human involvement with the environment.   (UNIT: 7; AREA: CORE)

biomass   The total mass of living material present at the various trophic levels in a food chain.   (UNIT: 7; AREA: EC)

biome   A major geographical grouping of similar ecosystems, usually named for the climax flora in the region (e.g., Northeast Deciduous Forest).   (UNIT: 7; AREA: CORE)

biosphere   The portion of the earth in which living things exist, including all land and water environments.   (UNIT: 7; AREA: CORE)

biotic factor   Any of several conditions associated with life and living things that affect the survival of living things in the environment.   (UNIT: 7; AREA: CORE)

birth   In placental mammals, a stage of embryonic development in which the baby passes through the vaginal canal to the outside of the mother's body.   (UNIT: 4; AREA: RD)

blastula   In certain animals, a stage of embryonic development in which the embryo resembles a hollow ball of undifferentiated cells.   (UNIT: 4; AREA: RD)

blood   The complex fluid tissue that functions to transport nutrients and respiratory gases to all parts of the body.   (UNIT: 3; AREA: CORE)

blood typing   An application of the study of immunity in which the blood of a person is characterized by its antigen composition.   (UNIT: 3; AREA: HP)

bone   A structure that provides mechanical support and protection for bodily organs, and levers for the body's locomotive activities.   (UNIT: 3; AREA: CORE)

Bowman's capsule   A cup-shaped portion of the nephron responsible for the filtration of soluble blood components.   (UNIT: 3; AREA: CORE)

brain   An organ of the central nervous system that is responsible for regulating conscious and much unconscious activity in the body.   (UNIT: 3; AREA: CORE)

breathing   A mechanical process by which air is forced into the lung by means of muscular contraction of the diaphragm and rib muscles.   (UNIT: 3; AREA: CORE)

bronchiole   One of several subdivisions of the bronchi that penetrate the lung interior and terminate in alveoli.   (UNIT: 3; AREA: CORE)

bronchitis   A disorder of the human respiratory system in which the bronchi become inflamed.   (UNIT: 3; AREA: HP)

bronchus   One of the two major subdivisions of the breathing tube; the bronchi are ringed with cartilage and conduct air from the trachea to the lung interior.   (UNIT: 3; AREA: CORE)

Bryophyta   A phylum of the Plant Kingdom that consists of organisms lacking vascular tissues (e.g., moss).   (UNIT: 1; AREA: CORE)

budding   A type of asexual reproduction in which mitosis is followed by unequal cytoplasmic division.   (UNIT: 4; AREA: CORE)

bulb   A type of vegetative propagation in which a plant bulb produces new bulbs that may be established as independent organisms with identical characteristics.   (UNIT: 4; AREA: CORE)

cambium   The lateral meristem tissue in woody plants responsible for annual growth in stem diameter.   (UNIT: 4; AREA: CORE)

cancer   Any of a number of conditions characterized by rapid, abnormal, and uncontrolled division of affected cells.   (UNIT: 4; AREA: CORE)

capillary   A very small, thin-walled blood vessel that connects an artery to a vein and through which all absorption into the blood fluid occurs.   (UNIT: 3; AREA: CORE)

carbohydrate   An organic compound composed of carbon, hydrogen, and oxygen in a $1:2:1$ ratio (e.g., $C_6H_{12}O_6$).   (UNIT: 1; AREA: CORE)

carbon-14   A radioactive isotope of carbon used to trace the movement of carbon in various biochemical reactions, and also used in the "carbon dating" of fossils.   (UNIT: 2; AREA: BC)

carbon-fixation reactions   A set of biochemical reactions in photosynthesis in which hydrogen atoms are combined with carbon and oxygen atoms to form PGAL and glucose.   (UNIT: 2; AREA: BC)

carbon-hydrogen-oxygen cycle   A process by which these three elements are made available for use by other organisms through the chemical reactions of respiration and photosynthesis.   (UNIT: 7; AREA: CORE)

carboxyl group A chemical group having the formula —COOH and found as part of all amino acid and fatty acid molecules. (UNIT: 1; AREA: BC)

cardiac muscle A type of muscle tissue in the heart and arteries that is associated with the rhythmic nature of the pulse and heartbeat. (UNIT: 3; AREA: CORE)

cardiovascular disease In human beings, any disease of the circulatory organs. (UNIT: 3; AREA: HP)

carnivore A heterotrophic organism that consumes animal tissue as its primary source of nutrition. (See **secondary consumer**.) (UNIT: 7; AREA: CORE)

carrier An individual who, though not expressing a particular recessive trait, carries this gene as part of his/her heterozygous genotype. (UNIT: 5; AREA: CORE)

carrier protein A specialized molecule embedded in the cell membrane that aids the movement of materials across the membrane. (UNIT: 2; AREA: CORE)

cartilage A flexible connective tissue found in many flexible parts of the body (e.g., knee); common in the embryonic stages of development. (UNIT: 3; AREA: CORE)

catalyst Any substance that speeds up or slows down the rate of a chemical reaction. (See **enzyme**.) (UNIT: 1; AREA: CORE)

cell plate A structure that forms during cytoplasmic division in plant cells and serves to separate the cytoplasm into two roughly equal parts. (UNIT: 4; AREA: CORE)

cell theory A scientific theory that states, "All cells arise from previously existing cells" and "Cells are the unit of structure and function of living things." (UNIT: 1; AREA: CORE)

cell wall A cell organelle that surrounds and gives structural support to plant cells; cell walls are composed of cellulose. (UNIT: 1; AREA: CORE)

central nervous system The portion of the vertebrate nervous system that consists of the brain and the spinal cord. (UNIT: 3; AREA: CORE)

centriole A cell organelle found in animal cells that functions in the process of cell division. (UNIT: 1; AREA: CORE)

centromere The area of attachment of two chromatids in a double-stranded chromosome. (UNIT: 4; AREA: CORE)

cerebellum The portion of the human brain responsible for the coordination of muscular activity. (UNIT: 3; AREA: CORE)

cerebral hemorrhage A disorder of the human regulatory system in which a broken blood vessel in the brain may result in severe dysfunction or death. (UNIT: 3; AREA: HP)

cerebral palsy A disorder of the human regulatory system in which the motor and speech centers of the brain are impaired. (UNIT: 3; AREA: HP)

cerebrum The portion of the human brain responsible for thought, reasoning, sense interpretation, learning, and other conscious activities. (UNIT: 3; AREA: CORE)

cervix A structure that bounds the lower end of the uterus and through which sperm must pass in order to fertilize the egg. (UNIT: 4; AREA: RD)

chemical digestion The process by which nutrient molecules are converted by chemical means into a form usable by the cells. (UNIT: 2; AREA: CORE)

chemosynthesis A type of autotrophic nutrition in which certain bacteria use the energy of chemical oxidation to convert inorganic raw materials to organic food molecules. (UNIT: 2; AREA: CORE)

chitin   A polysaccharide substance that forms the exoskeleton of the grasshopper and other arthropods. (UNIT: 2; AREA: CORE)

chlorophyll   A green pigment in plant cells that absorbs sunlight and makes possible certain aspects of the photosynthetic process. (UNIT: 2; AREA: CORE)

chloroplast   A cell organelle found in plant cells that contains chlorophyll and functions in photosynthesis. (UNIT: 1; AREA: CORE)

Chordata   A phylum of the Animal Kingdom whose members (chordates) have internal skeletons made of cartilage and/or bone (e.g., human being). (UNIT: 1; AREA: CORE)

chorion   A membrane that surrounds all other embryonic membranes in many animal species, protecting them from mechanical damage. (UNIT: 4; AREA: RD)

chromatid   One strand of a double-stranded chromosome. (UNIT: 4; AREA: CORE)

chromosome mutation   An alteration in the structure of a chromosome involving many genes. (See **nondisjunction; translocation; addition; deletion**.) (UNIT: 5; AREA: CORE)

cilia   Small, hairlike structures in paramecia and other unicellular organisms that aid in nutrition and locomotion. (UNIT: 2; AREA: CORE)

classification   A technique by which scientists sort, group, and name organisms for easier study. (UNIT: 1; AREA: CORE)

cleavage   A series of rapid mitotic divisions that increase cell number in a developing embryo without corresponding increase in cell size. (UNIT: 4; AREA: CORE)

climax community   A stable, self-perpetuating community that results from an ecological succession. (UNIT: 7; AREA: CORE)

cloning   A technique of genetic investigation in which undifferentiated cells of an organism are used to produce new organisms with the same set of traits as the original cells. (UNIT: 5; AREA: MG)

closed transport system   A type of circulatory system in which the transport fluid is always enclosed within blood vessels (e.g., earthworm, human). (UNIT: 2; AREA: CORE)

clot   A structure that forms as a result of enzyme-controlled reactions following the rupturing of a blood vessel and serves as a plug to prevent blood loss. (UNIT: 3; AREA: HP)

codominance   A type of intermediate inheritance that results from the simultaneous expression of two dominant alleles with contrasting effects. (UNIT: 5; AREA: CORE)

codon   See **triplet codon**. (UNIT: 5; AREA: MG)

Coelenterata   A phylum of the Animal Kingdom whose members (coelenterates) have bodies that resemble a sack (e.g., hydra, jellyfish). (UNIT: 1; AREA: CORE)

coenzyme   A chemical substance or chemical subunit that functions to aid the action of a particular enzyme. (See **vitamin**.) (UNIT: 1; AREA: CORE)

cohesion   A force binding water molecules together that aids in the upward conduction of materials in the xylem. (UNIT: 2; AREA: CORE)

commensalism   A type of symbiosis in which one organism in the relationship benefits and the other is neither helped nor harmed. (UNIT: 7; AREA: EC)

common ancestry A concept central to the science of evolution which postulates that all organisms share a common ancestry whose closeness varies with the degree of shared similarity. (UNIT: 6, AREA: CORE)

community A level of biological organization that includes all of the species populations inhabiting a particular geographic area. (UNIT: 7; AREA: CORE)

comparative anatomy The study of similarities in the anatomical structures of organisms, and their use as an indicator of common ancestry and as evidence of organic evolution. (UNIT: 6; AREA: CORE)

comparative biochemistry The study of similarities in the biochemical makeups of organisms, and their use as an indicator of common ancestry and as evidence of organic evolution. (UNIT: 6; AREA: CORE)

comparative cytology The study of similarities in the cell structures of organisms, and their use as an indicator of common ancestry and as evidence of organic evolution. (UNIT: 6; AREA: CORE)

comparative embryology The study of similarities in the patterns of embryological development of organisms, and their use as an indicator of common ancestry and as evidence of organic evolution. (UNIT: 6; AREA: CORE)

competition A condition that arises when different species in the same habitat attempt to use the same limited resources. (UNIT: 7; AREA: CORE)

complete protein A protein that contains all eight essential amino acids. (UNIT: 3; AREA: HP)

compound A substance composed of two or more different kinds of atom (e.g., water: H_2O). (UNIT: 1; AREA: CORE)

compound light microscope A tool of biological study capable of producing a magnified image of a biological specimen by using a focused beam of light. (UNIT: 1; AREA: CORE)

conditioned behavior A type of response that is learned, but that becomes automatic with repetition. (UNIT: 3; AREA: CORE)

conservation of resources The development and application of practices to protect valuable and irreplaceable soil and mineral resources—a positive aspect of human involvement with the environment. (UNIT: 7; AREA: CORE)

constipation A disorder of the human digestive tract in which fecal matter solidifies and becomes difficult to egest. (UNIT: 3; AREA: HP)

consumer Any heterotrophic animal organism (e.g., human being). (UNIT: 7; AREA: CORE)

coronary artery An artery that branches off the aorta to feed the heart muscle. (UNIT: 3; AREA: HP)

coronary thrombosis A disorder of the human transport system in which the heart muscle becomes damaged as a result of blockage of the coronary artery. (UNIT: 3; AREA HP)

corpus luteum A structure resulting from the hormone-controlled transformation of the ovarian follicle that produces the hormone progesterone. (UNIT: 4; AREA: RD)

corpus luteum stage A stage of the menstrual cycle in which the cells of the follicle are transformed into the corpus luteum under the influence of the hormone LH. (UNIT: 4; AREA: RD)

cotyledon   A portion of the plant embryo that serves as a source of nutrition for the young plant before photosynthesis begins.   (UNIT: 4; AREA: CORE)

cover-cropping   A proper agricultural practice in which a temporary planting (cover crop) is used to limit soil erosion between seasonal plantings of main crops.   (UNIT: 7; AREA: CORE)

crop   A portion of the digestive tract of certain animals that stores food temporarily before digestion.   (UNIT: 2; AREA: CORE)

cross-pollination   A type of pollination in which pollen from one flower pollinates flowers of a different plant of the same species.   (UNIT: 4; AREA: CORE)

crossing-over   A pattern of inheritance in which linked genes may be separated during synapsis in the first meiotic division, when sections of homologous chromosomes may be exchanged.   (UNIT: 5; AREA: CORE)

cuticle   A waxy coating that covers the upper epidermis of most leaves and acts to help the leaf retain water.   (UNIT: 2; AREA: CORE)

cutting   A technique of plant propagation in which vegetative parts of the parent plant are cut and rooted to establish new plant organisms with identical characteristics. (UNIT: 4; AREA: CORE)

cyclosis   The circulation of the cell fluid (cytoplasm) within the cell interior.   (UNIT: 2; AREA: CORE)

cyton   The "cell body" of the neuron, which generates the nerve impulse.   (UNIT: 2; AREA: CORE)

cytoplasm   The watery fluid that provides a medium for the suspension of organelles within the cell.   (UNIT: 1; AREA: CORE)

cytoplasmic division   The separation of daughter nuclei into two new daughter cells. (UNIT: 4; AREA: CORE)

cytosine   A nitrogenous base found in both DNA and RNA molecules.   (UNIT: 5; AREA: CORE)

daughter cell   A cell that results from mitotic cell division.   (UNIT: 4; AREA: CORE)

daughter nucleus   One of two nuclei that form as a result of mitosis.   (UNIT: 4; AREA: CORE)

deamination   A process by which amino acids are broken down into their component parts for conversion into urea.   (UNIT: 3; AREA: HP)

death   The irreversible cessation of bodily functions and cellular activities.   (UNIT: 4; AREA: RD)

deciduous   A term relating to broadleaf trees which shed their leaves in the fall. (UNIT: 7; AREA: EC)

decomposer   Any saprophytic organism that derives its energy from the decay of plant and animal tissues (e.g., bacteria of decay, fungus); the final stage of a food chain. (UNIT: 7; AREA: CORE)

decomposition bacteria   In the nitrogen cycle, bacteria that break down plant and animal protein and produce ammonia as a by-product.   (UNIT: 7; AREA: EC)

dehydration synthesis   A chemical process in which two organic molecules may be joined after removing the atoms needed to form a molecule of water as a by-product.   (UNIT: 1; AREA: CORE)

deletion   A type of chromosome mutation in which a section of a chromosome is separated and lost.   (UNIT: 5; AREA: MG)

dendrite   A cytoplasmic extension of a neuron that serves to detect an environmental stimulus and carry an impulse to the cell body of the neuron.   (UNIT: 2; AREA: CORE)

denitrifying bacteria   In the nitrogen cycle, bacteria that convert excess nitrate salts into gaseous nitrogen.   (UNIT: 7; AREA: EC)

deoxygenated blood   Blood that has released its transported oxygen to the body tissues.   (UNIT: 3; AREA: HP)

deoxyribonucleic acid (DNA)   A nucleic acid molecule known to be the chemically active agent of the gene; the fundamental hereditary material of living organisms.   (UNIT: 5; AREA: CORE)

deoxyribose   A five-carbon sugar that is a component part of the nucleotide unit in DNA only.   (UNIT: 5; AREA: CORE)

desert   A terrestrial biome characterized by sparse rainfall, extreme temperature variation, and a climax flora that includes cactus.   (UNIT: 7; AREA: EC)

diabetes   A disorder of the human regulatory system in which insufficient insulin production leads to elevated blood sugar concentrations.   (UNIT: 3; AREA: HP)

diarrhea   A disorder of the human digestive tract in which the large intestine fails to absorb water from the waste matter, resulting in watery feces.   (UNIT: 3; AREA: HP)

diastole   The lower pressure registered during blood pressure testing. (See **systole**.)   (UNIT: 3; AREA: HP)

differentiation   The process by which embryonic cells become specialized to perform the various tasks of particular tissues throughout the body.   (UNIT: 4; AREA: RD)

diffusion   A form of passive transport by which soluble substances are absorbed or released by cells.   (UNIT: 2; AREA: CORE)

digestion   The process by which complex foods are broken down by mechanical or chemical means for use by the body.   (UNIT: 2; AREA: CORE)

dipeptide   A chemical unit composed of two amino acid units linked by a peptide bond.   (UNIT: 1; AREA: CORE)

diploid chromosome number   The number of chromosomes found characteristically in the cells (except gametes) of sexually reproducing species.   (UNIT: 4; AREA: CORE)

disaccharidase   Any disaccharide-hydrolyzing enzyme.   (UNIT: 3; AREA: CORE)

disaccharide   A type of carbohydrate known also as a "double sugar"; all disaccharides have the molecular formula $C_{12}H_{22}O_{11}$.   (UNIT: 1; AREA: CORE)

disjunction   The separation of homologous chromosome pairs at the end of the first meiotic division.   (UNIT: 4; AREA: CORE)

disposal problems   Problems, due to technological oversight, that result when commercial and technological activities produce solid and/or chemical wastes that must be disposed of.   (UNIT: 7; AREA: CORE)

dissecting microscope   A tool of biological study that magnifies the image of a biological specimen up to 20 times normal size for purposes of gross dissection.   (UNIT: 1; AREA: CORE)

dominance A pattern of genetic inheritance in which the effects of a dominant allele mask those of a recessive allele. (UNIT: 5; AREA: CORE)

dominant allele (gene) An allele (gene) whose effect masks that of its recessive allele. (UNIT: 5; AREA: CORE)

double-stranded chromosome The two-stranded structure that results from chromosomal replication. (UNIT: 4; AREA: CORE)

Down's syndrome In human beings, a condition, characterized by mental and physical retardation, that may be caused by the nondisjunction of chromosome number 21. (UNIT: 5; AREA: CORE)

Drosophila The common fruit fly, an organism that has served as an object of genetic research in the development of the gene-chromosome theory. (UNIT: 5; AREA: CORE)

ductless gland See **endocrine gland**. (UNIT: 2; AREA: CORE)

ecology The science that studies the interactions of living things with each other and with the nonliving environment. (UNIT: 7; AREA: CORE)

ecosystem The basic unit of study in ecology, including the plant and animal community in interaction with the nonliving environment. (UNIT: 7; AREA: CORE)

ectoderm An embryonic tissue that differentiates into skin and nerve tissue in the adult animal. (UNIT: 4; AREA: RD)

effector An organ specialized to produce a response to an environmental stimulus: effectors may be muscles or glands. (UNIT: 2; AREA: CORE)

egestion The process by which undigested food materials are eliminated from the body. (UNIT: 2; AREA: CORE)

electron microscope A tool of biological study that uses a focused beam of electrons to produce an image of a biological specimen magnified up to 25,000 times normal size. (UNIT: 1; AREA: CORE)

element The simplest form of matter; an element is a substance (e.g., nitrogen) made up of a single type of atom. (UNIT: 1; AREA: CORE)

embryo An organism in the early stages of development following fertilization. (UNIT: 4; AREA: CORE)

embryonic development A series of complex processes by which animal and plant embryos develop into adult organisms. (UNIT: 4; AREA: CORE)

emphysema A disorder of the human respiratory system in which lung tissue deteriorates, leaving the lung with diminished capacity and efficiency. (UNIT: 3; AREA: HP)

emulsification A process by which fat globules are mechanically digested to form fat droplets. (UNIT: 3; AREA: CORE)

endocrine ("ductless") gland A gland (e.g., thyroid, pituitary) specialized for the production of hormones and their secretion directly into the bloodstream; such glands lack ducts. (UNIT: 2; AREA: CORE)

endoderm An embryonic tissue that differentiates into the digestive and respiratory tract lining in the adult animal. (UNIT: 4; AREA: RD)

endoplasmic reticulum (ER) A cell organelle known to function in the transport of cell products from place to place within the cell. (UNIT: 1; AREA: CORE)

environmental laws   Federal, state, and local legislation enacted in an attempt to protect environmental resources—a positive aspect of human involvement with the environment.   (UNIT: 7; AREA: CORE)

enzymatic hydrolysis   An enzyme-controlled reaction by which complex food molecules are broken down chemically into simpler subunits.   (UNIT: 2; AREA: CORE)

enzyme   An organic catalyst that controls the rate of metabolism of a single type of substrate; enzymes are protein in nature.   (UNIT: 1; AREA: CORE)

enzyme-substrate complex   A physical association between an enzyme molecule and its substrate within which the substrate is metabolized.   (UNIT: 1; AREA: CORE)

epicotyl   A portion of the plant embryo that specializes to become the upper stem, leaves, and flowers of the adult plant.   (UNIT: 4; AREA: CORE)

epidermis   The outermost cell layer in a plant or animal.   (UNIT: 2; AREA: CORE)

epiglottis   In a human being, a flap of tissue that covers the upper end of the trachea during swallowing and prevents inhalation of food.   (UNIT: 3; AREA: CORE)

esophagus   A structure in the upper portion of the digestive tract that conducts the food from the pharynx to the midgut.   (UNIT: 2; AREA: CORE)

essential amino acid   An amino acid that cannot be synthesized by the human body, but must be obtained by means of the diet.   (UNIT: 3; AREA: HP)

estrogen   A hormone, secreted by the ovary, that regulates the production of female secondary sex characteristics.   (UNIT: 3; AREA: HP)

evolution   Any process of gradual change through time.   (UNIT: 6; AREA: CORE)

excretion   The life function by which living things eliminate metabolic wastes from their cells.   (UNIT: 1; AREA: CORE)

exoskeleton   A chitinous material that covers the outside of the bodies of most arthropods and provides protection for internal organs and anchorage for muscles.   (UNIT: 2; AREA: CORE)

exploitation of organisms   Systematic removal of animals and plants with commercial value from their environments, for sale—a negative aspect of human involvement with the environment.   (UNIT: 7; AREA: CORE)

extensor   A skeletal muscle that extends (opens) a joint.   (UNIT: 3; AREA: CORE)

external development   Embryonic development that occurs outside the body of the female parent (e.g., birds).   (UNIT: 4; AREA: CORE)

external fertilization   Fertilization that occurs outside the body of the female parent (e.g., fish).   (UNIT: 4; AREA: RD)

extracellular digestion   Digestion that occurs outside the cell.   (UNIT: 2; AREA: CORE)

Fallopian tube   See **oviduct**.   (UNIT: 4; AREA: RD)

fatty acid   An organic molecule that is a component of certain lipids.   (UNIT: 1; AREA: CORE)

fauna   The animal species comprising an ecological community.   (UNIT: 7; AREA: EC)

feces   The semisolid material that results from the solidification of undigested foods in the large intestine.   (UNIT: 3; AREA: CORE)

fertilization The fusion of gametic nuclei in the process of sexual reproduction. (UNIT: 4; AREA: CORE)

filament The portion of the stamen that supports the anther. (UNIT: 4; AREA: CORE)

flagella Microscopic, whiplike structures found on certain cells that aid in locomotion and circulation. (UNIT: 2; AREA: CORE)

flexor A skeletal muscle that flexes (closes) a joint. (UNIT: 3; AREA: CORE)

flora The plant species comprising an ecological community. (UNIT: 7; AREA: EC)

flower The portion of a flowering plant that is specialized for sexual reproduction. (UNIT: 4; AREA: CORE)

fluid-mosaic model A model of the structure of the cell membrane in which large protein molecules are thought to be embedded in a bilipid layer. (UNIT: 2; AREA: CORE)

follicle One of many areas within the ovary that serve as sites for the periodic maturation of ova. (UNIT: 4; AREA: RD)

follicle stage The stage of the menstrual cycle in which an ovum reaches its final maturity under the influence of the hormone FSH. (UNIT: 4; AREA: RD)

follicle-stimulating hormone (FSH) A pituitary hormone that regulates the maturation of, and the secretion of estrogen by, the ovarian follicle. (UNIT: 3; AREA: HP)

food chain A series of nutritional relationships in which food energy is passed from producer to herbivore to carnivore to decomposer; a segment of a food web. (UNIT: 7; AREA: CORE)

food web A construct showing a series of interrelated food chains and illustrating the complex nutritional interrelationships that exist in an ecosystem. (UNIT: 7; AREA: CORE)

fossil The preserved direct or indirect remains of an organism that lived in the past, as found in the geologic record. (UNIT: 6; AREA: CORE)

fraternal twins In human beings, twin offspring that result from the simultaneous fertilization of two ova by two sperm; such twins are not genetically identical. (UNIT: 4; AREA: RD)

freshwater biome An aquatic biome made up of many separate freshwater systems that vary in size and stability and may be closely associated with terrestrial biomes. (UNIT: 7; AREA: EC)

fruit Any plant structure that contains seeds; a mechanism of seed dispersal. (UNIT: 4; AREA: CORE)

Fungi One of the five biological kingdoms; it includes organisms unable to manufacture their own organic foods (e.g., mushroom). (UNIT: 1; AREA: CORE)

gall bladder An accessory organ that stores bile. (UNIT: 3; AREA: CORE)

gallstones A disorder of the human digestive tract in which deposits of hardened cholesterol lodge in the gallbladder. (UNIT: 3; AREA: HP)

gamete A specialized reproductive cell produced by organisms of sexually reproducing species. (See **sperm; ovum; pollen; ovule**.) (UNIT: 4; AREA: CORE)

gametogenesis The process of cell division by which gametes are produced. (See **meiosis; spermatogenesis; oogenesis**.) (UNIT: 4; AREA: CORE)

ganglion An area of bunched nerve cells that acts as a switching point for nerve impulses traveling from receptors and to effectors. (UNIT: 2; AREA: CORE)

garden pea The research organism used by Mendel in his early scientific work in genetic inheritance. (UNIT: 5; AREA: CORE)

gastric cecum A gland in the grasshopper that secretes digestive enzymes. (UNIT: 2; AREA: CORE)

gastrula A stage of embryonic development in animals in which the embryo assumes a tube-within-a-tube structure and distinct embryonic tissues (ectoderm, mesoderm, endoderm) begin to differentiate. (UNIT: 4; AREA: RD)

gastrulation The process by which a blastula becomes progressively more indented, forming a gastrula. (UNIT: 4; AREA: RD)

gene A unit of heredity; a discrete portion of a chromosome thought to be responsible for the production of a single type of polypeptide; the "factor" responsible for the inheritance of a genetic trait. (UNIT: 5; AREA: CORE)

gene frequency The proportion (percentage) of each allele for a particular trait that is present in the gene pool of a population. (UNIT: 5; AREA: MG)

gene linkage A pattern of inheritance in which genes located along the same chromosome are prevented from assorting independently, but are linked together in their inheritance. (UNIT: 5; AREA: CORE)

gene mutation An alteration of the chemical nature of a gene that changes its ability to control the production of a polypeptide chain. (UNIT: 5; AREA: CORE)

gene pool The sum total of all the inheritable genes for the traits in a given sexually reproducing population. (UNIT: 5; AREA: MG)

gene-chromosome theory A theory of genetic inheritance that is based on current understanding of the relationships between the biochemical control of traits and the process of cell division. (UNIT: 5; AREA: CORE)

genetic counseling Clinical discussions concerning inheritance patterns that are designed to inform prospective parents of the potential for expression of a genetic disorder in their offspring. (UNIT: 5; AREA: MG)

genetic engineering The use of various techniques to move genes from one organism to another. (UNIT: 5; AREA: MG)

genetic screening A technique for the detection of human genetic disorders in which bodily fluids are analyzed for the presence of certain marker chemicals. (UNIT: 5; AREA: MG)

genotype The particular combination of genes in an allele pair. (UNIT: 5; AREA: CORE)

genus A level of biological classification that represents a subdivision of the phylum level; having fewer organisms with great similarity (e.g., *Drosophila*, *Paramecium*). (UNIT: 1; AREA: CORE)

geographic isolation The separation of species populations by geographical barriers, facilitating the evolutionary process. (UNIT: 6; AREA: CORE)

geologic record A supporting item of evidence of organic evolution, supplied within the earth's rock and other geological deposits. (UNIT: 6; AREA: CORE)

germination The growth of the pollen tube from a pollen grain; the growth of the embryonic root and stem from a seed. (UNIT: 4; AREA: CORE)

gestation   The period of prenatal development of a placental mammal; human gestation requires approximately 9 months.   (UNIT: 4; AREA: RD)

gizzard   A portion of the digestive tract of certain organisms, including the earthworm and the grasshopper in which food is ground into smaller fragments.   (UNIT: 2; AREA: CORE)

glomerulus   A capillary network lying within Bowman's capsule of the nephron.   (UNIT: 3; AREA: CORE)

glucagon   A hormone, secreted by the islets of Langerhans, that regulates the release of blood sugar from stored glycogen.   (UNIT: 3; AREA: HP)

glucose   A monosaccharide produced commonly in photosynthesis and used by both plants and animals as a "fuel" in the process of respiration.   (UNIT: 1; AREA: CORE)

glycerol   An organic compound that is a component of certain lipids.   (UNIT: 1; AREA: CORE)

glycogen   A polysaccharide synthesized in animals as a means of storing glucose; glycogen is stored in the liver and in the muscles.   (UNIT: 3; AREA: HP)

goiter   A disorder of the human regulatory system in which the thyroid gland enlarges because of a deficiency of dietary iodine.   (UNIT: 3; AREA: HP)

Golgi complex   Cell organelles that package cell products and move them to the plasma membrane for secretion.   (UNIT: 1; AREA: CORE)

gonad   An endocrine gland that produces the hormones responsible for the production of various secondary sex characteristics. (See **ovary; testis**.)   (UNIT: 3; AREA: HP)

gout   A disorder of the human excretory system in which uric acid accumulates in the joints, causing severe pain.   (UNIT: 3; AREA: HP)

gradualism   A theory of the time frame required for organic evolution which assumes that evolutionary change is slow, gradual, and continuous.   (UNIT: 6; AREA: CORE)

grafting   A technique of plant propagation in which the stems of desirable plants are attached (grafted) to rootstocks of related varieties to produce new plants for commercial purposes.   (UNIT: 4; AREA: CORE)

grana   The portion of the chloroplast within which chlorophyll molecules are concentrated.   (UNIT: 2; AREA: BC)

grassland   A terrestrial biome characterized by wide variation in temperature and a climax flora that includes grasses.   (UNIT: 7; AREA: EC)

growth   A process by which cells increase in number and size, resulting in an increase in size of the organism.   (UNIT: 1; AREA: CORE)

growth-stimulating hormone (GSH)   A pituitary hormone regulating the elongation of the long bones of the body.   (UNIT: 3; AREA: HP)

guanine   A nitrogenous base found in both DNA and RNA molecules.   (UNIT: 5; AREA: CORE)

guard cell   One of a pair of cells that surround the leaf stomate and regulate its size.   (UNIT: 2; AREA: CORE)

habitat   The environment or set of ecological conditions within which an organism lives.   (UNIT: 7; AREA: CORE)

Hardy-Weinberg principle A hypothesis, advanced by G. H. Hardy and W. Weinberg, which states that the gene pool of a population should remain stable as long as a set of "ideal" conditions is met. (UNIT: 5; AREA: MG)

heart In human beings, a four-chambered muscular pump that facilitates the movement of blood throughout the body. (UNIT: 3; AREA: CORE)

helix Literally a spiral; a term used to describe the "twisted ladder" shape of the DNA molecule. (UNIT: 5; AREA: CORE)

hemoglobin A type of protein specialized for the transport of respiratory oxygen in certain organisms, including earthworms and human beings. (UNIT: 2; AREA: CORE)

herbivore A heterotrophic organism that consumes plant matter as its primary source of nutrition. (See **primary consumer**.) (UNIT: 7; AREA: CORE)

hermaphrodite An animal organism that produces both male and female gametes. (UNIT: 4; AREA: CORE)

heterotroph An organism that typically carries on heterotrophic nutrition. (UNIT: 2; AREA: CORE)

heterotroph hypothesis A scientific hypothesis devised to explain the probable origin and early evolution of life on earth. (UNIT: 6; AREA: CORE)

heterotrophic nutrition A type of nutrition in which organisms must obtain their foods from outside sources of organic nutrients. (UNIT: 2; AREA: CORE)

heterozygous A term used to refer to an allele pair in which the alleles have different contrasting effects (e.g., *Aa, RW*). (UNIT: 5; AREA: CORE)

high blood pressure A disorder of the human transport system in which systolic and diastolic pressures register higher than normal because of narrowing of the artery opening. (UNIT: 3; AREA: HP)

histamine A chemical product of the body that causes irritation and swelling of the mucous membranes. (UNIT: 3; AREA: HP)

homeostasis The condition of balance and dynamic stability that characterizes living systems under normal conditions. (UNIT: 1; AREA: CORE)

homologous chromosomes A pair of chromosomes that carry corresponding genes for the same traits. (UNIT: 4; AREA: CORE)

homologous structures Structures present within different species that can be shown to have had a common origin, but that may or may not share a common function. (UNIT: 6; AREA: CORE)

homozygous A term used to refer to an allele pair in which the alleles are identical in terms of effect (e.g., *AA, aa*). (UNIT: 5; AREA: CORE)

hormone A chemical product of an endocrine gland which has a regulatory effect on the cell's metabolism. (UNIT: 2; AREA: CORE)

host The organism that is harmed in a parasitic relationship. (UNIT: 7; AREA: EC)

hybrid A term used to describe a heterozygous genotype. (See **heterozygous**.) (UNIT: 5; AREA: CORE)

hybridization A technique of plant/animal breeding in which two varieties of the same species are crossbred in the hope of producing offspring with the favorable traits of both varieties. (UNIT: 5; AREA: CORE)

hydrogen bond   A weak electrostatic bond that holds together the twisted strands of DNA and RNA molecules.   (UNIT: 5; AREA: CORE)

hydrolysis   The chemical process by which a complex food molecule is split into simpler components through the addition of a molecule of water to the bonds holding it together.   (UNIT: 3; AREA: CORE)

hypocotyl   A portion of the plant embryo that specializes to become the root and lower stem of the adult plant.   (UNIT: 4; AREA: CORE)

hypothalamus   An endocrine gland whose secretions affect the pituitary gland.   (UNIT: 3; AREA: HP)

identical twins   In human beings, twin offspring resulting from the separation of the embryonic cell mass of a single fertilization into two separate masses; such twins are genetically identical.   (UNIT: 4; AREA: RD)

importation of organisms   The introduction of nonactive plants and animals into new areas where they compete strongly with native species—a negative aspect of human involvement with the environment.   (UNIT: 7; AREA: CORE)

***in vitro* fertilization**   A laboratory technique in which fertilization is accomplished outside the mother's body using mature ova and sperm extracted from the parents' bodies.   (UNIT: 4; AREA: RD)

inbreeding   A technique of plant/animal breeding in which a "purebred" variety is bred only with its own members, so as to maintain a set of desired characteristics.   (UNIT: 5; AREA: CORE)

independent assortment   A pattern of inheritance in which genes on different, nonhomologous chromosomes are free to be inherited randomly and regardless of the inheritance of the others.   (UNIT: 5; AREA: CORE)

ingestion   The mechanism by which an organism takes in food from its environment.   (UNIT: 2; AREA: CORE)

inorganic compound   A chemical compound that lacks the element carbon or hydrogen (e.g., table salt: NaCl).   (UNIT: 1; AREA: CORE)

insulin   A hormone, secreted by the islets of Langerhans, that regulates the storage of blood sugar as glycogen.   (UNIT: 3; AREA: HP)

intercellular fluid (ICF)   The fluid that bathes cells and fills intercellular spaces.   (UNIT: 3; AREA: CORE)

interferon   A substance, important in the fight against human cancer, that may now be produced in large quantities through techniques of genetic engineering.   (UNIT: 5; AREA: MG)

intermediate inheritance   Any pattern of inheritance in which the offspring expresses a phenotype different from the phenotypes of its parents and usually representing a form intermediate between them.   (UNIT: 5; AREA: CORE)

internal development   Embryonic development that occurs within the body of the female parent.   (UNIT: 4; AREA: CORE)

internal fertilization   Fertilization that occurs inside the body of the female parent.   (UNIT: 4; AREA: CORE)

interneuron   A type of neuron, located in the central nervous system, that is responsible for the interpretation of impulses received from sensory neurons.   (UNIT: 3; AREA: CORE)

intestine A portion of the digestive tract in which chemical digestion and absorption of digestive end products occur. (UNIT: 2; AREA: CORE)

intracellular digestion A type of chemical digestion carried out within the cell. (UNIT: 2; AREA: CORE)

iodine A chemical stain used in cell study; an indicator used to detect the presence of starch. (See **staining**.) (UNIT: 1; AREA: CORE)

islets of Langerhans An endocrine gland, located within the pancreas, that produces the hormones insulin and glucagon. (UNIT: 3; AREA: HP)

karyotype An enlarged photograph of the paired homologous chromosomes of an individual cell that is used in the detection of certain genetic disorders involving chromosome mutation. (UNIT: 5; AREA: MG)

karyotyping A technique for the detection of human genetic disorders in which a karyotype is analyzed for abnormalities in chromosome structure or number. (UNIT: 5; AREA: MG)

kidney The excretory organ responsible for maintaining the chemical composition of the blood. (See **nephron**.) (UNIT: 3; AREA: CORE)

kidney failure A disorder of the human excretory system in which there is a general breakdown of the kidney's ability to filter blood components. (UNIT: 3; AREA: HP)

kingdom A level of biological classification that includes a broad grouping of organisms displaying general structural similarity; five kingdoms have been named by scientists. (UNIT: 1; AREA: CORE)

lacteal A small extension of the lymphatic system, found inside the villus, that absorbs fatty acids and glycerol resulting from lipid hydrolysis. (UNIT: 3; AREA: CORE)

lactic acid fermentation A type of anaerobic respiration in which glucose is converted to two lactic acid molecules. (UNIT: 2; AREA: CORE)

large intestine A portion of the digestive tract in which undigested foods are solidified by means of water absorption to form feces. (UNIT: 3; AREA: CORE)

lateral meristem A plant growth region located under the epidermis or bark of a stem. (See **cambium**.) (UNIT: 4; AREA: CORE)

Latin The language used in biological classification for naming organisms by means of binomial nomenclature. (UNIT: 1; AREA: CORE)

lenticel A small pore in the stem surface that permits the absorption and release of respiratory gases within stem tissues. (UNIT: 2; AREA: CORE)

leukemia A disorder of the human transport system in which the bone marrow produces large numbers of abnormal white blood cells. (See **cancer**.) (UNIT: 3; AREA: HP)

lichen A symbiosis of alga and fungus that frequently acts as a pioneer species on bare rock. (UNIT: 7; AREA: EC)

limiting factor Any abiotic or biotic condition that places limits on the survival of organisms and on the growth of species populations in the environment. (UNIT: 7; AREA: CORE)

lipase Any lipid-hydrolyzing enzyme. (UNIT: 7; AREA: CORE)

lipid An organic compound composed of carbon, hydrogen, and oxygen in which hydrogen and oxygen are *not* in a 2 : 1 ratio (e.g., a wax, plant oil); many lipids are constructed of a glycerol and three fatty acids. (UNIT: 1; AREA: CORE)

liver An accessory organ that stores glycogen, produces bile, destroys old red blood cells, deaminates amino acids, and produces urea. (UNIT: 3; AREA: CORE)

lock-and-key model A theoretical model of enzyme action that attempts to explain the concept of enzyme specificity. (UNIT: 1; AREA: CORE)

lung The major organ of respiratory gas exchange. (UNIT: 3; AREA: CORE)

lutinizing hormone (LH) A pituitary hormone that regulates the conversion of the ovarian follicle into the corpus luteum. (UNIT: 4; AREA: RD)

lymph Intercellular fluid (ICF) that has passed into the lymph vessels. (UNIT: 3; AREA: CORE)

lymph node One of a series of structures in the body that act as reservoirs of lymph and also contain white blood cells as part of the body's immune system. (UNIT: 3; AREA: CORE)

lymph vessel One of a branching series of tubes that collect ICF from the tissues and redistribute it as lymph. (UNIT: 3; AREA: CORE)

lymphatic circulation The movement of lymph throughout the body. (UNIT: 3; AREA: CORE)

lymphocyte A type of white blood cell that produces antibodies. (UNIT: 3; AREA: CORE)

lysosome A cell organelle that houses hydrolytic enzymes used by the cell in the process of chemical digestion. (UNIT: 1; AREA: CORE)

Malpighian tubules In arthropods (e.g., grasshopper), an organ specialized for the removal of metabolic wastes. (UNIT: 2; AREA: CORE)

maltase A specific enzyme that catalyzes the hydrolysis (and dehydration synthesis) of maltose. (UNIT: 1; AREA: CORE)

maltose A type of disaccharide; a maltose molecule is composed of two units of glucose joined together by dehydration synthesis. (UNIT: 1; AREA: CORE)

marine biome An aquatic biome characterized by relatively stable conditions of moisture, salinity, and temperature. (UNIT: 7; AREA: EC)

marsupial mammal See **nonplacental mammal**. (UNIT: 4; AREA: CORE)

mechanical digestion Any of the processes by which foods are broken apart physically into smaller particles. (UNIT: 2; AREA: CORE)

medulla The portion of the human brain responsible for regulating the automatic processes of the body. (UNIT: 3; AREA: CORE)

meiosis The process by which four monoploid nuclei are formed from a single diploid nucleus. (UNIT: 4; AREA: CORE)

meningitis A disorder of the human regulatory system in which the membranes of the brain or spinal cord become inflamed. (UNIT: 3; AREA: HP)

menstrual cycle A hormone-controlled process responsible for the monthly release of mature ova. (UNIT: 4; AREA: RD)

menstruation The stage of the menstrual cycle in which the lining of the uterus breaks down and is expelled from the body via the vaginal canal. (UNIT: 4; AREA: RD)

meristem   A plant tissue specialized for embryonic development. (See **apical meristem; lateral meristem; cambium.**)   (UNIT: 4; AREA: CORE)

mesoderm   An embryonic tissue that differentiates into muscle, bone, the excretory system, and most of the reproductive system in the adult animal.   (UNIT: 4; AREA: RD)

messenger RNA (m-RNA)   A type of RNA that carries the genetic code from the nuclear DNA to the ribosome for transcription.   (UNIT: 5; AREA· MG)

metabolism   All of the chemical processes of life considered together; the sum total of all the cell's chemical activity.   (UNIT: 1; AREA: CORE)

methylene blue   A chemical stain used in cell study. (See **staining.**)   (UNIT: 1; AREA: CORE)

microdissection instruments   Tools of biological study that are used to remove certain cell organelles from within cells for examination.   (UNIT: 1; AREA: CORE)

micrometer (μm)   A unit of linear measurement equal in length to 0.001 millimeter (0.000001 meter), used for expressing the dimensions of cells and cell organelles. (UNIT: 1; AREA: CORE)

mitochondrion   A cell organelle that contains the enzymes necessary for aerobic respiration.   (UNIT: 1; AREA: CORE)

mitosis   A precise duplication of the contents of a parent cell nucleus, followed by an orderly separation of these contents into two new, identical daughter nuclei.   (UNIT: 4; AREA: CORE)

mitotic cell division   A type of cell division that results in the production of two daughter cells identical to each other and to the parent cell.   (UNIT: 4; AREA: CORE)

Monera   One of the five biological kingdoms; it includes simple unicellular forms lacking nuclear membranes (e.g., bacteria),   (UNIT: 1; AREA: CORE)

monohybrid cross   A genetic cross between two organisms both heterozygous for a trait controlled by a single allele pair. The phenotypic ratio resulting is 3 : 1; the genotypic ratio is 1 : 2 : 1.   (UNIT: 5; AREA: CORE)

monoploid chromosome number   The number of chromosomes commonly found in the gametes of sexually reproducing species.   (UNIT: 4; AREA: CORE)

monosaccharide   A type of carbohydrate known also as a "simple sugar"; all monosaccharides have the molecular formula $C_6H_{12}O_6$.   (UNIT: 1; AREA: CORE)

motor neuron   A type of neuron that carries "command" impulses from the central nervous system to an effector organ.   (UNIT: 3; AREA: CORE)

mucus   A protein-rich mixture that bathes and moistens the respiratory surfaces (UNIT: 2; AREA: CORE)

multicellular   Having a body that consists of large groupings of specialized cells (e.g., human being).   (UNIT: 1; AREA: CORE)

multiple alleles   A pattern of inheritance in which the existence of more than two alleles is hypothesized, only two of which are present in the genotype of any one individual. (UNIT: 5; AREA: CORE)

muscle   A type of tissue specialized to produce movement of body parts.   (UNIT: 2; AREA: CORE)

mutagenic agent   Any environmental condition that initiates or accelerates genetic mutation.   (UNIT: 5; AREA: CORE)

mutation   Any alteration of the genetic material, either the chromosome or the gene, in an organism.   (UNIT: 5; AREA: CORE)

mutualism   A type of symbiosis beneficial to both organisms in the relationship.   (UNIT: 7; AREA: EC)

nasal cavity   A series of channels through which outside air is admitted to the body interior and is warmed and moistened before entering the lung.   (UNIT: 3; AREA: CORE)

natural selection   A concept, central to Darwin's theory of evolution, to the effect that the individuals best adapted to their environment tend to survive and to pass their favorable traits on to the next generation.   (UNIT: 6; AREA: CORE)

negative feedback   A type of endocrine regulation in which the effects of one gland may inhibit its own secretory activity, while stimulating the secretory activity of another gland.   (UNIT: 3; AREA: HP)

nephridium   An organ found in certain organisms, including the earthworm, specialized for the removal of metabolic wastes.   (UNIT: 2; AREA: CORE)

nephron   The functional unit of the kidney. (See **glomerulus; Bowman's capsule**.)   (UNIT: 3; AREA: CORE)

nerve   A structure formed from the bundling of neurons carrying sensory or motor impulses.   (UNIT: 3; AREA: CORE)

nerve impulse   An electrochemical change in the surface of the nerve cell.   (UNIT: 2; AREA: CORE)

nerve net   A network of "nerve" cells in coelenterates such as the hydra.   (UNIT: 2; AREA: CORE)

neuron   A cell specialized for the transmission of nerve impulses.   (UNIT: 2; AREA: CORE)

neurotransmitter   A chemical substance secreted by a neuron that aids in the transmission of the nerve impulse to an adjacent neuron.   (UNIT: 2; AREA: CORE)

niche   The role that an organism plays in its environment.   (UNIT: 7; AREA: EC)

nitrifying bacteria   In the nitrogen cycle, bacteria that absorb ammonia and convert it into nitrate salts.   (UNIT: 7; AREA: EC)

nitrogen cycle   The process by which nitrogen is recycled and made available for use by other organisms.   (UNIT: 7; AREA: CORE)

nitrogen-fixing bacteria   A type of bacteria responsible for absorbing atmospheric nitrogen and converting it to nitrate salts in the soil.   (UNIT: 7; AREA: EC)

nitrogenous base   A chemical unit composed of carbon, hydrogen, and nitrogen that is a component part of the nucleotide unit.   (UNIT: 5; AREA: CORE)

nitrogenous waste   Any of a number of nitrogen-rich compounds that result from the metabolism of proteins and amino acids in the cell. (See **ammonia; urea; uric acid**.)   (UNIT: 2; AREA: CORE)

nondisjunction   A type of chromosome mutation in which the members of one or more pairs of homologous chromosomes fail to separate during the disjunction phase of the first meiotic division.   (UNIT: 5; AREA: CORE)

nonplacental mammal   A species of mammal in which internal development is accomplished without the aid of a placental connection (marsupial mammals).   (UNIT: 4; AREA: CORE)

nucleic acid An organic compound composed of repeating units of nucleotide. (UNIT: 1; AREA: CORE)

nucleolus A cell organelle located within the nucleus that is known to function in protein synthesis. (UNIT: 1; AREA: CORE)

nucleotide The repeating unit making up the nucleic acid polymer (e.g., DNA, RNA). (UNIT: 5; AREA: CORE)

nucleus A cell organelle that contains the cell's genetic information in the form of chromosomes. (UNIT: 1; AREA: CORE)

nutrition The life function by which living things obtain food and process it for their use. (UNIT: 1; AREA: CORE)

omnivore A heterotrophic organism that consumes both plant and animal matter as sources of nutrition. (UNIT: 7; AREA: CORE)

one gene-one polypeptide A scientific hypothesis concerning the role of the individual gene in protein synthesis. (UNIT: 5; AREA: MG)

oogenesis A type of meiotic cell division in which one ovum and three polar bodies are produced from each primary sex cell. (UNIT: 4; AREA: CORE)

open transport system A type of circulatory system in which the transport fluid is *not* always enclosed within blood vessels (e.g., grasshopper). (UNIT: 2; AREA: CORE)

oral cavity In human beings, the organ used for the ingestion of foods. (UNIT: 3; AREA: CORE)

oral groove The ingestive organ of the paramecium. (UNIT: 2; AREA: CORE)

organ transplant An application of the study of immunity in which an organ or tissue of a donor is transplanted into a compatible recipient. (UNIT: 3; AREA: HP)

organelle A small, functional part of a cell specialized to perform a specific life function (e.g., nucleus, mitochondrion). (UNIT: 1; AREA: CORE)

organic compound a chemical compound that contains the elements carbon and hydrogen (e.g., carbohydrate, protein). (UNIT: 1; AREA: CORE)

organic evolution The mechanism thought to govern the changes in living species over geologic time. (UNIT: 6; AREA: CORE)

osmosis A form of passive transport by which water is absorbed or released by cells. (UNIT: 2; AREA: CORE)

ovary A female gonad that secretes the hormone estrogen, which regulates female secondary sex characteristics; the ovary also produces ova, which are used in reproduction. (UNIT: 3; AREA: HP)

overcropping A negative aspect of human involvement with the environment in which soil is overused for the production of crops, leading to exhaustion of soil nutrients. (UNIT: 7; AREA: CORE)

overgrazing The exposure of soil to erosion due to the loss of stabilizing grasses when it is overused by domestic animals—a negative aspect of human involvement with the environment. (UNIT: 7; AREA: CORE)

overhunting A negative aspect of human involvement with the environment in which certain species have been greatly reduced or made extinct by uncontrolled hunting practices. (UNIT: 7; AREA: CORE)

oviduct   A tube that serves as a channel for conducting mature ova from the ovary to the uterus; the site of fertilization and the earliest stages of embryonic development. (UNIT: 4; AREA: RD)

ovulation   The stage of the menstrual cycle in which the mature ovum is released from the follicle into the oviduct.   (UNIT: 4; AREA: RD)

ovule   A structure located within the flower ovary that contains a monoploid egg nucleus and serves as the site of fertilization.   (UNIT: 4; AREA: CORE)

ovum   A type of gamete produced as a result of oogenesis in female animals; the egg, the female sex cell.   (UNIT: 4; AREA: CORE)

oxygen-18   A radioactive isotope of oxygen that is used to trace the movement of this element in biochemical reaction sequences.   (UNIT: 2; AREA: BC)

oxygenated blood   Blood that contains a high percentage of oxyhemoglobin.   (UNIT: 3; AREA: HP)

Oxyhemoglobin   Hemoglobin that is loosely bound to oxygen for purposes of oxygen transport.   (UNIT: 3; AREA: CORE)

pH   A chemical unit used to express the concentration of hydrogen ion (H^+), or the acidity, of a solution.   (UNIT: 1; AREA: CORE)

palisade layer   A cell layer found in most leaves that contains high concentrations of chloroplasts.   (UNIT: 2; AREA: CORE)

pancreas   An accessory organ which produces enzymes that complete the hydrolysis of foods to soluble end products; also the site of insulin and glucagon production. (UNIT: 3; AREA: CORE)

parasitism   A type of symbiosis from which one organism in the relationship benefits, while the other (the "host") is harmed, but not ordinarily killed.   (UNIT: 7; AREA: EC)

parathormone   A hormone of the parathyroid gland that regulates the metabolism of calcium in the body.   (UNIT: 3; AREA: HP)

parathyroid gland   An endocrine gland whose secretion, parathormone, regulates the metabolism of calcium in the body.   (UNIT: 3; AREA: HP)

passive immunity   A temporary immunity produced as a result of the injection of pre-formed antibodies.   (UNIT: 3; AREA: HP)

passive transport   Any process by which materials are absorbed into the cell interior from an area of high concentration to an area of low concentration, without the expenditure of cell energy (e.g., osmosis, diffusion).   (UNIT: 2; AREA: CORE)

penis   A structure that permits internal fertilization through direct implantation of sperm into the female reproductive tract.   (UNIT: 4; AREA: RD)

peptide bond   A type of chemical bond that links the nitrogen atom of one amino acid with the terminal carbon atom of a second amino acid in the formation of a dipeptide.   (UNIT: 1; AREA: BC)

peripheral nerves   Nerves in the earthworm and grasshopper that branch from the ventral nerve cord to other parts of the body.   (UNIT: 2; AREA: CORE)

peripheral nervous system   A major subdivision of the nervous system that consists of all the nerves of all types branching through the body. (See **autonomic nervous system; somatic nervous system.**)   (UNIT: 3; AREA: CORE)

peristalsis   A wave of contraction of the smooth muscle lining the digestive tract that causes ingested food to pass along the food tube.   (UNIT: 3; AREA: CORE)

petal   An accessory part of the flower that is thought to attract pollinating insects. (UNIT: 4; AREA: CORE)

phagocyte   A type of white blood cell that engulfs and destroys bacteria.   (UNIT: 3; AREA: CORE)

phagocytosis   The process by which the ameba surrounds and ingests large food particles for intracellular digestion.   (UNIT: 2; AREA: CORE)

pharynx   The upper part of the digestive tube that temporarily stores food before digestion.   (UNIT: 2; AREA: CORE)

phenotype   The observable trait that results from the action of an allele pair.   (UNIT: 5; AREA: CORE)

phenylketonuria (PKU)   A genetically related human disorder in which the homozygous combination of a particular mutant gene prevents the normal metabolism of the amino acid phenylalanine.   (UNIT: 5; AREA: MG)

phloem   A type of vascular tissue through which water and dissolved sugars are transported in plants from the leaf downward to the roots for storage.   (UNIT: 2; AREA: CORE)

phosphate group   A chemical group made up of phosphorus and oxygen that is a component part of the nucleotide unit.   (UNIT: 5; AREA: CORE)

phosphoglyceraldehyde (PGAL)   An intermediate product formed during photosynthesis that acts as the precursor of glucose formation.   (UNIT: 2; AREA: BC)

photochemical reactions   A set of biochemical reactions in photosynthesis in which light is absorbed and water molecules are split. (See **photolysis**.)   (UNIT: 2; AREA: BC)

photolysis   The portion of the photochemical reactions in which water molecules are split into hydrogen atoms and made available to the carbon fixation reactions. (UNIT: 2; AREA: BC)

photosynthesis   A type of autotrophic nutrition in which green plants use the energy of sunlight to convert carbon dioxide and water into glucose.   (UNIT: 2; AREA: CORE)

phylum   A level of biological classification that is a major subdivision of the kingdom level, containing fewer organisms with greater similarity (e.g., Chordata).   (UNIT: 1; AREA: CORE)

pinocytosis   A special type of absorption by which liquids and particles too large to diffuse through the cell membrane may be taken in by vacuoles formed at the cell surface.   (UNIT: 2; AREA: CORE)

pioneer autotrophs   The organisms supposed by the heterotroph hypothesis to have been the first to evolve the ability to carry on autotrophic nutrition.   (UNIT: 6; AREA: CORE)

pioneer species   In an ecological succession, the first organisms to inhabit a barren environment.   (UNIT: 7; AREA: EC)

pistil   The female sex organ of the flower. (See **stigma; style; ovary**.)   (UNIT: 4; AREA: CORE)

pituitary gland   An endocrine gland that produces hormones regulating the secretions of other endocrine glands; the "master gland."   (UNIT: 3; AREA: HP)

placenta   In placental mammals, a structure composed of both embryonic and maternal tissues that permits the diffusion of soluble substances to and from the fetus for nourishment and the elimination of fetal waste.   (UNIT: 4; AREA: CORE)

placental mammal   A mammal species in which embryonic development occurs internally with the aid of a placental connection to the female parent's body.   (UNIT: 4; AREA: CORE)

Plant   One of the five biological kingdoms; it includes multicellular organisms whose cells are bounded by cell walls and which are capable of photosynthesis (e.g., maple tree).   (UNIT: 1; AREA: CORE)

plasma   The liquid fraction of blood, containing water and dissolved proteins.   (UNIT: 3; AREA: CORE)

plasma membrane   A cell organelle that encloses the cytoplasm and other cell organelles and regulates the passage of materials into and out of the cell.   (UNIT: 1; AREA: CORE)

platelet   A cell-like component of the blood that is important in clot formation.   (UNIT: 3; AREA: CORE)

polar body   One of three nonfunctional cells produced during oogenesis that contain monoploid nuclei and disintegrate soon after completion of the process.   (UNIT: 4; AREA: CORE)

polio   A disorder of the human regulatory system in which viral infection of the central nervous system may result in severe paralysis.   (UNIT: 3; AREA: HP)

pollen   The male gamete of the flowering plant.   (UNIT: 4; AREA: CORE)

pollen tube   A structure produced by the germinating pollen grain that grows through the style to the ovary and carries the sperm nucleus to the ovule for fertilization.   (UNIT: 4; AREA: CORE)

pollination   The transfer of pollen grains from anther to stigma.   (UNIT: 4; AREA: CORE)

pollution control   The development of new procedures to reduce the incidence of air, water, and soil pollution—a positive aspect of human involvement with the environment.   (UNIT: 7; AREA: CORE)

polyploidy   A type of chromosome mutation in which an entire set of homologous chromosomes fail to separate during the disjunction phase of the first meiotic division.   (UNIT: 5; AREA: CORE)

polysaccharide   A type of carbohydrate composed of repeating units of monosaccharide that form a polymeric chain.   (UNIT: 1; AREA: CORE)

polyunsaturated fat   A type of fat in which many bonding sites are unavailable for the addition of hydrogen atoms.   (UNIT: 3; AREA: HP)

population   All the members of a particular species in a given geographical location at a given time.   (UNIT: 5; AREA: MG)

population control   The use of various practices to slow the rapid growth in the human population—a positive aspect of human interaction with the environment.   (UNIT: 7; AREA: CORE)

population genetics   A science that studies the genetic characteristics of a sexually reproducing species and the factors that affect its gene frequencies.   (UNIT: 5; AREA: MG)

postnatal development   The growth and maturation of an individual from birth, through aging, to death.   (UNIT: 4; AREA: RD)

prenatal development   The embryonic development that occurs before birth within the uterus. (See **gestation**.)   (UNIT: 4; AREA: RD)

primary consumer   Any herbivorous organism that receives food energy from the producer level (e.g., mouse); the second stage of a food chain. (UNIT: 7; AREA: CORE)

primary sex cell   The diploid cell that undergoes meiotic cell division to produce monoploid gametes. (UNIT: 4; AREA: CORE)

producer   Any autotrophic organism capable of trapping light energy and converting it to the chemical bond energy of food (e.g., green plants); the organisms forming the basis of the food chain. (UNIT: 7; AREA: CORE)

progesterone   A hormone produced by the corpus luteum and/or placenta that has the effect of maintaining the uterine lining and suppressing ovulation during gestation. (UNIT: 4; AREA: RD)

protease   Any protein-hydrolyzing enzyme. (UNIT: 3; AREA: CORE)

protein   A complex organic compound composed of repeating units of amino acid. (UNIT: 1; AREA: CORE)

Protista   One of the five biological kingdoms; it includes simple unicellular forms whose nuclei are surrounded by nuclear membranes (e.g., ameba, paramecium). (UNIT: 1; AREA: CORE)

pseudopod   A temporary, flowing extension of the cytoplasm of an ameba that is used in nutrition and locomotion. (UNIT: 2; AREA: CORE)

pulmonary artery   One of two arteries that carry blood from the heart to the lungs for reoxygenation. (UNIT: 3; AREA: HP)

pulmonary circulation   Circulation of blood from the heart through the lungs and back to the heart. (UNIT: 3; AREA: HP)

pulmonary vein   One of four veins that carry oxygenated blood from the lungs to the heart. (UNIT: 3; AREA: HP)

pulse   Rhythmic contractions of the artery walls that help to push the blood fluid through the capillary networks of the body. (UNIT: 3; AREA: CORE)

punctuated equilibrium   A theory of the time frame required for evolution which assumes that evolutionary change occurs in "bursts" with long periods of relative stability intervening. (UNIT: 6; AREA: CORE)

pyramid of biomass   A construct used to illustrate the fact that the total biomass available in each stage of a food chain diminishes from producer level to consumer level. (UNIT: 7; AREA: EC)

pyramid of energy   A construct used to illustrate the fact that energy is lost at each trophic level in a food chain, being most abundant at the producer level. (UNIT: 7; AREA: CORE)

pyruvic acid   An intermediate product in the aerobic or anaerobic respiration of glucose. (UNIT: 2; AREA: BC)

receptor   An organ specialized to receive a particular type of environmental stimulus. (UNIT: 2; AREA: CORE)

recessive allele (gene)   An allele (gene) whose effect is masked by that of its dominant allele. (UNIT: 5; AREA: CORE)

recombinant DNA   DNA molecules that have been moved from one cell to another in order to give the recipient cell a genetic characteristic of the donor cell. (UNIT: 5; AREA: MG)

recombination The process by which the members of segregated allele pairs are randomly recombined in the zygote as a result of fertilization. (UNIT: 5; AREA: CORE)

rectum The portion of the digestive tract in which digestive wastes are stored until they can be released to the environment. (UNIT: 2; AREA: CORE)

red blood cell Small, nonnucleated cells in the blood that contain hemoglobin and carry oxygen to bodily tissues. (UNIT: 3; AREA: CORE)

reduction division See **meiosis**. (UNIT: 4; AREA: CORE)

reflex A simple, inborn, involuntary response to an environmental stimulus. (UNIT: 3; AREA: CORE)

reflex arc The complete path, involving a series of three neurons (sensory, interneuron, and motor), working together, in a reflex action. (UNIT: 3; AREA: CORE)

regeneration A type of asexual reproduction in which new organisms are produced from the severed parts of a single parent organism; the replacement of lost or damaged tissues. (UNIT: 4; AREA: CORE)

regulation The life process by which living things respond to changes within and around them, and by which all life processes are coordinated. (UNIT: 1; AREA: CORE)

replication An exact self-duplication of the chromosome during the early stages of cell division; the exact self-duplication of a molecule of DNA. (UNIT: 4; AREA: CORE)

reproduction The life process by which new cells arise from preexisting cells by cell division. (UNIT: 1; AREA: CORE)

reproductive isolation The inability of species varieties to interbreed and produce fertile offspring, because of variations in behavior or chromosome structure. (UNIT: 6; AREA: CORE)

respiration The life function by which living things convert the energy of organic foods into a form more easily used by the cell. (UNIT: 1; AREA: CORE)

response The reaction of an organism to an environmental stimulus. (UNIT: 2; AREA: CORE)

rhizoid A rootlike fiber produced by fungi that secrete hydrolytic enzymes and absorb digested nutrients. (UNIT: 2; AREA: CORE)

ribonucleic acid (RNA) A type of nucleic acid that operates in various ways to facilitate protein synthesis. (UNIT: 5; AREA: MG)

ribose A five-carbon sugar found as a component part of the nucleotides of RNA molecules only. (UNIT: 5; AREA: MG)

ribosomal RNA (r-RNA) The type of RNA that makes up the ribosome. (UNIT: 5; AREA: MG)

ribosome A cell organelle that serves as the site of protein synthesis in the cell. (UNIT: 1; AREA: CORE)

root A plant organ specialized to absorb water and dissolved substances from the soil, as well as to anchor the plant to the soil. (UNIT: 2; AREA: CORE)

root hair A small projection of the growing root that serves to increase the surface area of the root for absorption. (UNIT: 2; AREA: CORE)

roughage A variety of undigestible carbohydrates that add bulk to the diet and facilitate the movement of foods through the intestine. (UNIT: 3; AREA: HP)

runner A type of vegetative propagation in which an above-ground stem (runner) produces roots and leaves and establishes new organisms with identical characteristics. (UNIT: 4; AREA: CORE)

saliva A fluid secreted by salivary glands that contains hydrolytic enzymes specific to the digestion of starches. (UNIT: 3; AREA: CORE)

salivary gland The gland that secretes saliva, important in the chemical digestion of certain foods. (UNIT: 2; AREA: CORE)

salt A chemical composed of a metal and a nonmetal joined by means of an ionic bond (e.g., sodium chloride). (UNIT: 1; AREA: CORE)

saprophyte A heterotrophic organism that obtains its nutrition from the decomposing remains of dead plant and animal tissues (e.g., fungus, bacteria). (UNIT: 7; AREA: CORE)

saturated fat A type of fat molecule in which all available bonding sites on the hydrocarbon chains are taken up with hydrogen atoms. (UNIT: 3; AREA: HP)

scrotum A pouch extending from the wall of the lower abdomen that houses the testes at a temperature optimum for sperm production. (UNIT: 4; AREA: RD)

secondary consumer Any carnivorous animal that derives its food energy from the primary consumer level (e.g., a snake); the third level of a food chain. (UNIT: 7; AREA: CORE)

secondary sex characteristics The physical features, different in males and females, that appear with the onset of sexual maturity. (UNIT: 4; AREA: RD)

seed A structure that develops from the fertilized ovule of the flower and germinates to produce a new plant. (UNIT: 4; AREA: CORE)

seed dispersal Any mechanism by which seeds are distributed in the environment so as to widen the range of a plant species. (See **fruit**.) (UNIT: 4; AREA: CORE)

segregation The random separation of the members of allele pairs that occurs during meiotic cell division. (UNIT: 5; AREA: CORE)

self-pollination A type of pollination in which the pollen of a flower pollinates another flower located on the same plant organism. (UNIT: 4; AREA: CORE)

sensory neuron A type of neuron specialized for receiving environmental stimuli, which are detected by receptor organs. (UNIT: 3; AREA: CORE)

sepal An accessory part of the flower that functions to protect the bud during development. (UNIT: 4; AREA: CORE)

sessile A term that relates to the "unmoving" state of certain organisms, including the hydra. (UNIT: 2; AREA: CORE)

seta One of several small, chitinous structures (setae) that aid the earthworm in its locomotor function. (UNIT: 2; AREA: CORE)

sex chromosomes A pair of homologous chromosomes carrying genes that determine the sex of an individual; these chromosomes are designated as X and Y. (UNIT: 5; AREA: CORE)

sex determination A pattern of inheritance in which the conditions of maleness and femaleness are determined by the inheritance of a pair of sex chromosomes (XX = female; XY = male). (UNIT: 5; AREA: CORE)

sex linkage A pattern of inheritance in which certain nonsex genes are located on the X sex chromosome, but have no corresponding alleles on the Y sex chromosome. (UNIT: 5; AREA: CORE)

sex-linked trait   A genetic trait whose inheritance is controlled by the genetic pattern of sex linkage (e.g., color blindness). (UNIT: 5; AREA: CORE)

sexual reproduction   A type of reproduction in which new organisms are formed as a result of the fusion of gametes from two parent organisms. (UNIT: 4; AREA: CORE)

shell   An adaptation for embryonic development in many terrestrial, externally developing species that protects the developing embryo from drying and physical damage (e.g., birds). (UNIT: 4; AREA: RD)

sickle-cell anemia   A genetically related human disorder in which the homozygous combination of a mutant gene leads to the production of abnormal hemoglobin and crescent-shaped red blood cells. (UNIT: 5; AREA: MG)

skeletal muscle   A type of muscle tissue associated with the voluntary movements of skeletal levers in locomotion. (UNIT: 3; AREA: CORE)

small intestine   In human beings, the longest portion of the food tube, in which final digestion and absorption of soluble end products occur. (UNIT: 3; AREA: CORE)

smooth muscle   See **visceral muscle**. (UNIT: 3; AREA: CORE)

somatic nervous system   A subdivision of the peripheral nervous system that is made up of nerves associated with voluntary actions. (UNIT: 3; AREA: HP)

speciation   The process by which new species are thought to arise from previously existing species. (UNIT: 6; AREA: CORE)

species   A biological grouping of organisms so closely related that they are capable of interbreeding and producing fertile offspring (e.g., human being). (UNIT: 1; AREA: CORE)

species preservation   The establishment of game lands and wildlife refuges that have permitted the recovery of certain endangered species—a positive aspect of human involvement with the environment. (UNIT: 1; AREA: CORE)

sperm   A type of gamete produced as a result of spermatogenesis in male animals; the male reproductive cell. (UNIT: 4; AREA: CORE)

spermatogenesis   A type of meiotic cell division in which four sperm cells are produced for each primary sex cell. (UNIT: 4; AREA: CORE)

spinal cord   The part of the central nervous system responsible for reflex action, as well as impulse conduction between the peripheral nervous system and the brain. (UNIT: 3; AREA: CORE)

spindle apparatus   A network of fibers that form during cell division and to which centromeres attach during the separation of chromosomes. (UNIT: 4; AREA: CORE)

spiracle   One of several small pores in arthropods, including the grasshopper, that serve as points of entry of respiratory gases from the atmosphere to the tracheal tubes. (UNIT: 2; AREA: CORE)

spongy layer   A cell layer found in most leaves that is loosely packed and contains many air spaces to aid in gas exchange. (UNIT: 2; AREA: CORE)

spore   A specialized asexual reproductive cell produced by certain plants. (UNIT: 4; AREA: CORE)

sporulation   A type of asexual reproduction in which spores released from special spore cases on the parent plant germinate and grow into new adult organisms of the species. (UNIT: 4; AREA: CORE)

staining   A technique of cell study in which chemical stains are used to make cell parts more visible for microscopic study.   (UNIT: 1; AREA: CORE)

stamen   The male reproductive structure in a flower. (See **anther; filament**.)   (UNIT: 4; AREA: CORE)

starch   A type of polysaccharide produced and stored by plants.   (UNIT: 1; AREA: CORE)

stem   A plant organ specialized to support the leaves and flowers of a plant, as well as to conduct materials between the roots and the leaves.   (UNIT: 2; AREA: CORE)

stigma   The sticky upper portion of the pistil, which serves to receive pollen.   (UNIT: 4; AREA: CORE)

stimulus   Any change in the environment to which an organism responds.   (UNIT: 2; AREA: CORE)

stomach   A muscular organ that acts to liquefy food and that produces gastric protease for the hydrolysis of protein.   (UNIT: 3; AREA: CORE)

stomate   A small opening that penetrates the lower epidermis of a leaf and through which respiratory and photosynthetic gases diffuse.   (UNIT: 2; AREA: CORE)

strata   The layers of sedimentary rock that contain fossils, whose ages may be determined by studying the patterns of sedimentation.   (UNIT: 6; AREA: CORE)

stroke   A disorder of the human regulatory system in which brain function is impaired because of oxygen starvation of brain centers.   (UNIT: 3; AREA: HP)

stroma   An area of the chloroplast within which the carbon-fixation reactions occur; stroma lie between pairs of grana.   (UNIT: 2; AREA: BC)

style   The portion of the pistil that connects the stigma to the ovary.   (UNIT: 4; AREA: CORE)

substrate   A chemical that is metabolized by the action of a specific enzyme.   (UNIT: 1; AREA: CORE)

succession   A situation in which an established ecological community is gradually replaced by another until a climax community is established.   (UNIT: 7; AREA: CORE)

survival of the fittest   The concept, frequently associated with Darwin's theory of evolution, that in the intraspecies competition among naturally occurring species the organisms best adapted to the particular environment will survive.   (UNIT: 6; AREA: CORE)

sweat glands   In human beings, the glands responsible for the production of perspiration.   (UNIT: 3; AREA: CORE)

symbiosis   A term which refers to a variety of biotic relationships in which organisms of different species live together in close physical association.   (UNIT: 7; AREA: EC)

synapse   The gap that separates the terminal branches of one neuron from the dendrites of an adjacent neuron.   (UNIT: 2; AREA: CORE)

synapsis   The intimate, highly specific pairing of homologous chromosomes that occurs in the first meiotic division, forming tetrads.   (UNIT: 4; AREA: CORE)

synthesis   The life function by which living things manufacture the complex compounds required to sustain life.   (UNIT: 1; AREA: CORE)

systemic circulation   The circulation of blood from the heart through the body tissues (except the lungs) and back to the heart.   (UNIT: 3; AREA: HP)

systole The higher pressure registered during blood pressure testing. (See **diastole**.) (UNIT: 3; AREA: HP)

taiga A terrestrial biome characterized by long, severe winters and climax flora that includes coniferous trees. (UNIT: 7; AREA: EC)

Tay-Sachs A genetically related human disorder in which fatty deposits in the cells, particularly of the brain, inhibit proper functioning of the nervous system. (UNIT: 5; AREA: MG)

technological oversight A term relating to human activities that adversely affect environmental quality due to failure to adequately assess the environmental impact of a technological development. (UNIT: 7; AREA: CORE)

teeth Structures located in the mouth that are specialized to aid in the mechanical digestion of foods. (UNIT: 3; AREA: CORE)

temperate deciduous forest A terrestrial biome characterized by moderate climatic conditions and climax flora that includes deciduous trees. (UNIT: 7; AREA: EC)

template A pattern or design provided by the DNA molecule for the synthesis of protein molecules. (UNIT: 5; AREA: MG)

tendon A type of connective tissue that attaches a skeletal muscle to a bone. (UNIT: 3; AREA: CORE)

tendonitis A disorder of the human locomotor system in which the junction between a tendon and a bone becomes irritated and inflamed. (UNIT: 3; AREA: HP)

tentacle A grasping structure in certain organisms, including the hydra, that contains stinging cells and is used for capturing prey. (UNIT: 2; AREA: CORE)

terminal branch A cytoplasmic extension of the neuron that transmits a nerve impulse to adjacent neurons via the secretion of neurotransmitters. (UNIT: 2; AREA: CORE)

terrestrial biome A biome that comprises primarily land ecosystems, the characteristics of which are determined by the major climate zone of the earth. (UNIT: 7; AREA: EC)

test cross A genetic cross accomplished for the purpose of determining the genotype of an organism expressing a dominant phenotype; the unknown is crossed with a homozygous recessive. (UNIT: 5; AREA: CORE)

testis A gonad in human males that secretes the hormone testosterone, which regulates male secondary sex characteristics; the testis also produces sperm cells for reproduction. (UNIT: 3; AREA: HP)

testosterone A hormone secreted by the testis that regulates the production of male secondary sex characteristics. (UNIT: 3; AREA: HP)

tetrad A grouping of four chromatids that results from synapsis. (UNIT: 4; AREA: CORE)

thymine A nitrogenous base found only in DNA. (UNIT: 5; AREA: CORE)

thyroid gland An endocrine gland that regulates the body's general rate of metabolism through secretion of the hormone thyroxin. (UNIT: 3; AREA: HP)

thyroid-stimulating hormone (TSH) A pituitary hormone that regulates the secretions of the thyroid gland. (UNIT: 3; AREA: HP)

thyroxin A thyroid hormone that regulates the body's general metabolic rate. (UNIT: 3; AREA: HP)

tongue A structure that aids in the mechanical digestion of foods. (UNIT: 3; AREA: CORE)

trachea A cartilage-ringed tube that conducts air from the mouth to the bronchi. (UNIT: 3; AREA: CORE)

tracheal tube An adaptation in arthropods (e.g., grasshopper) which functions to conduct respiratory gases from the environment to the moist internal tissues. (UNIT: 2; AREA: CORE)

Tracheophyta A phylum of the Plant Kingdom whose members (tracheophytes) contain vascular tissues and true roots, stems, and leaves (e.g., geranium, fern, bean, maple tree, corn). (UNIT: 1; AREA: CORE)

transfer RNA (t-RNA) A type of RNA that functions to transport specific amino acids from the cytoplasm to the ribosome for protein synthesis. (UNIT: 5; AREA: MG)

translocation A type of chromosome mutation in which a section of a chromosome is transferred to a nonhomologous chromosome. (UNIT: 5; AREA: MG)

transpiration The evaporation of water from leaf stomates. (UNIT: 2; AREA: CORE)

transpiration pull A force that aids the upward conduction of materials in the xylem by means of the evaporation of water (transpiration) from leaf surfaces. (UNIT: 2; AREA: CORE)

transport The life function by which substances are absorbed, circulated, and released by living things. (UNIT: 1; AREA: CORE)

triplet codon A group of three nitrogenous bases that provide information for the placement of amino acids in the synthesis of proteins. (UNIT: 5; AREA: CORE)

tropical forest A terrestrial biome characterized by a warm, moist climate and a climax flora that includes many species of broadleaf trees. (UNIT: 7; AREA: EC)

tropism A plant growth response to an environmental stimulus. (UNIT: 2; AREA: CORE)

tuber A type of vegetative propagation in which an underground stem (tuber) produces new tubers, each of which is capable of producing new organisms with identical characteristics. (UNIT: 4; AREA: CORE)

tundra A terrestrial biome characterized by permanently frozen soil and climax flora that includes lichens and mosses. (UNIT: 7; AREA: EC)

tympanum A receptor organ in arthropods (e.g., grasshopper) which is specialized to detect vibrational stimuli. (UNIT: 2; AREA: CORE)

ulcer A disorder of the human digestive tract in which a portion of its lining erodes and becomes irritated. (UNIT: 3; AREA: HP)

ultracentrifuge A tool of biological study that uses very high speeds of centrifugation to separate cell parts for examination. (UNIT: 1; AREA: CORE)

umbilical cord In placental mammals, a structure containing blood vessels that connects the placenta to the embryo. (UNIT: 4; AREA: CORE)

unicellular Having a body that consists of a single cell (e.g., paramecium). (UNIT: 1; AREA: CORE)

uracil A nitrogenous base that is a component part of the nucleotides of RNA molecules only. (UNIT: 5; AREA: MG)

urea   A type of nitrogenous waste with moderate solubility and moderate toxicity. (UNIT: 2; AREA: CORE)

ureter   In human beings, a tube that conducts urine from the kidney to the urinary bladder.   (UNIT: 3; AREA: CORE)

urethra   In human beings, a tube that conducts urine from the urinary bladder to the exterior of the body.   (UNIT: 3; AREA: CORE)

uric acid   A type of nitrogenous waste with low solubility and low toxicity.   (UNIT: 2; AREA: CORE)

urinary bladder   An organ responsible for the temporary storage of urine.   (UNIT: 3; AREA: CORE)

urine   A mixture of water, salts, and urea excreted from the kidney.   (UNIT: 3; AREA: CORE)

use and disuse   A term associated with the evolutionary theory of Lamarck, since proved incorrect.   (UNIT: 6; AREA: CORE)

uterus   In female placental mammals, the organ within which embryonic development occurs.   (UNIT: 4; AREA: CORE)

vaccination   An inoculation of dead or weakened disease organisms that stimulates the body's immune system to produce active immunity.   (UNIT: 3; AREA: HP)

vacuole   A cell organelle that contains storage materials (e.g., starch, water) housed inside the cell.   (UNIT: 1; AREA: CORE)

vagina   In female placental mammals, the portion of the reproductive tract into which sperm are implanted during sexual intercourse and through which the baby passes during birth.   (UNIT: 4; AREA: RD)

variation   A concept, central to Darwin's theory of evolution, that refers to the range of adaptation which can be observed in all species.   (UNIT: 6; AREA: CORE)

vascular tissues   Tubelike plant tissues specialized for the conduction of water and dissolved materials within the plant. (See **xylem; phloem**.)   (UNIT: 2; AREA: CORE)

vegetative propagation   A type of asexual reproduction in which new plant organisms are produced from the vegetative (nonfloral) parts of the parent plant.   (UNIT: 4; AREA: CORE)

vein (human)   A relatively thin-walled blood vessel that carries blood from capillary networks back toward the heart.   (UNIT: 3; AREA: CORE)

vein (plant)   An area of vascular tissues located in the leaf that aid the upward transport of water and minerals through the leaf and the transport of dissolved sugars to the stem and roots.   (UNIT: 2; AREA: CORE)

vena cava   One of two major arteries that return blood to the heart from the body tissues.   (UNIT: 3; AREA: HP)

ventral nerve cord   The main pathway for nerve impulses between the brain and peripheral nerves of the grasshopper and earthworm.   (UNIT: 2; AREA: CORE)

ventricle   One of two thick-walled, muscular chambers of the heart that pump blood out to the lungs and body   (UNIT: 3; AREA: CORE)

villi Microscopic projections of the lining of the small intestine that absorb the soluble end products of digestion. (See **lacteal.**) (UNIT: 3; AREA: CORE)

visceral muscle A type of muscle tissue associated with the involuntary movements of internal organs (e.g., peristalsis in the small intestine). (UNIT: 3; AREA: CORE)

vitamin a type of nutrient that acts as a coenzyme in various enzyme-controlled reactions. (UNIT: 1; AREA: CORE)

water cycle The mechanism by which water is made available to living things in the environment through the processes of precipitation, evaporation, runoff, and percolation. (UNIT: 7; AREA: CORE)

water pollution A type of technological oversight that involves the addition of some unwanted factor (e.g., sewage, heavy metals, heat, toxic chemicals) to our water resources. (UNIT: 7; AREA: CORE)

Watson-Crick model A model of DNA structure devised by J. Watson and F. Crick that hypothesizes a "twisted ladder" arrangement for the DNA molecule. (UNIT: 5; AREA: CORE)

white blood cell A type of blood cell that functions in disease control. (See **phagocyte; lymphocyte.**) (UNIT: 3; AREA: CORE)

xylem A type of vascular tissue through which water and dissolved minerals are transported upward through a plant from the root to the stems and leaves. (UNIT: 2; AREA: CORE)

yolk A food substance, rich in protein and lipid, found in the eggs of many animal species. (UNIT: 4; AREA: CORE)

yolk sac The membrane that surrounds the yolk food supply of the embryos of many animal species. (UNIT: 4; AREA: RD)

zygote The single diploid cell that results from the fusion of gametes in sexual reproduction; a fertilized egg. (UNIT: 4; AREA: CORE)

SELECTED ANSWERS

UNIT 1

SET 1.1.	SET 1.2.	SET 1.3.	SET 1.4.	SET 1.5.	SET 1.6.	SET 1.7.	UNIT 1 REVIEW
1. 4	1. 4	1. 4	1. 1	1. 1	1. 2	1. 4	1. 3
3. 3	3. 2	3. 4	3. 3	3. 4	3. 2	3. 4	3. 4
5. 2	5. 4	5. 3	5. 4	5. 1	5. 1	5. 3	5. 1
7. 1	7. 1	7. 3	7. 2	7. 1	7. 1	7. 1	7. 3
9. 4	9. 4	9. 3	9. 2		9. 4	9. 2	9. 2
11. 3			11. 3			11. 4	11. 1
13. 1							13. 1
							15. 1
							17. 4
							19. 3
							21. 2
							23. 4
							25. 1
							27. 2
							29. 1
							31. 2
							33. 4

UNIT 2

SET 2.1.	SET 2.2.	SET 2.3.	SET 2.4.	SET 2.5.	SET 2.6.	SET 2.7.	SET 2.8
1. 3	1. 1	1. 1	1. 3	1. 4	1. 2	1. 3	1. 4
3. 3	3. 4	3. 4	3. 2	3. 2	3. 3	3. 1	3. 3
5. 2	5. 2	5. 1	5. 2	5. 3	5. 4	5. 2	5. 1
7. 3	7. 1	7. 1	7. 1	7. 1	7. 3	7. 1	7. 3
9. 4	9. 2	9. 1	9. 2	9. 3	9. 3	9. 1	
11. 1		11. 4	11. 1	11. 4		11. 1	
		13. 1	13. 3			13. 3	

UNIT 2 REVIEW

1. 1	19. 1	37. 1
3. 3	21. 4	39. 3
5. 3	23. 1	41. 3
7. 3	25. 4	43. 2
9. 3	27. 3	45. 4
11. 3	29. 3	47. 1
13. 1	31. 2	49. 4
15. 4	33. 1	51. 4
17. 3	35. 2	53. 2

UNIT 3

SET 3.1.	SET 3.2.	SET 3.3.	SET 3.4.	SET 3.5.	SET 3.6.	UNIT 3 REVIEW	
1. 1	**1.** 2	**1.** 1	**1.** 2	**1.** 3	**1.** 4	**1.** 2	**31.** 4
3. 3	**3.** 2	**3.** 4	**3.** 1	**3.** 4	**3.** 4	**3.** 4	**33.** 4
5. 2	**5.** 4	**5.** 3	**5.** 4	**5.** 3	**5.** 1	**5.** 4	**35.** 3
7. 4	**7.** 1	**7.** 2	**7.** 1	**7.** 3	**7.** 4	**7.** 4	**37.** 3
9. 3	**9.** 4	**9.** 3	**9.** 2	**9.** 3		**9.** 2	**39.** 3
11. 2	**11.** 2			**11.** 1		**11.** 2	**41.** 4
13. 1	**13.** 2			**13.** 2		**13.** 1	**43.** 2
15. 1	**15.** 1			**15.** 2		**15.** 4	**45.** 1
17. 2						**17.** 2	**47.** 2
						19. 2	**49.** 3
						21. 2	**51.** 5
						23. 3	**53.** 2
						25. 4	**55.** 3
						27. 3	**57.** 2
						29. 4	**59.** 3

UNIT 4

SET 4.1.	SET 4.2.	SET 4.3.	SET 4.4.	SET 4.5.	UNIT 4 REVIEW	
1. 4	**1.** 1	**1.** 2	**1** 3	**1.** 3	**1.** 3	**29.** 1
3. 3	**3.** 2	**3.** 3	**3** 3	**3.** 1	**3.** 1	**31.** 4
5. 3	**5.** 1	**5.** 1	**5.** 4	**5.** 2	**5.** 2	**33.** 1
7. 1	**7.** 4	**7.** 2	**7** 3	**7.** 4	**7.** 3	**35.** 4
9. 4	**9.** 2	**9.** 4	**9.** 1	**9.** 4	**9.** 3	**37.** 3
11. 3	**11.** 3	**11.** 1	**11.** 1	**11.** 3	**11.** 4	**39.** 3
13. 2		**13.** 1		**13.** 2	**13.** 3	**41.** 5
15. 4		**15.** 2			**15.** 1	**43.** 4
17. 1		**17.** 1			**17.** 4	**45.** 1
19. 2					**19.** 4	**47.** 3
					21. 4	**49.** 1
					23. 3	**51.** 3
					25. 4	**53.** 1
					27. 4	**55.** 4

UNIT 5

SET 5.1.	SET 5.2.	SET 5.3.	SET 5.4.	SET 5.5.	UNIT 5 REVIEW	
1. 3	**1.** 2	**1.** 2	**1** 1	**1.** 1	**1.** 4	**35.** 3
3. 4	**3.** 2	**3.** 3	**3.** 1	**3.** 1	**3.** 2	**37.** 1
5. 2	**5.** 2	**5.** 1	**5.** 1	**5.** 1	**5.** 1	**39.** 4
7. 3	**7.** 4	**7.** 2	**7** 2	**7.** 1	**7.** 3	**41.** 1
9. 3	**9.** 4	**9.** 3	**9** 3		**9.** 1	**43.** 4
11. 1	**11.** 3	**11.** 2	**11** 1		**11.** 4	**45.** 1
13. 3	**13.** 1	**13.** 2	**13** 1		**13.** 1	**47.** 4
15. 3	**15.** 2	**15.** 4	**15** 2		**15.** 4	**49.** 3
			17 4		**17.** 1	**51.** 4
			19 3		**19.** 3	**53.** 4
			21 3		**21.** 4	**55.** 3
			23 3		**23.** 1	**57.** 2
			25 4		**25.** 1	**59.** 2
			27 2		**27.** 4	**61.** 3
			29 1		**29.** 1	**63.** 4
			31 1		**31.** 2	**65.** 4
					33. 3	

UNIT 6

SET 6.1.	SET 6.2.	SET 6.3.	SET 6.4.	UNIT 6 REVIEW
1. 1	**1.** 3	**1.** 1	**1.** 4	**1.** 2
3. 1	**3.** 2	**3.** 3	**3.** 1	**3.** 1
5. 2	**5.** 4	**5.** 4	**5.** 4	**5.** 4
7. 2	**7.** 2	**7.** 2		**7.** 3
9. 1	**9.** 1	**9.** 4		**9.** 1
11. 1	**11.** 3	**11.** 4		**11.** 1
				13. 1
				15. 1
				17. 3
				19. 3
				21. 4
				23. 2
				25. 4
				27. 3
				29. 4
				31. 3
				33. 2

UNIT 7

SET 7.1.	SET 7.2.	SET 7.3.	SET 7.4.	SET 7.5.	SET 7.6.	UNIT 7 REVIEW	
1. 2	**1.** 2	**1.** 4	**1.** 4	**1.** 3	**1.** 4	**1.** 2	**35.** 2
3. 3	**3.** 3	**3.** 2	**3.** 4	**3.** 5	**3.** 3	**3.** 1	**37.** 1
5. 3	**5.** 4	**5.** 2	**5.** 4	**5.** 2	**5.** 4	**5.** 1	**39.** 4
7. 1	**7.** 4	**7.** 1	**7.** 1	**7.** 1	**7.** 3	**7.** 1	**41.** 1
9. 3	**9.** 2	**9.** 2	**9.** 2	**9.** 1	**9.** 2	**9.** 3	**43.** 2
11. 1	**11.** 2		**11.** 4	**11.** 1	**11.** 1	**11.** 4	**45.** 2
13. 2	**13.** 1				**13.** 2	**13.** 1	**47.** 1
15. 4					**15.** 4	**15.** 1	**49.** 1
17. 1						**17.** 3	**51.** 4
19. 4						**19.** 3	**53.** 2
						21. 3	**55.** 1
						23. 2	**57.** 1
						25. 4	**59.** 4
						27. 3	**61.** 2
						29. 3	**63.** 1
						31. 4	**65.** 1
						33. 1	

LABORATORY SKILLS

A. 1.1. 2
A. 2.2. C
A. 3.1. 4
A. 4.1. 1 This part is used to provide general focus.
  3 This part is used to take in light passing through the object.
A. 4.2. 1
A. 5.2. 4
A. 6.2. 4
A. 7.1. 1
A. 7.3. 1
A. 8.2. 3
A. 9.2. 8.9 cm
A.10.1. 4
A.10.3. This structure functions in the manufacture of pollen.
A.11.2. 3
A.12.1. Height of Bean Plants (cm)
A.13.1. 4
A.13.3. 2
A.13.5. 3
A.14.2. 4
A.14.4. 2
B.1. 3
B.3. 1
B.5 3
C.1. Gradualism proposes that evolutionary change is continuous and slow, whereas punctuated equilibrium proposes that species undergo long periods without change, and then undergo much change quickly.
C.3. Gradualism is supported by fossil evidence which shows transitional forms.
C.4. PANCREAS—This organ produces digestive enzymes used in the digestive function.
  EPIGLOTTIS—This structure prevents swallowed food from entering the trachea.
  VILLUS—This structure permits the absorption of digested foods from the intestine to the blood stream.
  PLATELET—This structure promotes blood clotting by releasing enzymes into the blood.
D.1. 4
D.3. 2
D.5. 3

Directions (1–59): For *each* statement or question, select the word or expression that, of those given, best completes the statement or answers the question. Record your answer on the separate answer paper in accordance with the directions on the front page of this booklet.

1 In plants, glucose is converted to cellulose, and in human muscle cells, glucose is converted to glycogen. These processes are examples of which life activity?

1 regulation   3 synthesis
2 respiration   4 excretion

2 Which organelles outside the cell nucleus contain genetic material?

1 lysosomes and cell walls
2 chloroplasts and mitochondria
3 endoplasmic reticula and cell membranes
4 vacuoles and Golgi complex

3 The diagrams below represent unicellular organisms. In the past, they were difficult to classify as either plants or animals.

Currently, these organisms are classified as

1 coelenterates   3 tracheophytes
2 annelids   4 protists

4 Which technique enabled scientists in the 1800's to identify cell organelles?

1 electron microscopy   3 staining
2 ultracentrifugation   4 dissection

5 Which substance is an inorganic compound?

1 water   3 maltase
2 glucose   4 insulin

6 Which chemical formula represents a carbohydrate?

(1) CH_4 (3) $C_{12}H_{22}O_{11}$
(2) $C_3H_7O_2N$ (4) CO_2

7 The equation below summarizes the process that produces the flashing light of a firefly. The molecule luciferin is broken down, and energy is released in the form of heat and light.

$$\text{luciferin} \xrightarrow[\text{luciferase}]{\text{ATP}} \text{heat + light}$$

In this process, luciferase functions as

1 a reactant
2 a substrate
3 an inorganic catalyst
4 an enzyme

8 Which statement best describes the process of digestion in fungi and bacteria?

1 It occurs as a result of the dehydration synthesis of foods within these organisms.
2 It results in the production of starch and protein molecules.
3 It occurs as a result of the enzymatic hydrolysis of foods outside these organisms.
4 It occurs within a highly specialized digestive system.

9 If a grasshopper's gastric caeca stopped functioning, which activity would be affected first?

1 chemical breakdown of food in the digestive tube
2 transmission of impulses by the nerve cord
3 taking in of air by the spiracles
4 release of hormones into the transport system

10 An iodine test of a tomato plant leaf revealed that starch was present at 5:00 p.m. on a sunny afternoon in July. When a similar leaf from the same tomato plant was tested with iodine at 6:00 a.m. the next morning, the test indicated that less starch was present in this leaf than in the leaf tested the day before. This reduction in starch content occurred because starch was

1 changed into cellulose
2 transported out of the leaves through the stomates
3 conducted downward toward the roots through vessels
4 digested into simple sugars

11 The diagram below illustrates the transport of oxygen and carbon dioxide.

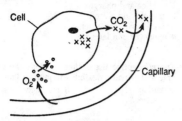

In which organisms does this type of transport occur?
1 hydra and ameba
2 human and frog
3 grasshopper and chicken
4 paramecium and earthworm

12 Which process requires cellular energy?
  1 diffusion   3 active transport
  2 passive transport   4 osmosis

13 The diagram below represents a cross section of a leaf.

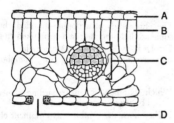

Food manufactured in the leaf is conducted to the rest of the plant by structure
  (1) A   (3) C
  (2) B   (4) D

14 What is a direct result of aerobic respiration?
1 The potential energy of glucose is transferred to ATP molecules.
2 The enzymes for anaerobic respiration are produced and stored in lysosomes.
3 Lactic acid is produced in muscle tissue.
4 Alcohol is produced by yeast and bacteria.

15 Hemoglobin is found in the blood of humans and earthworms, but not in the blood of grasshoppers. Which conclusion is best supported by this statement?
1 The human and the earthworm have lungs, but the grasshopper does not have lungs.
2 The human and the earthworm transport far more oxygen with their blood than the grasshopper transports with its blood.
3 The human and the earthworm have open circulatory systems, but the grasshopper has a closed circulatory system.
4 The human and the earthworm are adapted for anaerobic respiration, but the grasshopper is adapted for aerobic respiration.

16 In a branch of a cherry tree, gases are exchanged between the environment and the cells through
  1 lenticels   3 xylem
  2 cambium   4 phloem

17 Some metabolic wastes produced by plant cells are
1 used by cells for the synthesis of chitin
2 used by centrioles to produce spindle fibers
3 stored in nuclei for DNA replication
4 stored in vacuoles, where they do not harm the cell

18 Nitrogenous wastes such as ammonia, urea, and uric acid all result from
1 dehydration synthesis
2 protein metabolism
3 aerobic respiration
4 carbohydrate metabolism

19 The growth of roots on plant cuttings may be stimulated by
1 a decrease in enzymes
2 the use of auxins
3 the use of endocrine secretions
4 a decrease in soil minerals

20 In humans, certain glands produce chemicals that are distributed by the circulatory system and influence various target organs. These glands are classified as
  1 intestinal glands   3 gastric glands
  2 salivary glands   4 endocrine glands

21 A change in the environment that initiates the transmission of an electrochemical charge along a neuron is known as
1 a reflex   3 an impulse
2 a response   4 a stimulus

22 Which specialized structures for locomotion are correctly paired with the organism that possesses those structures?
1 setae—earthworm
2 chitinous appendages—paramecium
3 flagella—grasshopper
4 pseudopods—human

23 Which process is illustrated in the diagrams below?

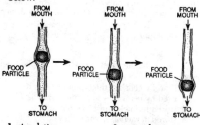

1 circulation   3 peristalsis
2 absorption   4 ingestion

24 Which statement best describes arteries?
1 They have thick walls and transport blood away from the heart.
2 They have thick walls and transport blood toward the heart.
3 They have thin walls and transport blood away from the heart.
4 They have thin walls and transport blood toward the heart.

25 To receive necessary nutrients and eliminate wastes, all human body cells must be
1 surrounded by cilia
2 endocrine in nature
3 able to carry on phagocytosis
4 surrounded by a transport medium

26 The nephron is the structural unit of the human
1 lung   3 kidney
2 liver   4 intestine

27 In the human respiratory system, bronchioles directly connect the
1 trachea and pharynx
2 bronchi and alveoli
3 nasal cavity and trachea
4 epiglottis and larynx

28 The connective tissue that cushions the vertebrae and provides flexibility to joints is known as
1 tendon   3 cartilage
2 muscle   4 bone

29 The diagrams below represent a cell process.

Diagram 1   Diagram 2   Diagram 3

If the cell in diagram 1 contains 4 chromosomes, what is the total number of chromosomes in each cell in diagram 3?
(1) 8   (3) 16
(2) 2   (4) 4

30 The chromatids of a double-stranded chromosome are held together at a region known as the
1 polar body   3 centriole
2 centromere   4 Golgi complex

31 Which statement is true regarding plants produced by vegetative propagation?
1 They normally exhibit only dominant characteristics.
2 They normally have the monoploid number of chromosomes.
3 They normally obtain most of their nourishment from the seed.
4 They are normally genetically identical to the parent.

32 After gametogenesis takes place, which process restores the diploid chromosome number of the species for the next generation?
1 oogenesis   3 fertilization
2 mitosis   4 meiosis

33 In the diagram below, which number indicates a structure that transmits impulses from a receptor to an interneuron?

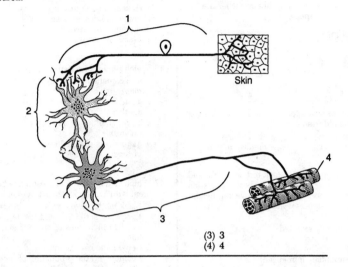

Skin

(1) 1
(2) 2

(3) 3
(4) 4

34 In many aquatic vertebrates, reproduction involves external fertilization. What is a characteristic of this type of fertilization?
1 Gametes fuse outside the body of the female.
2 Gametes fuse in the moist reproductive tract of the female.
3 Offspring produced have twice as many chromosomes as each of the parents.
4 Offspring produced have only half the number of chromosomes as each of the parents.

35 In a rabbit, the embryo normally develops within the
1 placenta
2 uterus
3 yolk sac
4 umbilical cord

36 The pistil of the flower includes the
1 stigma, anther, and filament
2 stamen, stigma, and anther
3 stigma, style, and ovary
4 petals, sepals, and pollen grains

37 In a plant, which structure enables sperm nuclei to reach the ovule?
1 stigma
2 pollen tube
3 stamen
4 seed coat

38 In screech owls, red feathers are dominant over gray feathers. If two heterozygous red-feathered owls are mated, what percentage of their offspring would be expected to have red feathers?
(1) 25%
(2) 50%
(3) 75%
(4) 100%

39 Two genes for two different traits located on the same chromosome are said to be
1 homozygous
2 independently assorted
3 mutagenic agents
4 linked

40 Diagram A below illustrates the chromosomes of a normal female insect. Diagram B illustrates the chromosomes found in an abnormal female insect of the same species.

Diagram A   Diagram B

The chromosomal alteration in diagram B most likely resulted from

1 codominance   3 sex linkage
2 crossing-over   4 nondisjunction

41 Identical twins were separated at birth and raised by different families. The best explanation for any differences between the twins in height, weight, and IQ scores is that the genes regulating these traits were

1 independently assorted
2 codominant
3 linked
4 environmentally influenced

42 In humans, normal color vision (N) is dominant over color blindness (n). A man and woman with normal color vision produced two colorblind sons and two daughters with normal color vision. The parental genotypes must be

(1) $X^N Y$ and $X^N X^N$ (3) $X^N Y$ and $X^N X^n$
(2) $X^n Y$ and $X^N X^N$ (4) $X^n Y$ and $X^n X^n$

43 If the pattern of inheritance for a trait is complete dominance, then an organism heterozygous for the trait would normally express

1 the recessive trait, only
2 the dominant trait, only
3 a blend of the recessive and dominant traits
4 a phenotype unlike that of either parent

44 Although genetic mutations may occur spontaneously in organisms, the incidence of such mutations may be increased by

1 radioactive substances in the environment
2 lack of vitamins in the diet
3 a long exposure to humid climates
4 a short exposure to freezing temperatures

45 In addition to a phosphate group, a DNA nucleotide could contain

1 thymine and deoxyribose
2 uracil and deoxyribose
3 thymine and ribose
4 uracil and ribose

46 Which phrase best defines evolution?

1 an adaptation of an organism to its environment
2 a sudden replacement of one community by another
3 a geographic or reproductive isolation of organisms
4 a process of change in organisms over a period of time

47 Which statement would most likely be in agreement with Lamarck's theory of evolution?

1 Black moths have evolved in an area because they were better adapted to the environment and had high rates of survival and reproduction.
2 Geographic barriers may lead to reproductive isolation and the production of new species.
3 Giraffes have long necks because their ancestors stretched their necks reaching for food, and this trait was passed on to their offspring.
4 Most variations in animals and plants are due to random chromosomal and gene mutation.

48 Organisms with favorable variations reproduce more successfully than organisms with less favorable variations. This statement best describes the concept of

1 overproduction
2 use and disuse
3 inheritance of acquired characteristics
4 survival of the fittest

49 How does natural selection operate to cause change in a population?

1 The members of the population are equally able to survive any environmental change.
2 The members of the population differ so that only some survive when the environment changes.
3 The members of the population do not adapt to environmental changes.
4 All the members of the population adapt to environmental changes.

50 Evolution is often represented as a tree similar to the one shown in the diagram below.

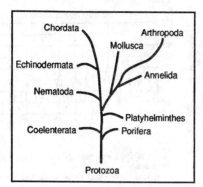

Chordata
Arthropoda
Mollusca
Echinodermata
Annelida
Nematoda
Platyhelminthes
Coelenterata
Porifera
Protozoa

This diagram suggests that

1 different groups of organisms may have similar characteristics because of common ancestry
2 because of biochemical differences, no two groups of organisms could have a common ancestor
3 evolution is a predictable event that happens every few years, adding new groups of organisms to the tree
4 only the best adapted organisms will survive from generation to generation

51 The concept that evolution is the result of long periods of stability interrupted by geologically brief periods of significant change is known as

1 gradualism
2 natural selection
3 geographic isolation
4 punctuated equilibrium

52 According to the heterotroph hypothesis, which gas given off by autotrophic activity made the evolution of aerobes possible?

1 oxygen   3 carbon dioxide
2 hydrogen   4 nitrogen

53 All the red foxes inhabiting a given forest constitute a

1 population   3 biome
2 community   4 biosphere

54 The diagram below shows living and nonliving factors and the interaction between them.

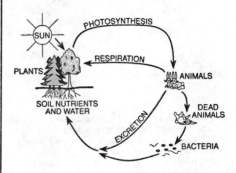

The diagram best represents

1 a species   3 a community
2 a population   4 an ecosystem

55 Which statement best describes an energy pyramid?

1 There is more energy at the consumer level than at the producer level.
2 There is more energy at the producer level than at the consumer level.
3 There is more energy at the secondary-consumer level than at the primary-consumer level.
4 There is more energy at the decomposer level than at the consumer level.

56 The exchange of useful chemicals between organisms and their abiotic environment is an example of

1 a material cycle   3 a limiting factor
2 competition   4 succession

57 In an ecosystem, the more living requirements that two different species have in common, the more intense will be their

1 ecological succession
2 competition
3 energy requirements
4 evolution

58 The diagrams below represent four members of a food chain.

Toads

Predaceous insects

Plants

Herbivorous insects

Which sequence best represents the transfer of energy between these organisms?

1 toads → predaceous insects → herbivorous insects → plants
2 predaceous insects → herbivorous insects → plants → toads
3 plants → herbivorous insects → predaceous insects → toads
4 plants → herbivorous insects → toads → predaceous insects

59 Based on the graph below, which conditions most likely existed during the period from 1860 to 1865?

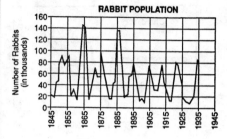

1 Plenty of food was available and there were few predators.
2 Food was scarce and there were few predators.
3 Plenty of food was available and there were many predators.
4 No conclusion can be drawn from the information given.

Part II

This part consists of five groups, each containing ten questions. Choose two of these five groups. Be sure that you answer all ten questions in each group chosen. Record the answers to these questions in accordance with the directions on the front page of this booklet.   [20]

Group 1 — Biochemistry

If you choose this group, be sure to answer questions 60–69.

Base your answers to questions 60 through 63 on the chemical equation below which represents a metabolic activity and on your knowledge of biology.

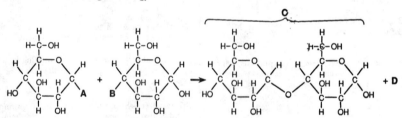

60 Which chemical substance is labeled C?
1 a lipid
2 a dipeptide
3 a disaccharide
4 a nucleotide

61 Which substance is represented by letter D?
1 water
2 salt
3 ammonia
4 carbon dioxide

62 Molecule C belongs in the general class of substances known as
1 vitamins
2 minerals
3 inorganic acids
4 organic compounds

63 Which structural group is represented at A and B?

(1) $-C\underset{OH}{\overset{O}{\lessgtr}}$

(2) $-OH$

(3) $-N\underset{H}{\overset{H}{\lessgtr}}$

(4) $-\overset{H}{\underset{H}{C}}-H$

Base your answers to questions 64 through 66 on the information below and on your knowledge of biology.

A solution of an enzyme normally found in the human body was added to a flask containing a solution of proteins in distilled water, and then the flask was stoppered. This mixture was then maintained at a temperature of 27°C and a pH of 7 for 48 hours. When the mixture was analyzed, the presence of amino acids was noted.

64 Which substances would most likely be present in the solution in the flask after 48 hours?
1 amino acids, only
2 amino acids and polypeptides, only
3 polypeptides, amino acids, and enzyme molecules
4 polysaccharides, amino acids, and enzyme molecules

65 One way to speed up the production of amino acids in the flask would be to
1 increase the temperature from 27°C to 37°C
2 increase the pH from 7 to 12
3 place the flask in bright light
4 decrease the amount of enzyme added

66 The enzymatic solution most likely contained
1 carbohydrases   3 lipases
2 maltases   4 proteases

Directions (67–68): For each phrase in questions 67 and 68 select the photosynthetic reactions, *chosen from the list below*, that are best described by that phrase. Then record the *number* of the reactions on the separate answer sheet.

Photosynthetic Reactions

(1) Photochemical reactions, only
(2) Carbon-fixation reactions, only
(3) Both photochemical and carbon-fixation reactions

67 The reactions in which the radioactive isotope carbon-14 can be used to trace the chemical pathway of the carbon in carbon dioxide

68 The reactions in which photolysis occurs

_____

69 Which intermediate substance is produced in both alcohol and lactic acid fermentation?
1 oxygen   3 nitrogen
2 pyruvic acid   4 starch

If you choose this group, be sure to answer questions 70–79.

Base your answers to questions 70 through 73 on the diagram below and on your knowledge of biology.

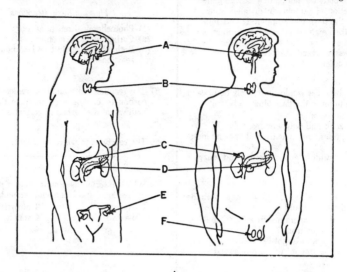

70 Which two glands produce gametes?
   (1) *A* and *B*   (3) *C* and *F*
   (2) *B* and *E*   (4) *E* and *F*

71 Which gland is associated with the malfunction known as goiter?
   (1) *E*   (3) *C*
   (2) *B*   (4) *D*

72 Which gland produces adrenalin?
   (1) *A*   (3) *C*
   (2) *B*   (4) *D*

73 Removing part of gland *D* would most likely result in
   1 a decrease in the secretions of other glands
   2 a decrease in the blood calcium level
   3 an increase in the growth rate of the individual
   4 an increase in the blood sugar level

Directions (74–76): For each statement in questions 74 through 76 select the part of the human transport system, *chosen from the list below*, that is best described by that statement. Then record its *number* on the separate answer sheet.

Parts of the Human Transport System

    (1) Pulmonary circulation
    (2) Systemic circulation
    (3) Coronary circulation
    (4) Lymphatic circulation

74 Cardiac muscle tissue is supplied with nutrients and oxygen.

75 The concentration of carbon dioxide in the blood decreases, and the concentration of oxygen increases.

76 Oxygen is delivered to the liver from the heart.

77 Which of the blood types in the ABO system may safely be given to a person with AB blood?
   (1) O or AB, only   (3) B or AB, only
   (2) A or B, only   (4) A, B, AB, or O

78 The movement of blood from the legs toward the heart is hindered by gravity. The effect of gravity is counteracted by
   1 smooth muscle in the capillaries
   2 cilia lining the blood vessels
   3 valves in the veins
   4 lymph nodes near major vessels

79 Which disease is linked to smoking and results in a reduction in the number and elasticity of alveoli?
   1 emphysema   3 bronchitis
   2 asthma   4 meningitis

Group 3 — Reproduction and Development

If you choose this group, be sure to answer questions 80–89.

Base your answers to questions 80 through 84 on the diagram below of some events in the human female reproductive cycle and on your knowledge of biology.

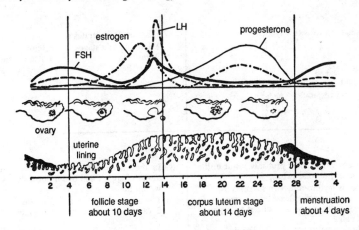

80 During which part of this cycle does the shedding of the thickened uterine lining occur?
   1 ovulation
   2 corpus luteum stage
   3 menstruation
   4 follicle stage

81 On or about which day does ovulation occur?
   (1) 8th day   (3) 20th day
   (2) 14th day   (4) 28th day

82 Which hormones are secreted by the ovaries?
   (1) progesterone and estrogen
   (2) FSH and progesterone
   (3) FSH and LH
   (4) LH and estrogen

83 What is the average length of this reproductive cycle?
   (1) 32 days   (3) 14 days
   (2) 28 days   (4) 4 days

84 The permanent cessation of this cycle is known as
   1 puberty   3 fertilization
   2 pregnancy   4 menopause

Base your answers to questions 85 through 87 on the diagrams below of some stages in the development of an embryo and on your knowledge of biology.

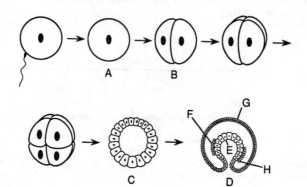

85 Which stage represents a zygote?
(1) A   (3) C
(2) B   (4) D

86 Which cell layer develops into the linings of the digestive and respiratory tracts?
(1) E   (3) G
(2) F   (4) H

87 Diagram D illustrates a
1 late blastula stage   3 fetus
2 gastrula   4 stage of meiosis

88 Which event would most probably result in the production of fraternal twins?
1 One egg is fertilized by two sperm cells.
2 Two egg cells are fertilized by one sperm cell.
3 Two egg cells are each fertilized by separate sperm cells.
4 Two eggs develop without fertilization.

89 In the human female reproductive system, the union of sperm and ovum normally takes place within the
1 ovary   3 uterus
2 vagina   4 oviduct

If you choose this group, be sure to answer questions 90–99.

Base your answers to questions 90 through 92 on the diagram below and on your knowledge of biology.

90 Structures 1, 2, and 3 make up a
1 nucleic acid   3 nucleolus
2 ribosome   4 nucleotide

91 If strand I represents a segment of a replicating DNA molecule with bases A–T–C–C–G–A, the complementary DNA strand would contain the bases
(1) T–A–G–G–C–T   (3) U–A–G–G–C–U
(2) T–U–G–G–C–T   (4) A–T–G–G–C–T

92 Structures 3 and 4 are held together by
1 weak peptide bonds
2 strong hydrogen bonds
3 weak hydrogen bonds
4 strong peptide bonds

Directions (93–94): For *each* statement in questions 93 and 94 select the genetic disorder, *chosen from the list below*, that is best described by that statement. Then record its *number* on the separate answer paper.

Genetic Disorders

(1) Phenylketonuria (PKU)
(2) Sickle-cell anemia
(3) Down's syndrome
(4) Tay-Sachs disease

93 The formation of abnormal hemoglobin results in severe pain due to obstructed blood vessels.

94 Fatty material accumulates because a specific enzyme cannot be synthesized, causing a deterioration of the nervous system.

95 Amino acid molecules are bonded together in a specific sequence on cell structures known as
1 ribosomes   3 mitochondria
2 vacuoles   4 centromeres

96 The core of a virus may contain either DNA or RNA. To identify which nucleic acid is present, a biochemist could chemically analyze the virus for the presence of
1 guanine   3 cytosine
2 ribose   4 phosphate

97 The sum of all the genes for all the heritable traits in a population is called the gene
1 frequency   3 pool
2 combination   4 system

98 A change in the base sequence of DNA is known as

1 a gene mutation   3 nondisjunction
2 a karyotype   4 polyploidy

99 In a large, randomly mating population of deer mice, the gene frequencies have stayed constant for several generations. Over this time interval, the rate of evolution in this mouse population most likely has

1 decreased, only
2 increased, only
3 increased, then decreased
4 remained the same

Group 5 — Ecology

If you choose this group, be sure to answer questions 100–109.

Base your answers to questions 100 and 101 on the chart below and on your knowledge of biology.

Characteristics	Climax Flora	Dominant Fauna
long, severe winters; summers with thawing subsoil	conifers	moose and black bears

100 Which biome is most accurately described by the data in the chart?

1 taiga
2 tropical rain forest
3 tundra
4 temperate deciduous forest

101 The climax flora in this biome are characterized by

1 leaves that are shed in the fall
2 needle-like leaves
3 no leaves at all
4 no stomates or lenticels

Base your answers to questions 102 and 103 on the passage below and on your knowledge of biology.

Algae live inside the body cells of a species of hydra. The hydra uses the products of the alga's photosynthesis. Ammonia resulting from the hydra's metabolism is thought to contribute to the alga's nutrition.

102 The relationship between the hydra and the alga is best described as

1 commensalism   3 saprophytism
2 mutualism   4 parasitism

103 The ammonia is a part of which important ecological cycle?

1 oxygen cycle   3 carbon cycle
2 water cycle   4 nitrogen cycle

104 Releasing sterilized male insects of a certain species into the environment can lead to a reduction in population size of that species. This process is an example of

1 the use of biological control
2 the use of a biocide
3 the effect of pesticides on reproduction rate
4 a technique used for species preservation

105 Which abiotic factor limits the distribution of life in the oceans, but does not usually limit the distribution of life on land?

1 minerals   3 nitrogen
2 water   4 oxygen

356

Base your answers to questions 106 through 109 on the diagrams below and on your knowledge of biology. The diagrams show various stages of plant development on a volcanic island over a 300-year period after it was formed.

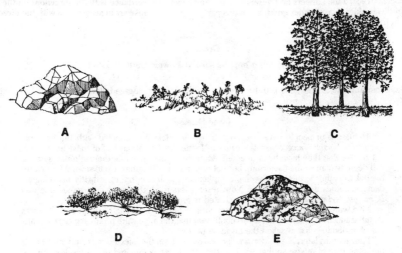

A   **B**   **C**

D   **E**

106 Pioneer organisms are the dominant flora in
  (1) *E*   (3) *C*
  (2) *B*   (4) *D*

107 If stage *C* illustrates the climax flora of the region, the dominant fauna would most likely include
  1 kangaroo rats and lizards
  2 snakes and monkeys
  3 squirrels and deer
  4 caribou and snowy owls

108 What is most likely the sequence of the stages of development that occurred on the island over the 300-year period?
  (1) *A → B → C → D → E*
  (2) *E → B → D → C → A*
  (3) *A → E → B → D → C*
  (4) *B → D → C → E → A*

109 Which type of organisms could most successfully colonize stage *A* of the volcanic island?
  1 ferns   3 woody shrubs
  2 lichens   4 flowering plants

Part III

This part consists of five groups. Choose three of these five groups. For those questions that are followed by four choices, record the answers on the separate answer paper in accordance with the directions on the front page of this booklet. For all other questions in this part, record your answers in accordance with the directions given in the question.   [15]

Group 1

If you choose this group, be sure to answer questions 110–114.

Base your answers to questions 110 through 114 on the reading passage below and on your knowledge of biology.

Lyme Disease

Thousands of people have been bitten by deer ticks and infected with the bacterial spirochete *Borrelia burgdorferi*, the cause of Lyme disease. About half of these people will not realize that they have been infected. After the initial infection, their immune systems will begin to control the bacterium, but not eliminate it altogether. Up to several years after the tick bite, the victims may develop complications such as crippling arthritis, neurological damage, and cardiac malfunctions. Now, researchers think they have determined one way *B. burgdorferi* manages to elude an activated immune system.

Five white-footed mice were infected with *B. burgdorferi*. The blood of the mice was sampled shortly thereafter, and it was confirmed that the mice were producing large quantities of antibodies that attacked the invading bacteria.

Four months later, *B. burgdorferi* were extracted from the infected mice and mixed with the same type of mouse antibodies. This time the bacteria initiated only a weak response, indicating that the antibodies were less able to recognize the bacteria. Since antibodies recognize a bacterium by binding to specific protein molecules on the bacterial surface, these surface molecules may somehow have changed over time. In this way, the bacteria are better able to escape early recognition by antibodies produced by the human immune system.

110 Shortly after the initial infection, the mice apparently
1 got rid of the bacteria
2 had no reaction to the infection
3 produced antibodies against the disease
4 suffered permanent neurological damage

111 The organisms that cause Lyme disease are able to
1 cause problems in plants as well as several species of animals
2 change their proteins, thus making recognition by the mouse's immune system more difficult
3 destroy mouse antibodies by chemically breaking them down into harmless end products
4 be transmitted directly from one mouse to another

112 According to the passage, which symptom of Lyme disease in humans might appear several years after the initial tick bite?
1 a severe rash   3 kidney failure
2 a high fever   4 joint inflammation

113 Which kingdom includes the organism that causes Lyme disease?
1 Monera   3 Fungi
2 Protista   4 Animal

114 The genus name of the organism that causes Lyme disease is
1 *spirochete*   3 *burgdorferi*
2 *Bacterium*   4 *Borrelia*

358

Group 2

If you choose this group, be sure to answer questions 115–119.

115 A student sees the image at the right when observing the letter "f" with the low-power objective lens of a microscope. Which diagram below most closely resembles the image the student will see after switching to high power?

(1)

(3)

(2)

(4)

116 A student views some cheek cells under low power. Before switching to high power, the student should
1 adjust the eyepiece
2 center the image being viewed
3 remove the slide from the stage
4 remove the coverslip

117 A student changes the objective of a microscope from 10× to 50×. If this is the only change made, what will happen to the field of view?
1 Its diameter will decrease.
2 Its diameter will increase.
3 Its brightness will increase.
4 Its brightness will remain the same.

118 When an onion cell is stained with iodine, which organelle becomes more visible under the compound light microscope?
1 mitochondrion   3 ribosome
2 lysosome   4 nucleus

119 The diagram below represents the field of view of a compound light microscope. Three unicellular organisms are located across the diameter of the field.

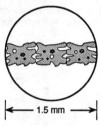

|←——— 1.5 mm ———→|

What is the approximate length of each unicellular organism?
(1) 250 μm   (3) 1,000 μm
(2) 500 μm   (4) 1,500 μm

Group 3

If you choose this group, be sure to answer questions 120–124.

Base your answers to questions 120 through 122 on the information below and on your knowledge of biology.

In a laboratory experiment, a student prepared a wet mount of the common aquarium plant elodea. After he used the low-power objective of a compound microscope to focus on the leaf cells, he switched to high power. In this field of view, he observed small green bodies moving along the boundary of each cell.

Directions (120–122): Your answers to questions 120 through 122 must be written in ink. Write your answers in the spaces provided on the separate answer paper.

120 Identify the small green bodies that the student observed within the elodea cells.

121 Identify one cell structure other than the small green bodies that the student should be able to observe in each cell under high power.

122 By which process do the small green bodies move within a cell?

123 A block of wood is measured, as shown in the diagram below.

What is the length of the wooden block in millimeters?
(1) 2.60 mm   (3) 260 mm
(2) 26.0 mm   (4) 2,600 mm

124 Iodine and Benedict's solution were both used to test for certain nutrients in a sample of food. If both tests were positive for the nutrients, the sample must contain
1 polysaccharide and simple sugar
2 protein and fat
3 carbohydrate and lipid
4 polysaccharide and protein

Group 4

If you choose this group, be sure to answer questions 125–129.

Base your answers to questions 125 through 127 on the graph below and on your knowledge of biology. The graph illustrates the growth curves for two types of bacteria (A and B) under differing pH values.

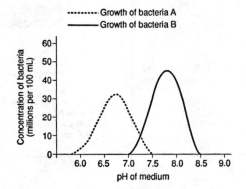

125 Bacteria A grows best in a medium that is
1 slightly acid   3 slightly basic
2 neutral   4 very basic

126 Which conclusion concerning bacteria A and bacteria B can correctly be drawn from the data provided in the graph?
1 They could not coexist in the same medium.
2 Their optimum pH values are different.
3 Bacteria A grows at a faster rate than bacteria B.
4 Bacteria A is larger than bacteria B.

127 A growth medium at pH 6.5 supports approximately what concentration of bacteria A?
(1) 15 million/100 mL   (3) 35 million/100 mL
(2) 25 million/100 mL   (4) 45 million/100 mL

Directions (128–129): Your answers must be written in complete sentences and must be written in ink. Write your answers in the spaces provided on the separate answer paper.

128 A student observed that on sunny days, green single-celled organisms in a lake were located several feet below the surface of the water. On cloudy days, she observed that these same organisms were only a few inches below the surface. Give a possible explanation for these observations.

129 The diagrams below show the setups for a particular experiment. Describe the relationship being studied in the experiment.

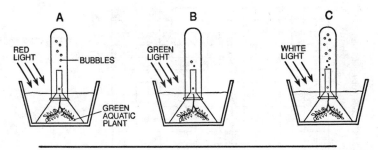

Group 5

If you choose this group, be sure to answer questions 130–134.

Base your answers to questions 130 through 133 on the information and data table below and on your knowledge of biology.

An experiment was set up to determine the effect of light intensity on the rate of photosynthesis in two cultures of the alga *Chlorella*, each grown in a different concentration of CO_2. The results are shown in the data table below.

Light Intensity (foot-candles)	Rate of Photosynthesis (bubbles of O_2 per minute)	
	Low CO_2 Concentration	High CO_2 Concentration
250	14	20
500	22	41
750	29	63
1,000	30	80
1,250	30	88
1,500	31	90
1,750	30	91
2,000	30	90

Directions (130–132): Using the information in the data table, construct a line graph on the grid provided on your answer paper, following the directions below. Pen or pencil may be used for your final answer.

The grid on the next page is provided for practice purposes only. Be sure your final answer appears *on your answer paper*.

130 Mark an appropriate scale on each axis.

131 Plot the data for the low CO_2 concentration on the graph. Surround each point with a small triangle and connect the points.

Example:

132 Plot the data for the high CO_2 concentration on the graph. Surround each point with a small circle and connect the points.

Example: ⊙–⊙

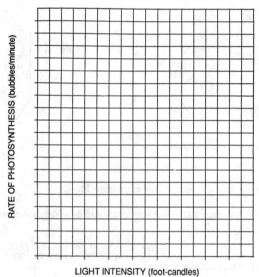

KEY

△ LOW CO_2 CONCENTRATION

⊙ HIGH CO_2 CONCENTRATION

RATE OF PHOTOSYNTHESIS (bubbles/minute)

LIGHT INTENSITY (foot-candles)

Directions (133): Your answer to question 133 must be written in ink. Write your answer in the space provided on the separate answer paper.

133 Using one or more complete sentences, state one factor in the experiment that affects the rate of photosynthesis from 250 to 1,000 foot-candles of light intensity and state this factor's effect on that rate.

134 An investigation was conducted using three groups of laboratory rats, *X*, *Y*, and *Z*, to determine the relative effects of glucose and adrenaline on the rate of heartbeat. The experimental conditions for each group of rats were kept the same except for the type of solution injected, as shown in the data table.

Data Table

Group	Solution Injected
X	1 mL adrenaline in distilled water
Y	1 mL glucose in distilled water
Z	1 mL distilled water, only

According to the data table, which group of rats functioned as the control?
(1) *X*, only   (3) *Z*, only
(2) *Y*, only   (4) both *X* and *Y*

362

Part I
Answer all 59 questions in this part. [65]

Directions (1–59): For *each* statement or question, select the word or expression that, of those given, best completes the statement or answers the question. Record your answer on the separate answer paper in accordance with the directions on the front page of this booklet.

1 In the diagram at the right, the arrows show the direction of movement of various substances. Which of the cell's life activities are represented by the arrows?

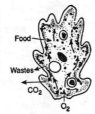

1 nutrition, reproduction, and regulation
2 excretion, transport, and respiration
3 growth, digestion, and locomotion
4 ingestion, regulation, and synthesis

2 Plants A and B are classified as members of the same species. Plants C and D are classified in the same genus as A and B, but not the same species as A and B. According to this information, which statement is correct?
1 Plant A has many characteristics in common with plant B.
2 Plant C cannot be the same species as plant D.
3 Plants A and B belong to a different kingdom than plants C and D.
4 Plants A, B, C, and D must all belong to different phyla.

3 Which statement describes an *exception* to the cell theory?
1 Cells arise from previously existing cells.
2 The cell is the basic unit of function in animals.
3 Mitochondria and chloroplasts can reproduce within the cell.
4 The cell is the basic unit of structure in plants.

4 The ultracentrifuge is an instrument that separates cellular components into distinct layers according to their relative
1 charges   3 acidities
2 solubilities   4 densities

5 Which organelles' activity contributes most directly to muscle contraction in an earthworm?
1 Golgi bodies   3 mitochondria
2 chloroplasts   4 lysosomes

6 Which of these elements is found in the *smallest* amount in living matter?
1 iodine   3 nitrogen
2 carbon   4 oxygen

7 A certain enzyme will hydrolyze egg white but not starch. Which statement best explains this observation?
1 Enzymes are specific in their actions.
2 Starch molecules are too large to be hydrolyzed.
3 Starch is composed of amino acids.
4 Egg white acts as a coenzyme for hydrolysis.

8 Which activity occurs in the process of photosynthesis?
1 Chemical energy from organic molecules is converted into light energy.
2 Organic molecules are obtained from the environment.
3 Organic molecules are converted into inorganic food molecules.
4 Light energy is converted into the chemical energy of organic molecules.

9 Which process is illustrated in the diagram below?

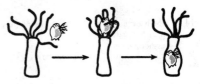

1 egestion   3 synthesis
2 ingestion   4 respiration

363

10 In the earthworm, the mixture of soil and food moving through the digestive tract is temporarily stored in the
1 mouth   3 ganglia
2 crop   4 setae

11 The process of osmosis would explain the net movement of water into a cell if the percentage of
1 water was 90% inside the cell and 95% outside the cell
2 protein was 30% inside the cell and 35% outside the cell
3 water was 95% inside the cell and 90% outside the cell
4 water and protein was equal inside and outside the cell

12 Vascular tissue in plants consists of
1 stomates and lenticels
2 xylem and phloem
3 spongy cells and xylem
4 lenticels and phloem

13 The diagram below represents a cross section of an earthworm.

The function of structure X is to provide a greater surface area for
1 transport of deoxygenated blood
2 transmission of nerve impulses
3 carbon dioxide absorption
4 nutrient absorption

14 In which organism is the transport of oxygen aided by hemoglobin?
1 grasshopper   3 earthworm
2 hydra   4 ameba

15 The products produced by yeast cells as a result of anaerobic respiration include ATP and
1 alcohol and oxygen
2 alcohol and carbon dioxide
3 water and oxygen
4 water and carbon dioxide

16 Which statement best describes one of the events taking place in the chemical reaction represented below?

$$H_2O + ATP \xrightarrow{\text{ATPase}} ADP + P + energy$$

1 Energy is being stored as a result of aerobic respiration.
2 Fermentation is taking place, resulting in the synthesis of ATP.
3 Energy is being released for metabolic activities.
4 Photosynthesis is taking place, resulting in the storage of energy.

17 The production of nitrogenous waste from excess amino acids is most directly associated with the process of
1 dehydration synthesis
2 glycogen storage
3 excretion
4 reproduction

18 Which organism eliminates water, urea, and mineral salts by means of nephridia?
1 human   3 grasshopper
2 hydra   4 earthworm

19 The diagram below represents a neuron.

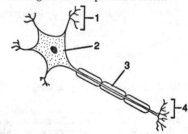

Which number indicates an area where a stimulus is detected and an electrochemical impulse is conducted to the cyton?
(1) 1   (3) 3
(2) 2   (4) 4

20 The diagrams below show two sequences of events.

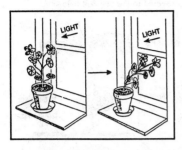 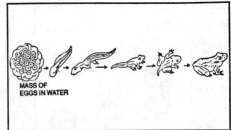

Both sequences of events most likely result from
1 impulse transmission
2 neurotransmitter secretion
3 artificial selection
4 hormonal control

---

21 The tympana and antennae of grasshoppers function primarily as
1 effectors for locomotion
2 glands for regulation
3 producers of hormones
4 receptors of stimuli

22 The diagrams below represent four different organisms.

The interaction of muscles and jointed appendages is responsible for movement in
(1) B, only
(2) B and D, only
(3) D, only
(4) A and C, only

23 Which body structures have walls one cell thick?
1 veins and arteries
2 trachea and bronchi
3 capillaries and alveoli
4 lymph vessels and stomach

24 Producing blood cells and providing anchorage sites for muscles are two functions of
1 skin
2 bones
3 cartilage
4 ligaments

25 The diagram below represents a portion of the esophagus.

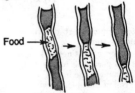

Food →

Which is a correct statement about the process shown in the diagram?
1 It transports nutrients within the digestive tract.
2 It must occur prior to mechanical digestion of food in the oral cavity.
3 It emulsifies fats for hydrolysis in the small intestine.
4 It increases water absorption by the esophagus.

26 Increased perspiration, a higher body temperature, and a rapidly beating heart are all possible responses to a stressful situation. These body responses are most likely a direct result of the interaction of the
1 digestive and endocrine systems
2 digestive and respiratory systems
3 nervous and endocrine systems
4 nervous and reproductive systems

27 The graph below shows the number of push-ups a student completed in each of four 2-minute trials (A–D) during a 15-minute exercise period.

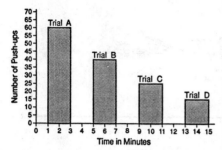

Time in Minutes

The concentration of lactic acid in the student's muscle tissue was most likely greatest during trial

(1) A   (3) C
(2) B   (4) D

28 Which statement best describes the human respiratory system?

1 It is composed of a network of moist passage-ways that permit air to flow from the external environment to the lungs.
2 Each cell of the human body is in direct contact with the external environment, and gas exchange occurs by diffusion.
3 The external body surface is kept moist to allow for gas exchange.
4 Gases diffuse across membranes on both the external and internal surfaces of the body.

29 The members of a certain species of grass in a lawn are genetically identical. The best explanation for this observation is that the species most probably reproduces

1 by an asexual method
2 after pollination by the wind
3 after pollination by a particular species of bee
4 by identical sperm fertilizing the eggs

30 In humans, which cell is produced most directly by mitotic cell division?

1 a sperm cell   3 an egg cell
2 a skin cell   4 a zygote

31 What is a major function of the blood vessel represented in the diagram below?

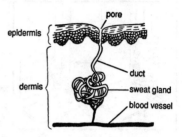

1 releasing carbon dioxide into the sweat gland
2 transporting oxygen away from the sweat gland
3 transporting wastes to the sweat gland
4 filtering starch out of the sweat gland

32 During meiotic cell division, the process in which homologous pairs of chromosomes separate and move apart is known as

1 internal fertilization
2 regeneration
3 binary fission
4 disjunction

33 In which region of the flower represented below do diploid cells change to monoploid male gametes?

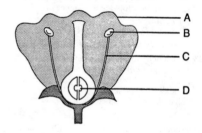

(1) A   (3) C
(2) B   (4) D

34 In the diagram of a dissected seed below, which letter indicates the epicotyl?

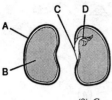

(1) A   (3) C
(2) B   (4) D

35 Which process is represented by the diagram below?

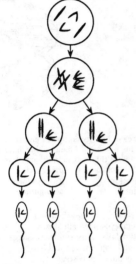

1 fertilization
2 gametogenesis
3 binary fission
4 vegetative propagation

36 An adaptation for reproduction in most terrestrial vertebrates is
1 internal fertilization
2 regeneration
3 mitosis
4 vegetative propagation

37 Which is a true statement about the process illustrated below?

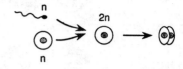

1 It is the beginning of embryonic development and occurs only in a freshwater environment.
2 It is the beginning of regeneration and occurs only within the female.
3 It is the beginning of embryonic development and occurs within the female or in water.
4 It is the beginning of ovule formation and occurs on the stigma of flowers.

38 The principles of dominance, segregation, and independent assortment resulted from studies by Mendel of the inheritance of traits in
1 four-o'clock flowers   3 fruit flies
2 roan cattle   4 pea plants

39 In the diagram below of two homologous chromosomes, what do r and R represent?

1 two different alleles
2 two gametes that can form a zygote
3 two identical alleles
4 two chromosomes in a hybrid pea plant

40 In certain rats, black fur is dominant over white fur. If two rats, both heterozygous for fur color, are mated, their offspring would be expected to have
1 four different genotypes and two different colors
2 two different genotypes and three different colors
3 three different genotypes and two different colors
4 three different genotypes and three different colors

367

41 Which pair of gametes can unite to produce a zygote that will develop into a normal human male embryo?

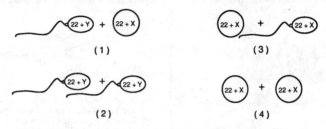

(1)

(3)

(2)

(4)

42 The diagrams below represent paired double-stranded chromosomes that contain genes indicated by letters.

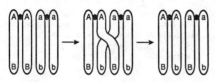

When does the process illustrated by the diagrams occur?
1 in meiosis, after disjunction of homologous chromosomes
2 in mitosis, after replication of chromosomes
3 in meiosis, during synapsis of homologous chromosomes
4 in mitosis, while chromosomes are attaching to spindle fibers

43 Mutagenic agents are substances that
1 increase the rate of gene mutations
2 decrease the rate of gene mutations
3 have no effect upon the rate of gene mutations
4 cause gene mutations but not other chromosomal changes

44 Which breeding method results in the production of offspring with the same genotype as the parents?
1 cross-pollination
2 vegetative propagation
3 inbreeding
4 hybridization

45 In the diagram of a polymer at the right, the repeating subunits are known as
1 amino acids
2 polysaccharides
3 nucleotides
4 fatty acids

46 Which area of biology compares and attempts to explain the structural changes that have taken place in living things over millions of years, as well as those changes occurring today?
1 classification   3 physiology
2 reproduction   4 evolution

47 In the early stages of development, both chicken and pig embryos have gill slits, two-chambered hearts, and tails. This similarity suggests that chickens and pigs most probably
1 have a common ancestry
2 carry on anaerobic respiration as adults
3 use gills for breathing during embryonic development
4 have inadequate circulation

48 The concept that, due to a need, organisms acquired the ability to move from an aquatic environment onto the land is most closely associated with a theory proposed by
1 Weismann   3 Miller
2 Lamarck   4 Mendel

368

49 Charles Darwin proposed that organisms produce many more offspring than can possibly survive on the limited amount of resources available to them. According to Darwin, the offspring most likely to survive are those that
1 are born first and grow fastest
2 are largest and most aggressive
3 are best adapted to the environment
4 have no natural predators

50 The graph below shows the results of an investigation related to evolution.

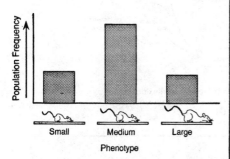

This graph was most likely developed from data involving a study of the
1 transmission of acquired characteristics
2 concept of punctuated equilibrium
3 concept of gradualism
4 variation within a species

51 According to the heterotroph hypothesis, which change contributed most directly to the evolution of aerobic organisms?
1 the appearance of organisms able to carry on photosynthesis
2 an increase in fermentation by organisms in the soil
3 a decrease in the intensity of light from the Sun
4 an increase in the concentration of hydrogen gas in the atmosphere

Base your answer to question 52 on the information and statement below.

Information

The Galapagos Islands in the Pacific were probably never connected to South America. However, in the various habitats on the islands, there are about 14 species of finchlike birds that appear to be related to the finches on the South American mainland. Although the Galapagos finches vary in beak structure, there is a close resemblance between these species in plumage, calls, nests, and eggs. These species do not interbreed and do not compete for food.

Statement

Isolation from the South American mainland and different habitats on the Galapagos Islands are important factors in the production of new species.

52 What is the relationship between the statement and the information given?
1 The statement is supported by the information given.
2 The statement is not supported by the information given.
3 The statement is contradicted by the information given.
4 No relevant information is given regarding the statement.

53 In the four diagrammatic statements below, the arrows should be read as "influences." For example, A ⟹ B would be read "A influences B," and B ⟲ B would read "B influences itself."
Which diagrammatic statement best defines ecology?

(1)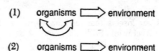

(2)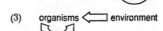

(3)

(4) organisms ⟺ environment

369

54 Which statement is best supported by the diagram below of the carbon-oxygen cycle?

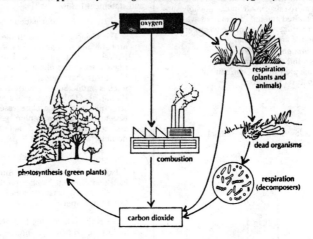

1 Decomposers add oxygen to the atmosphere and remove carbon dioxide.
2 Combustion adds oxygen to the atmosphere and removes carbon dioxide.
3 Producers generate oxygen and utilize carbon dioxide.
4 Consumers generate oxygen and utilize carbon dioxide.

---

55 An owl cannot entirely digest the animals it preys upon. Therefore, each day it expels from its mouth a pellet composed of fur, bones, and sometimes cartilage. By examining owl pellets, ecologists would be able to determine the
1 consumers that owls prefer
2 autotrophs that owls prefer
3 organisms that feed on owls
4 saprophytes that affect owls

56 If birds eat insects that feed on corn, which pyramid level would birds occupy?

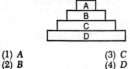

(1) A   (3) C
(2) B   (4) D

57 In an attempt to prevent certain species from becoming extinct, humans have
1 placed all endangered species in zoos
2 increased the trapping of predators
3 increased wildlife management and habitat protection
4 attempted to mate organisms from different species to create new and stronger organisms

58 All the interacting populations in a given area represent an ecological unit known as a
1 biosphere
2 community
3 world biome
4 saprophytic relationship

370

59 Which is an abiotic factor that functions as a limiting factor for the autotrophs in the ecosystem below?

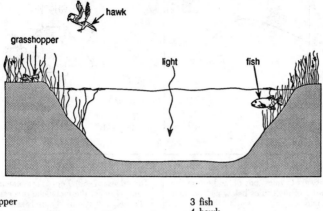

1 grasshopper   3 fish
2 light   4 hawk

Part II

This part consists of five groups, each containing ten questions. Choose two of these five groups. Be sure that you answer all ten questions in each group chosen. Record the answers to these questions in accordance with the directions on the front page of this booklet.   [20]

Group 1 — Biochemistry

If you choose this group, be sure to answer questions 60–69.

Base your answers to questions 60 through 62 on the model of a biochemical reaction below and on your knowledge of biology.

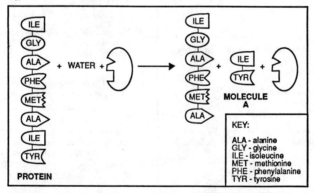

60 The process represented in the diagram is known as
1 dehydration synthesis   3 photolysis
2 carbon fixation   4 hydrolysis

61 Molecule A can best be described as a
1 dipeptide   3 starch
2 disaccharide   4 fat

62 The bond that exists between alanine and phenylalanine is known as
1 an ionic bond   3 a hydrogen bond
2 a peptide bond   4 a phosphate bond

63 According to the graph below, at what temperature will the denaturation of lipase begin?

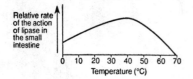

Relative rate of the action of lipase in the small intestine

Temperature (°C)

1 below 0°C
2 between 0°C and 38°C
3 at 40°C
4 at 68°C

64 Which compound has the structural formula shown below?

CH_2OH

(1) starch   (3) ATP
(2) PGAL   (4) glucose

Base your answers to questions 65 and 66 on the structural formulas of molecules below and on your knowledge of biology.

Molecule A

Molecule B

65 The portion of molecule A represented in box X is known as
1 a nitrogenous base
2 an amino group
3 a hydrocarbon chain
4 a carboxyl group

66 How many molecules of A normally combine with one molecule of B to form a single fat molecule?
(1) 5   (3) 3
(2) 6   (4) 4

Directions (67–68): For each statement in questions 67 and 68 select the metabolic process, chosen from the list below, that is most closely associated with that statement. Then record its number on the separate answer paper.

Metabolic Processes

(1) $2ATP + C_6H_{12}O_6 \xrightarrow{enzymes}$
     $4ATP + 2$ lactic acid

(2) $6CO_2 + 12H_2O \xrightarrow[chlorophyll]{light, enzymes}$
     $C_6H_{12}O_6 + 6O_2 + 6H_2O$

(3) $C_{12}H_{22}O_{11} + H_2O \xrightarrow{enzymes}$
     $C_6H_{12}O_6 + C_6H_{12}O_6$

67 This process occurs in humans only when certain cells do not receive an adequate supply of oxygen.

68 Part of this process takes place in structures known as grana.

---

69 What is the net gain in ATP following the completion of aerobic cellular respiration of one molecule of glucose in a brain cell?
(1) 30   (3) 36
(2) 2   (4) 4

If you choose this group, be sure to answer questions 70–79.

Directions (70–72): For *each* phrase in questions 70 through 72 select the transport pathway, *chosen from the list below*, that is most closely related to that phrase. Then record its *number* on the separate answer paper.

Transport Pathways

(1) Coronary circulation
(2) Systemic circulation
(3) Lymphatic circulation
(4) Pulmonary circulation

70 Carries blood from the heart to the lungs and from the lungs to the heart

71 Contains nodes that filter foreign substances such as bacteria from transport fluid

72 Carries blood from the heart to the digestive and reproductive structures of the body

73 The nerves that directly control the muscles used in writing are
1 part of the autonomic nervous system
2 regulated by the hypothalamus
3 part of the somatic nervous system
4 regulated by the medulla

74 Structures 1, 2, and 3 in the diagram at the right are connected to striated muscles by connective tissue. An inflammation of this connective tissue is known as
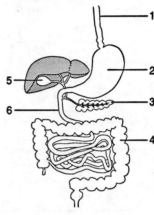
1 tendinitis
2 gout
3 polio
4 angina pectoris

Directions (75–78): For *each* statement in questions 75 through 78 select the organ, *indicated in the diagram below*, that is most closely associated with that statement. Then record its *number* on the separate answer paper. [A number may be used more than once or not at all.]

75 This organ stores bile.

76 Gastric juice is produced in this organ.

77 The chemical digestion of protein begins within this organ..

78 Materials to be egested are stored in this organ.

79 An example of the maintenance of homeostasis in humans is the action of glucagon and insulin in regulating the
1 temperature of the body
2 concentration of blood sugar
3 excretion of urine from the bladder
4 secretion of thyroxin

Group 3 — Reproduction and Development

If you choose this group, be sure to answer questions 80–89.

Base your answers to questions 80 through 82 on the diagram below of the male reproductive system and on your knowledge of biology.

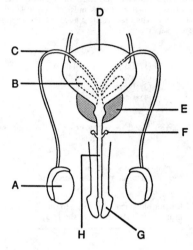

80 Which structures are glands that secrete a liquid for the transport of sperm?
(1) *A* and *D*
(2) *B* and *E*
(3) *C* and *H*
(4) *F* and *G*

81 A male sex hormone is produced within structure
(1) *A*
(2) *B*
(3) *E*
(4) *F*

82 Male gametes are produced within structure
(1) *A*
(2) *B*
(3) *F*
(4) *D*

Directions (83–85): For *each* statement in questions 83 through 85 select the part of the human female reproductive system, *indicated in the diagram below,* that is most closely associated with that statement. Then record its *number* on the separate answer paper. [A number may be used more than once or not at all.]

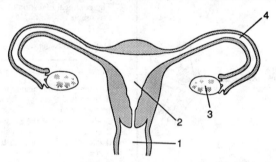

83 The wall of this structure breaks down during one of the stages of the menstrual cycle.

84 This structure produces chemicals that regulate the development of secondary sex characteristics such as the mammary glands.

85 The process of embryo implantation normally occurs within this structure.

374

86 Which is the correct sequence of stages in a normal menstrual cycle?
  1 corpus luteum stage → menstruation → ovulation → follicle stage
  2 ovulation → follicle stage → menstruation → corpus luteum stage
  3 follicle stage → ovulation → corpus luteum stage → menstruation
  4 menstruation → corpus luteum stage → ovulation → follicle stage

87 Different embryonic layers of tissue form during the process of
  1 fertilization   3 cleavage
  2 birth   4 gastrulation

88 Which adaptation for successful development is characteristic of all embryos?
  1 a shell for protection from predators
  2 a parent for protection from predators
  3 a sac for storage of wastes
  4 a mechanism for absorbing oxygen

89 Eggs that develop externally on land contain a membrane that collects and stores nitrogenous wastes until the egg hatches. This membrane is known as the
  1 amnion   3 allantois
  2 yolk sac   4 chorion

Group 4 — Modern Genetics
If you choose this group, be sure to answer questions 90–99.

90 The diagram below shows some steps involved in preparing tissue cultures of a plant.

Which technique is represented in the diagram?
  1 hybridization   3 cloning
  2 amniocentesis   4 karyotyping

Base your answers to questions 91 through 93 on the information below and on your knowledge of biology.

Two alleles for coloration, black and silver, exist in small minnows that inhabit a particular lake. Largemouth bass also live in this lake, and the minnows are a major portion of their diet. Bass recognize their prey to some extent by the degree to which the prey contrast with their background.

91 According to the Hardy-Weinberg principle, which factor will contribute to the maintenance of a stable gene pool in the minnow population?
  1 mutations
  2 random mating
  3 minnow migration
  4 increased predation

92 The percentage of each allele for coloration in minnows is known as the
  1 gene frequency   3 mutation rate
  2 genetic code   4 abiotic factor

93 In areas of aquatic vegetation, black minnows outnumber silver minnows, but in areas of open water, silver minnows outnumber black minnows. In time, the aquatic vegetation will increase and cover the entire lake. What change will most likely occur in the frequency of the minnows' alleles for color?

1 The frequency of the black allele will decrease.
2 The frequency of both alleles will change randomly.
3 The frequency of the black allele will increase.
4 The frequency of both alleles will remain the same.

Base your answers to questions 94 and 95 on the pedigree chart below and on your knowledge of biology. The pedigree chart represents the inheritance of color blindness through three generations.

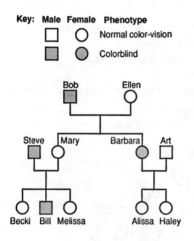

96 Genetic engineering has been utilized for the production of
1 salivary amylase
2 human growth hormone
3 hydrochloric acid
4 uric acid crystals

97 Which nitrogenous base is normally present in RNA molecules but *not* in DNA molecules?
1 adenine   3 thymine
2 cytosine   4 uracil

Base your answers to questions 98 and 99 on the "help-wanted advertisements" below and on your knowledge of biology.

Job A	**Accuracy and Speed** vital for this job in the field of translation. Applicants must demonstrate skills in transporting and positioning amino acids. Salary commensurate with experience.
Job B	**Executive Position** available. Must be able both to maintain genetic continuity through replication and to control cellular activity by regulation of enzyme production. Limited number of openings. All benefits.

94 Mary and Steve are expecting another child. What is the probability that the new baby will be colorblind?
(1) 0%   (3) 50%
(2) 25%   (4) 100%

95 Which is a true statement about the genotype of Alissa and Haley regarding color blindness?
1 Both carry one recessive allele.
2 Alissa is a carrier, and Haley is homozygous dominant.
3 Both are homozygous recessive.
4 Alissa is homozygous dominant, and Haley is a carrier.

98 Which "applicant" would qualify for job A?
(1) DNA
(2) messenger RNA
(3) recombinant DNA
(4) transfer RNA

99 Which "applicant" would qualify for job B?
(1) DNA   (3) transfer RNA
(2) messenger RNA (4) ADP

Group 5 — Ecology

If you choose this group, be sure to answer questions 100–109.

Base your answers to questions 100 through 102 on the chart below and on your knowledge of biology.

Symbiotic Relationships

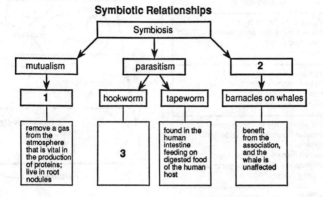

100 Which organisms are represented by box 1?
1 nitrifying bacteria
2 nitrogen-fixing bacteria
3 saprophytic bacteria
4 denitrifying bacteria

101 Which term belongs in box 2?
1 tropism   3 saprophytism
2 gradualism   4 commensalism

102 Which description belongs in box 3?
1 derives nourishment from human body fluids
2 feeds on dead animals, assisting in the recycling of nutrients
3 stalks, kills, and eats fish in deep ocean environments
4 carries on autotrophic nutrition in the tropical forest biome

---

103 Which group in the food web represented below would most likely have the greatest biomass?

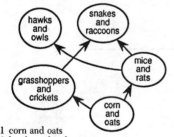

1 corn and oats
2 hawks and owls
3 mice and rats
4 snakes and raccoons

104 When animals excrete nitrogenous wastes into the soil, certain soil bacteria convert these wastes into nitrates, which are absorbed by plants. These soil bacteria function as
1 autotrophs
2 secondary consumers
3 decomposers
4 abiotic factors

105 Which organisms would most likely be the pioneer organisms on a newly formed volcanic island?
1 conifers   3 deciduous trees
2 lichens   4 tall grasses

Base your answers to questions 106 and 107 on the information in the chart below and on your knowledge of biology.

Biome	Characteristics
A	Moisture is a limiting factor Extreme daily temperature variations Climax flora includes many succulent plants
B	Heavy annual rainfall Constant warm temperature Climax flora includes many species of broad-leaved plants
C	Areas vary greatly in concentration of dissolved particles and velocity of currents Seasonal dieback of vegetation Concentration of dissolved gases is a limiting factor
D	Provides the most stable aquatic environment Provides a large amount of the world's food production Contains a relatively constant supply of nutrient materials and dissolved salts

106 Which letter most likely indicates a marine biome?
(1) A   (3) C
(2) B   (4) D

107 Which letter most likely indicates a tropical forest biome?
(1) A   (3) C
(2) B   (4) D

Base your answers to questions 108 and 109 on the diagram below and on your knowledge of biology.

Desert Community

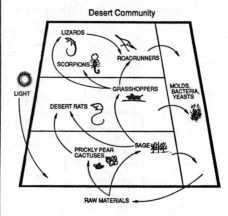

108 Which is an example of the nutritional pattern of a primary consumer?
1 grasshoppers → lizards
2 scorpions → bacteria
3 prickly pear cactuses → desert rats
4 lizards → roadrunners

109 A carnivore in this desert community is represented by the
1 lizard   3 yeast
2 sage   4 desert rat

378

Part III

This part consists of five groups. Choose three of these five groups. For those questions that are followed by four choices, record the answers on the separate answer paper in accordance with the directions on the front page of this booklet. For all other questions in this part, record your answers in accordance with the directions given in the question. [15]

Group 1

If you choose this group, be sure to answer questions 110–114.

Base your answers to questions 110 through 114 on the information and data table below and on your knowledge of biology.

When a culture of cells is exposed to gamma rays, chromosome damage results. This damage is very evident when the cells are stained and observed with a compound light microscope. The chromosome damage is primarily in the form of breaks and gaps, which are commonly referred to as chromosome aberrations. Investigations have shown that when the amino acid cysteine is added to the cell culture prior to gamma-ray exposure, the number of aberrations is reduced. The results of one investigation are shown in the data table below. In this investigation, each cell culture received the same amount of gamma-ray exposure.

Data Table

Cell Culture Tube Number	Amount of Cysteine Added (g)	Average Number of Chromosome Aberrations per Cell After Gamma-Ray Exposure
1	0.0	1.20
2	0.7	0.65
3	1.0	0.58
4	2.6	0.40
5	5.3	0.33
6	10.5	0.25
7	15.8	0.18

Directions (110–111): Using the information in the data table, construct a line graph on the grid provided on your answer paper, following the directions below. The grid on the next page is for practice purposes only. Be sure your final answer appears *on your answer paper.*

110 Mark an appropriate scale on each labeled axis.

111 Plot the data and connect the points. Surround each point with a circle

Example:

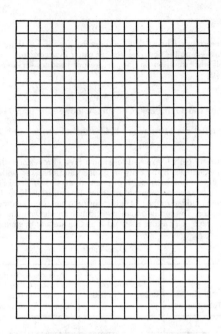

Amount of Cysteine Added (g)

112 A culture tube in which the average number of chromosome aberrations per cell is 0.30 would most likely contain approximately how many grams of added cysteine?
  (1) 7.0   (3) 0.3
  (2) 1.15   (4) 0.4

113 Chromosome aberrations that result from gamma-ray exposure would most likely cause
  1 the condition known as polyploidy
  2 mutations in cells
  3 an increase in the lifespan of cells
  4 a reduction in cysteine synthesis

Directions (114): Write your answer to question 114 in the space provided on the separate answer paper. Your answer must be written in ink.

114 Culture tube 1 was exposed to the same amount of gamma rays as the other six tubes, but no cysteine was added to this tube. Using one or more complete sentences, explain the role of culture tube 1 as a control in this investigation.

_____

Group 2

If you choose this group, be sure to answer questions 115–119.

Base your answers to questions 115 through 119 on the reading passage below and on your knowledge of biology.

Carnivorous Plants

Carnivorous plants make carbohydrates by the process of photosynthesis, as do other green plants. However, few nitrogenous minerals are available in the acid bog environment, and the roots of carnivorous plants are not efficient at absorbing them. In order to survive, these plants have evolved modified leaves that trap insects to supplement their nutrition. The modified leaves contain nectar glands which give off substances that attract and aid in the capture of insects. Once an insect is trapped, the leaves begin to produce digestive enzymes. Nutrients from the digested insects are then absorbed by the leaves.

Carnivorous plants have several types of traps. The Venus flytrap is an example of an active trap. It has colorful red-lined leaves that are hinged in the middle. When an insect lands and touches the sensitive hairs on the inner surface of the leaf, the leaf folds. Spines along the leaf's edges interlock, keeping the insect from escaping. Glands secrete enzymes that digest the insect's soft parts. When digestion is complete, the leaf reopens, allowing the undigested parts to blow away.

The largest of the carnivorous plants are the pitcher plants, which have pitfall traps. The leaves of these plants form a slender tube with a hood that prevents rain from entering. Nectar glands on the lip of the tube attract insects. The insects land on a slick area of the tube and fall into a pool of digestive juices at the bottom. Hairs inside the plant prevent the insect from crawling out.

The sundew is a flypaper trap. The attractive leaves of this plant are covered with hairs that secrete sticky droplets. The odor produced by this liquid lures insects to the plant. The insect then becomes entangled in the hairs on the leaf, where enzymes digest the soft parts of the insect.

These and other varieties of carnivorous plants grow in the marshes, swamps, and bogs of the eastern United States. These plants are becoming endangered or threatened species as wetlands are drained for commercial or residential development.

115 Which is a characteristic of some varieties of carnivorous plants?
   1 They carry on autotrophic nutrition similar to the fungi.
   2 They have modified leaves that trap insects.
   3 They have roots that absorb nutrients from dead insects.
   4 They trap all species of insects.

116 Carnivorous plants are similar to other green plants because they have the ability to
   1 secrete enzymes from leaf surfaces
   2 absorb digestive end products through the leaves
   3 produce carbohydrates from inorganic materials
   4 carry on heterotrophic nutrition

117 Nitrogen-containing minerals are inefficiently absorbed by which structures of carnivorous plants?
   1 roots   3 flowers
   2 leaves   4 stems

118 Which carnivorous plant is correctly paired with its adaptation for the capture of insects?
   1 sundew — spines on leaf edges
   2 pitcher plant — sensitive hairs
   3 pitcher plant — hairs with sticky droplets
   4 Venus flytrap — hinged leaves

119 Which statement is true of many carnivorous plant species of the United States?
   1 They are used in the chemical control of insects.
   2 They are becoming endangered species.
   3 They are poisonous to various species of mammals.
   4 They are adapted to habitats with a high pH.

Base your answers to questions 120 and 121 on the diagram below of a single-celled organism observed by a student using the low-power objective of a microscope.

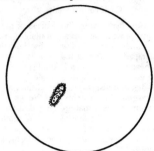

120 How should the student move the slide on the stage to center the single-celled organism in the field?
1 away from herself and to her right
2 away from herself and to her left
3 toward herself and to her right
4 toward herself and to her left

121 As the student observes the organism under the high-power objective, the organism swims out of focus. To bring it back into focus, the student should
1 open the diaphragm
2 turn the fine adjustment
3 turn the ocular
4 adjust the light source

122 The diagram below shows a portion of a compound microscope.

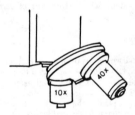

A student observes 12 onion epidermal cells along the diameter of the low-power field. How many of these cells would the student observe along the diameter of the high-power field?
(1) 48   (3) 3
(2) 40   (4) 24

123 To locate a specimen on a prepared slide with a compound microscope, a student should begin with the low-power objective rather than the high-power objective because the
1 field of vision is smaller under low power than under high power
2 field of vision is larger under low power than under high power
3 specimen does not need to be stained for observation under low power but must be stained for observation under high power
4 amount of the specimen that can be observed under low power is less than the amount that can be observed under high power

Directions (124): Write your answer to question 124 in the space provided on the separate answer paper. Your answer must be written in ink.

124 Using one or more complete sentences, explain how the light intensity in the high-power field of view of a compound microscope may be increased.

125 Two groups of 100 corn seeds were planted in two separate containers of soil and watered regularly. Group I was grown in light for 4 weeks and group II was grown in the dark for 2 weeks and then in the light for 2 weeks. The color of the seedlings was recorded after each 2-week period. Light was the only variable in the experiment. The results are summarized in the data table below.

	Group I		Group II	
	After 2 Weeks (in light)	After 2 More Weeks (in light)	After 2 Weeks (in darkness)	After 2 More Weeks (in light)
Number of Green Seedlings	75	75	0	80
Number of White Seedlings	25	25	100	20

This experiment demonstrates that
1 the environment interacts with genes in the expression of an inherited trait
2 water and fertilizer are important for seed germination
3 heat should have been provided along with carbon dioxide for proper growth
4 the principles of genetics apply only to plants and not to animals

126 To test for the presence of glucose, a student added the same amount of Benedict's solution to three test tubes, two of which contained unknown solutions. The third test tube contained water. The chart below shows the color results obtained after the solutions were heated in the three test tubes in a hot water bath.

Tube	Contents	Color After Heating
1	unknown solution plus Benedict's solution	Royal blue
2	unknown solution plus Benedict's solution	Red orange
3	water plus Benedict's solution	Royal blue

The student could correctly conclude that
1 none of the tubes contained glucose
2 tubes 1 and 2 contained glucose
3 tube 1 did not contain glucose, but tube 2 did
4 tube 2 did not contain glucose, but tube 1 did

Base your answers to questions 127 and 128 on the diagram below and on your knowledge of biology.

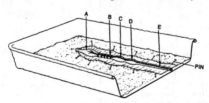

Directions (127): Write your answer to question 127 in the space provided on the separate answer paper. Your answer must be written in ink.

127 Select one of the lettered digestive structures from the diagram. Record the letter of this structure and the name of the structure it indicates.

128 Which statement best describes the relative positions of two structures in the diagram?
(1) *A* is anterior to *C*.
(2) *E* is dorsal to *B*.
(3) *C* is ventral to *A*.
(4) *D* is posterior to *E*.

Directions (129): Write your answer to question 129 in the space provided on the separate answer paper. Your answer must be written in ink.

129 Select one of the lettered parts from the diagram below of a human cheek cell. Record the letter of the part chosen. Using one or more complete sentences, state the function of this part.

Group 5

If you choose this group, be sure to answer questions 130–134.

130 An organism was kept at a temperature of 40°C for a period of 2 weeks. At the end of that time, the investigator determined that the organism was sterile. To support the hypothesis that high temperatures cause sterility, the investigator should be able to show that the
1 organism was not sterile before the experimental period began
2 high temperature did not alter the blood pressure of the organism
3 pituitary gland of the organism had not degenerated
4 organism was homozygous for temperature sensitivity

131 A student wants to test the hypothesis that a certain aquarium plant absorbs CO_2. The student placed a sprig of the aquarium plant in a test tube with water and exhaled CO_2 into the water. Which indicator should the student add to the test tube to help him test this hypothesis?
1 Benedict's solution
2 iodine solution
3 salt solution
4 bromthymol blue solution

Base your answer to question 132 on the diagram below and on your knowledge of biology.

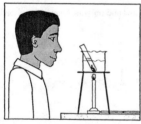

132 Which statement describes *two* unsafe laboratory practices represented in the diagram?
1 The flame is too high and the test tube is unstoppered.
2 The opening of the test tube is pointed toward the student and the student is not wearing goggles.
3 The test tube is unstoppered and the student is not wearing goggles.
4 The beaker has water in it and the flame is under the tripod.

133 What must a student do to obtain a volume of 12.5 milliliters of liquid in the graduated cylinder shown below?

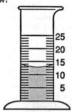

1 Add 0.5 mL of liquid.
2 Add 1.5 mL of liquid.
3 Remove 0.5 mL of liquid.
4 Remove 1.5 mL of liquid.

Base your answer to question 134 on the information below, which describes the purpose and procedure of two investigations, A and B.

Investigation A

Purpose: To observe the nucleus of an ameba

Procedure: Place a drop of water containing living amebas on the stage of a compound microscope. Add a drop of Benedict's solution and observe with the low-power objective and then with the high-power objective.

Investigation B

Purpose: To observe the nerve cord of an earthworm

Procedure: Place an earthworm dorsal side up on a slide. Cut through the skin and muscle of the dorsal surface and, using a compound microscope, observe the nerve cord on the upper surface of the intestine.

Directions (134): Write your answer in the space provided on the separate answer paper. Your answer must be written in ink.

134 Choose *one* of these investigations, and using one or more complete sentences, identify *one* error in the procedure. Then describe the proper procedure to correct this error.

Part I

Answer all 59 questions in this part. [65]

Directions (1–59): For *each* statement or question, select the word or expression that, of those given, best completes the statement or answers the question. Record your answer on the separate answer paper in accordance with the directions on the front page of this booklet.

1 In which group do all the organisms belong to the same kingdom?
1 yeast, mushroom, maple tree
2 paramecium, ameba, spirogyra
3 bacteria, ameba, spirogyra
4 bacteria, moss, geranium

2 Stained yeast were added to a paramecium culture, and some of the yeast were ingested by the paramecia. This activity is most closely associated with which life function?
1 synthesis   3 nutrition
2 regulation   4 growth

3 Which is the most specific term used to classify humans?
1 *sapiens*   3 *Homo*
2 Animal   4 Chordate

4 Which statement is *not* part of the cell theory?
1 Cells are the structural units in living things.
2 Cells are the functional units in living things.
3 New cells arise from preexisting cells.
4 New cells have nuclei identical to those of preexisting cells.

5 Which metabolic process is most closely associated with the organelle represented in the diagram below?

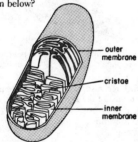

1 intracellular digestion
2 cellular respiration
3 synthesis of glycogen
4 hydrolysis of lipids

6 Which substances are inorganic compounds?
1 water and salts
2 proteins and carbohydrates
3 fats and oils
4 enzymes and hormones

7 Compound X increases the rate of the reaction shown below.

$$CO_2 + H_2O \xrightarrow{\ \ X\ \ } H_2CO_3$$

Compound X is most likely
1 an enzyme   3 an indicator
2 a lipid molecule   4 an ADP molecule

8 A green plant is kept in a brightly lighted area for 48 hours. What will most likely occur if the light intensity is then reduced slightly during the next 48 hours?
1 Photosynthesis will stop completely.
2 The rate at which nitrogen is used by the plant will increase.
3 The rate at which oxygen is released from the plant will decrease.
4 Glucose production inside each plant cell will increase.

9 Organisms that obtain and ingest organic molecules for their nutrition are classified as
1 autotrophs   3 algae
2 producers   4 heterotrophs

10 Which statement best describes extracellular digestion?
1 Large insoluble molecules are converted to small soluble molecules outside the cell.
2 Large insoluble molecules are converted to small soluble molecules within the cell.
3 Small soluble molecules are converted to large insoluble molecules outside the cell.
4 Small soluble molecules are converted to large insoluble molecules within the cell.

11 The end products of digestion enter the cells of a vertebrate by the process of
1 absorption   3 emulsification
2 osmosis   4 egestion

12 The diagram below represents an organism.

This organism is able to survive without a specialized transport system because
1 most of its cells are in direct contact with a terrestrial environment
2 it possesses a brain, which transmits impulses to control locomotion
3 most of its cells are in direct contact with a water environment
4 it possesses a cell wall, which keeps the internal body cells moist

13 Which structures in humans perform a function most similar to that performed by the aortic arches of the earthworm?
1 nephrons   3 alveoli
2 ventricles   4 neurons

14 In a bean plant, which reaction will release the greatest amount of energy?
1 aerobic respiration of a glucose molecule
2 anaerobic respiration of a glucose molecule
3 synthesis of a chlorophyll molecule
4 hydrolysis of a cellulose molecule

15 Which plant structures are specifically adapted for the absorption of materials from the substratum?
1 xylem tubes   3 palisade cells
2 root hairs   4 phloem tubes

16 The function of the contractile vacuole in many freshwater protozoans is to
1 store carbon dioxide
2 aid in mechanical digestion
3 eliminate wastes such as ammonia
4 eliminate excess water

17 Which structure of the woody plant represented in the diagram below most likely contains lenticels?

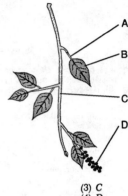

(1) A   (3) C
(2) B   (4) D

18 Which organism is correctly paired with the structures it uses for the excretion of nitrogenous wastes and carbon dioxide?
1 hydra — cell membranes
2 earthworm — alveoli
3 paramecium — nephridia
4 human — Malpighian tubules

19 In the reflex arc represented by the diagram below, which type of substance is normally secreted in the area indicated by letter X?

1 an antibody   3 a neurotransmitter
2 a pigment   4 an antigen

20 Which organism has a nervous system most similar to that of the grasshopper?
1 hydra   3 human
2 earthworm   4 paramecium

21 When a geranium plant is placed in a horizontal position, auxins accumulate on the side of the stem closest to the ground. As a result, what will most likely occur in the stem of the geranium?
1 Stomates will close.
2 Leaves will develop.
3 Cells will grow unequally.
4 Cell growth will stop.

22 In the diagram below, which structures aid the muscles of the earthworm as it moves from region A to region B?

1 setae
2 bones
3 long cilia
4 jointed appendages

23 In humans, structures that release digestive secretions directly into the small intestine include both the
1 salivary glands and the pancreas
2 gall bladder and the lacteals
3 villi and the salivary glands
4 pancreas and the gall bladder

24 Blood normally flows from the capillaries directly into
1 small arteries
2 small veins
3 lymph vessels
4 heart atria

25 The liquid that is derived from human blood plasma and is in direct contact with the cells of the body is known as
1 bile
2 cytoplasm
3 intercellular fluid
4 whole blood

26 The human trachea is a passageway that remains open due to the presence of
1 bones
2 ligaments
3 skeletal muscles
4 cartilaginous rings

27 In the diagram below, which structure is indicated by letter X?

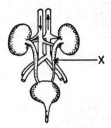

1 ureter
2 urinary bladder
3 artery
4 urethra

28 What is a function of the structure labeled Y in the diagram below?

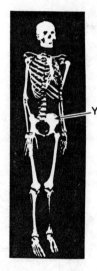

1 It serves as a site for the synthesis of hormones.
2 It supports and protects body structures.
3 It contracts to aid in locomotion.
4 It provides vitamins during periods of physical stress.

29 Which process is represented by the series of diagrams below?

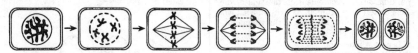

1 gametogenesis
2 fertilization

3 mitotic cell division
4 meiotic cell division

30 When leg muscles respond to a stimulus by moving the foot, the response depends most directly on the functioning of
1 bronchioles   3 capillaries
2 nephrons   4 neurons

31 If an organism reproduces asexually, its offspring will most likely be
1 genetically different from each other
2 produced from specialized cells known as gametes
3 genetically identical to the parent
4 produced as a result of fertilization

32 In flowers, pollination will most likely occur when
1 gravity causes pollen to fall from the stigma to the anthers
2 insects transfer pollen from an anther to a stigma
3 the wind transfers pollen from the pistil to the stamen
4 many specialized cells are released from the ovary

33 Which letters indicate structures where meiosis normally occurs?

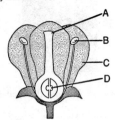

(1) A and D   (3) B and D
(2) B and C   (4) A and C

34 The function of the cotyledon in a seed is to
1 form the upper portion of the plant
2 form the lower portion of the plant
3 protect the ovary from drying out
4 provide nutrients for the germinating plant

35 Growth in higher plants most often takes place in regions of undifferentiated tissue known as
1 meristems   3 palisade layers
2 lenticels   4 vascular bundles

36 In most animal species with internal development, the embryo becomes implanted in the lining of the
1 stomach   3 ovary
2 liver   4 uterus

37 In humans, where does the exchange of respiratory gases take place between a mother's blood and the blood of her fetus?
1 amnion   3 placenta
2 yolk sac   4 umbilical cord

38 According to the gene-chromosome theory, which statement is true?
1 Genes are present only on human chromosomes.
2 Genes are arranged in a linear sequence on a chromosome.
3 Alleles are located on nonhomologous chromosomes.
4 Mutations occur mainly in sex cells.

39 When roan cattle are crossed, 25% of the offspring produced will have white coats, 50% will have roan coats, and 25% will have red coats. These results illustrate
1 polyploidy   3 codominance
2 crossing-over   4 sex linkage

40 In watermelon plants, the allele for solid green fruit (G) is dominant over the allele for striped fruit (g). Pollen from a flower of a homozygous green watermelon plant is used to pollinate a flower from a heterozygous green watermelon plant. What percent of the offspring of this cross will bear striped watermelons?

(1) 0%   (3) 50%
(2) 25%   (4) 100%

41 The diagram below shows a diploid cell with two homologous pairs of chromosomes.

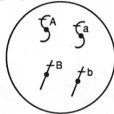

Due to independent assortment, what possible normal allelic combinations could be found in gametes produced from this cell by meiosis?

(1) Aa, Bb, AA, and bb   (3) AB and Ab, only
(2) AaBb and ABab   (4) AB, Ab, aB, and ab

42 The diagram below represents human gametes.

Which statement best describes the fertilized egg that would result if this sperm cell and egg cell unite?

(1) It would contain 44 autosomes and develop into a male.
(2) It would contain 44 autosomes and develop into a female.
(3) It would contain 46 sex chromosomes and develop into a female.
(4) It would contain 46 sex chromosomes and develop into a male.

43 Because x rays and ultraviolet light can change the chemical nature of DNA, they are known as

1 mutagenic agents   3 hydrolytic enzymes
2 growth regulators   4 toxic wastes

44 A green corn plant, when grown in reduced light for a period of time, will show a yellowing of the leaves. This yellowing is partly due to the

1 effect of pH on gene action
2 increase in polyploidy in the plant
3 effect of light on gene action
4 expression of recessive traits in reduced light

45 A DNA molecule is a polymer composed of subunits known as

1 disaccharides   3 ribose sugars
2 nucleotides   4 uracil molecules

46 The process of structural modification over a long period of time that helps to explain the diversity of living things is known as

1 metamorphosis   3 migration
2 succession   4 evolution

47 Which technique has been used by scientists to determine that the Earth is at least 4.5 billion years old?

1 radioactive dating of rocks found in the Earth's crust
2 comparing fossils found in the upper rock strata to fossils found in the lower rock strata
3 using the electron microscope to observe fossils of prehistoric microscopic life forms
4 using x rays to find fossils buried in the Earth's crust

48 Which statement would most likely have been used by Lamarck to explain the development of a long trunk in elephants?

1 Elephants stretched their trunks to reach a food supply, and this longer trunk was passed on.
2 A mutation occurred, and its frequency increased in later generations.
3 Elephants with longer trunks had a higher survival rate, and the longer trunk was passed on.
4 Elephants with short trunks were most likely sterile.

49 Which two factors provide the genetic basis for variation within a species?

1 asexual reproduction and geographic isolation
2 mutations and sexual reproduction
3 competition and protein synthesis
4 constant gene frequency and reproductive isolation

50 The diagram below represents a taxonomic tree showing the possible evolution of six species of finches.

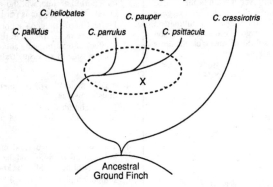

The most likely explanation for the branching pattern seen in the circled region labeled X is that

1 environmental changes resulted in extinction
2 speciation occurred as a result of inbreeding
3 no speciation occurred during this time
4 environmental changes influenced speciation

51 According to the heterotroph hypothesis, autotrophs developed after the evolution of heterotrophs partly because the primitive environment of the Earth *lacked*

1 methane   3 carbon dioxide
2 water   4 solar radiation

52 According to the heterotroph hypothesis, what contribution did ultraviolet light, heat, and lightning make to the formation of the first organic molecules?

1 They metabolized inorganic substances.
2 They synthesized radioactive materials.
3 They provided energy sources.
4 They hydrolyzed dissolved particles.

53 The fact that no organism exists as an entity separate and distinct from its environment is a major concept in the branch of biology known as

1 cytology   3 genetics
2 biochemistry   4 ecology

54 Which is an abiotic factor that would affect the ability of a species of tree to survive in a particular habitat?

1 availability of minerals in the soil
2 type and number of tree parasites present
3 climax vegetation of the area
4 type and number of herbivores present

55 Young rabbits that eat grass are sometimes eaten by raccoons, which also eat seeds and berries. Bacteria help to decompose the excretions of the raccoon. Which statement about these nutritional relationships is accurate?

1 Bacteria are scavengers.
2 Rabbits are secondary consumers.
3 Raccoons eat only producers.
4 Raccoons are both primary and secondary consumers.

56 The nitrogen cycle is most directly dependent upon the

1 process of transpiration in autotrophs
2 ability of consumers to move from place to place
3 metabolic activities of soil bacteria
4 evaporation of water from the Earth's surface

390

57 Which negative human environmental influence is matched with the positive influence that has helped to correct it?
  1 overhunting — pollution controls
  2 disposal problems — species preservation
  3 importation of organisms — energy conservation
  4 pesticide use — biological control of insects

58 Recycling solid wastes is an example of
  1 utilizing covercropping techniques
  2 preventing erosion in landfills
  3 eliminating sulfur pollutants from the atmosphere
  4 conserving natural resources without harming the environment

59 The type of climax vegetation that grows in a certain geographical area is most directly influenced by the
  1 climatic limitations of the area
  2 dominant biotic factors present
  3 number of consumers present
  4 complexity of the food webs in the area

Part II

This part consists of five groups, each containing ten questions. Choose two of these five groups. Be sure that you answer all ten questions in each group chosen. Record the answers to these questions in accordance with the directions on the front page of this booklet. [20]

Group 1 — Biochemistry

If you choose this group, be sure to answer questions 60–69.

Base your answers to questions 60 through 62 on the diagram below which represents some of the events that take place in a chloroplast and on your knowledge of biology.

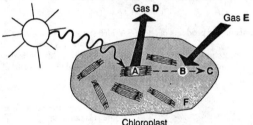

Chloroplast

60 If C represents PGAL, gas E would most likely represent
  (1) CO_2 (3) O_2
  (2) N_2 (4) H_2O

61 The isotope O^{18} has been used to trace the origin of
  1 structure A   3 gas E
  2 reaction B   4 gas D

62 Chlorophyll is most closely associated with the chemical events that involve the
  1 production of gas D   3 use of compound C
  2 absorption of gas E   4 contents of region F

Base your answers to questions 63 and 64 on the graph below and on your knowledge of biology.

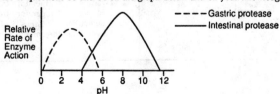

391

63 The most likely result of mixing both enzymes with their substrates in a single test tube is that
  1 only gastric protease would be active if the pH of the mixture was basic
  2 intestinal protease would be **more active** than gastric protease at pH 4
  3 both enzymes would exhibit some activity at pH 5
  4 gastric protease would be more active than intestinal protease at pH 6

64 Which is a true statement about the relationship between pH and enzyme action?
  1 All enzymes work best at a neutral pH.
  2 Adding more acid does not affect the rate of activity of an enzyme.
  3 Enzymes function only in a pH range of 4.0 to 5.5.
  4 The activity of an enzyme is affected by pH.

Base your answers to questions 65 and 66 on the equation below which represents a biochemical reaction and on your knowledge of biology.

65 To which class of compounds do the products of this equation belong?
  1 lipids   3 proteins
  2 carbohydrates   4 nucleic acids

66 Which statement concerning the water molecule in the equation is true?
  1 It is split by light into hydrogen and oxygen.
  2 It is split and becomes part of the product molecules.
  3 It serves as a medium in which the reacting molecules move.
  4 It helps cool the reacting molecules, slowing the reaction rate.

67 In the diagram below, which number indicates a carboxyl group?

  (1) 1   (3) 3
  (2) 2   (4) 4

68 In a lipid synthesis reaction, the greatest number of fatty acid molecules that could combine with one glycerol molecule is
  (1) 1   (3) 3
  (2) 2   (4) 6

69 Which action would most likely increase the rate of an enzyme-controlled reaction during an experiment?
  1 raising the temperature from 20°C to 30°C
  2 raising the temperature from 50°C to 100°C
  3 decreasing the intensity of light
  4 decreasing the amount of enzyme

Base your answers to questions 70 through 73 on the diagram below which represents endocrine glands of both human sexes and on your knowledge of biology.

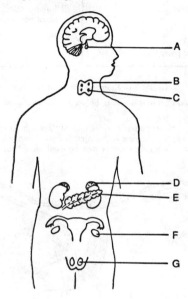

70 Which endocrine glands represented in the diagram secrete substances having the regulatory relationship shown in the graph below?

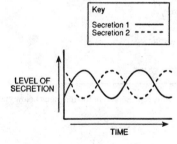

Key

Secretion 1 ——————
Secretion 2 - - - - -

LEVEL OF
SECRETION

TIME

(1) A and B   (3) E and F
(2) C and E   (4) F and G

71 Secretions classified as steroids are released by the structure labeled
(1) E   (3) C
(2) B   (4) D

72 A malfunction in the hormonal secretions from structure E may require a person to
1 eliminate all carbohydrates from the diet
2 increase the amount of iodine in the diet
3 receive daily injections of insulin
4 undergo surgery to remove the organ

73 A secretion responsible for the development of a deep voice and facial hair is synthesized by structure
(1) F   (3) C
(2) B   (4) G

74 Which blood type could a person with blood type O safely receive?
(1) A   (3) AB
(2) B   (4) O

75 A blood abnormality that results from uncontrolled production of abnormal white blood cells by the bone marrow is known as
1 leukemia   3 thrombosis
2 anemia   4 emphysema

76 In some individuals, substances such as pollen, mold, dust, or animal hair may cause an allergic response by stimulating the release of
1 glycogen   3 thyroxin
2 histamine   4 urea

393

Directions (77-79): For each phrase in questions 77 through 79 select the malfunction, chosen from the list below, that is best described by that phrase. Then record its number on the separate answer paper.

Malfunctions
(1) Polio
(2) Stroke
(3) Meningitis
(4) Cerebral palsy

77 A congenital disease characterized by abnormal motor functions

78 A clot in a cerebral blood vessel that may result in brain damage

79 Inflammation of the membranes surrounding the brain and spinal cord

Group 3 — Reproduction and Development
If you choose this group, be sure to answer questions 80-89.

Directions (80-83): For each statement in questions 80 through 83 select the part of the human female reproductive tract, chosen from the diagram below, that is best described by that statement. Then record its number on the separate answer paper. [A number may be used more than once or not at all.]

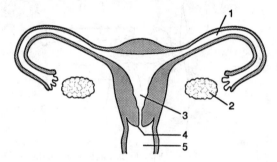

80 The process of meiotic cell division begins within this structure.

81 Fertilization normally occurs within this structure.

82 Sperm is deposited within this structure during mating.

83 This structure releases estrogen into the circulatory system.

Base your answers to questions 84 through 86 on the organisms represented in the diagrams below and on your knowledge of biology.

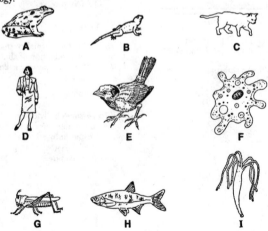

394

84 Which two organisms have the characteristics in the list below?

Number of eggs produced per year	greater than 500
Amount of parental care	none
Environment for deposited eggs	aquatic
Type of fertilization	external
Type of development	external

(1) A and B　　　　(3) F and H
(2) G and I　　　　(4) A and H

85 Which two organisms normally produce eggs that develop externally and have an allantois, amnion, chorion, and yolk sac?
(1) A and I　　　　(3) C and D
(2) B and E　　　　(4) F and G

86 Which two organisms typically reproduce asexually?
(1) A and H　　　　(3) E and G
(2) C and F　　　　(4) F and I

87 In human males, which structure is used to transport sperm out of the body?
1 urethra　　　　3 scrotum
2 ureter　　　　4 oviduct

88 An embryonic structure that functions both as a respiratory membrane and as a site for the storage of nitrogenous wastes is the
1 amnion　　　　3 allantois
2 yolk sac　　　　4 chorion

89 The process of cleavage leads most directly to the formation of a
1 zygote　　　　3 gamete
2 gonad　　　　4 blastula

Group 4 — Modern Genetics

If you choose this group, be sure to answer questions 90–99.

Base your answers to questions 90 through 92 on the diagram below which represents a biochemical process that occurs in a cell and on your knowledge of biology.

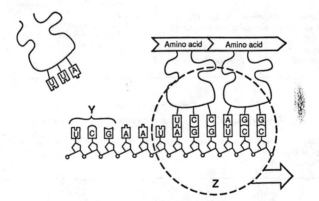

90 The organelle labeled Z represents a
1 ribosome　　　　3 mitochondrion
2 Golgi body　　　　4 nucleus

91 A change in the region labeled Y from U–C–G to U–G–C would most likely cause
1 the synthesis of a different polypeptide
2 polyploidy
3 the formation of recombinant DNA
4 crossing-over

92 The arrangement of the nitrogenous bases at region Y was determined by the
1 type of amino acids present in the cytoplasm
2 sequence of nucleotides in DNA
3 number of ATP molecules in the cytoplasm
4 concentration of enzyme in region Z

Base your answers to questions 93 and 94 on the information in the chart below and on your knowledge of biology. In the chart, X indicates that a component is present within a substance.

Components	Substance 1	Substance 2
Compound A	X	
Compound B		X
Cytosine	X	X
Guanine	X	X
Thymine	X	
Adenine	X	X
Uracil		X
Phosphate	X	X

93 Substances 1 and 2 are most likely
 (1) ATP and ICF
 (2) DNA and ATP
 (3) RNA and ATP
 (4) DNA and RNA

94 The correct name for compound A is
 1 ribose
 2 deoxyribose
 3 nucleotide
 4 polysaccharide

95 In a population of red squirrels, the sum of all the genes that can be passed on to the offspring is known as the
 1 gene frequency
 2 migration rate
 3 mutation frequency
 4 gene pool

96 Which genetic disorder can be detected by karyotyping?
 (1) PKU
 (2) Tay-Sachs
 (3) Down's syndrome
 (4) sickle-cell anemia

97 In recent research, a specific DNA code for an organic catalyst was removed from each of three different species of soil bacteria. Using these DNA codes, a single bacterium capable of synthesizing the three different organic catalysts was produced. Which technique was used to produce this new bacterium?
 1 hybridization
 2 translocation
 3 mutagen screening
 4 genetic engineering

98 Tay-Sachs disease is a genetic disease that occurs mainly in certain people of Eastern European descent. This gene was recently found in a population of people of Cajun descent in Louisiana. The most likely reason that the gene appeared in the Cajun population is that
 1 a chromosomal mutation occurred in the Cajun population
 2 polyploidy occurred in the Cajun population
 3 a person of Cajun descent mated with a person of Eastern European descent
 4 the environment of Louisiana has become more like that of Eastern Europe

99 Which scientists developed the double helix model of the DNA molecule?
 1 Watson and Crick
 2 Hardy and Weinberg
 3 Darwin and Lamarck
 4 Weismann and Miller

Group 5 — Ecology
If you choose this group, be sure to answer questions 100–109.

Base your answers to questions 100 through 103 on the diagrams below of four stages of succession and on your knowledge of biology.

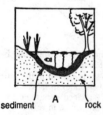 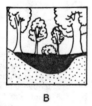  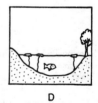

sediment   A   rock   B   C   D

396

100 The sequence of succession for typical communities in New York State would be
   (1) $A \rightarrow C \rightarrow B \rightarrow D$ (3) $B \rightarrow C \rightarrow A \rightarrow D$
   (2) $C \rightarrow A \rightarrow D \rightarrow B$ (4) $D \rightarrow A \rightarrow C \rightarrow B$

101 Which factor would most likely speed up the rate of succession?
   1 increased intensity of sunlight during winter
   2 erosion of the banks of the pond
   3 the presence of predator fish in the pond
   4 decreased transpiration in land plants

102 This series of diagrams best illustrates the succession of a
   1 freshwater ecosystem into a terrestrial ecosystem
   2 terrestrial community into a marine biome
   3 freshwater ecosystem into a taiga biome
   4 terrestrial community into a freshwater ecosystem

103 Which stage of succession would most likely experience the *least* variation in daily temperature?
   (1) A (3) C
   (2) B (4) D

104 The larvae of the tent caterpillar eat the leaves of deciduous trees. The tent caterpillars serve as food for several species of birds. Which biomass pyramid best represents these organisms?

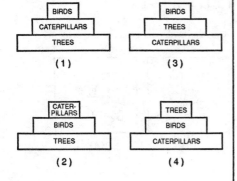

105 The plant represented in the diagram below is associated with the nitrogen cycle.

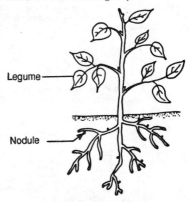

This plant is directly involved in the
   1 release of ammonia to the atmosphere
   2 decomposition of dead organisms
   3 conversion of atmospheric nitrogen to nitrates
   4 synthesis of atmospheric nitrogen from ammonia

106 A major land area characterized by permanently frozen subsoil is the
   1 grassland   3 taiga
   2 tundra   4 desert

Directions (107–109): For *each* statement in questions 107 through 109 select the nutritional term, *chosen from the list below,* that is best described by that statement. Then record its *number* on the separate answer paper.

Nutritional Terms
   (1) Scavenging
   (2) Mutualism
   (3) Parasitism
   (4) Saprophytism

107 Fish living deep in the ocean obtain energy by eating animal remains that sink to the bottom from upper water regions.

108 A mushroom grows on a dead tree in the woods.

109 Certain protozoans derive food and shelter from the digestive tract of termites and, in return, help digest food for the termites.

This part consists of five groups. Choose three of these five groups. For those questions that are followed by four choices, record the answers on the separate answer paper in accordance with the directions on the front page of this booklet. For all other questions in this part, record your answers in accordance with the directions given in the question.   [15]

Group 1

If you choose this group, be sure to answer questions 110–114.

110 The apparatus below was designed with the understanding that animals take in oxygen and release carbon dioxide, and that ordinary air contains very little carbon dioxide.

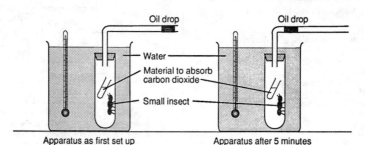

The apparatus can be used to measure the
1 amount of heat produced by the animal
2 rate of respiration of the animal
3 effect of carbon dioxide on the animal
4 amount of carbon dioxide absorbed by the animal

111 Which nutrient is correctly paired with the indicator used to detect its presence?
1 glucose — bromthymol blue   3 protein — pH paper
2 oil — Benedict's solution   4 starch — Lugol's iodine

112 Which diagram illustrates how a cross section of an earthworm should be prepared?

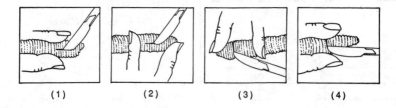

(1)   (2)   (3)   (4)

Base your answers to questions 113 and 114 on the information below and on your knowledge of biology.

A student cut a piece of potato into 8 cubes, each measuring 5 millimeters along each edge, and placed 7 of them in a beaker containing a 25% salt solution. One cube was removed from the beaker every 10 minutes for a period of 70 minutes and the average length of two sides was determined. The results are shown in the data table below.

Data Table

Time (min)	Average Length (mm)
0	5.0
10	4.5
20	4.0
30	3.5
40	3.0
50	2.5
60	2.5
70	2.5

113 Which line graph most accurately shows the relationship between time in the salt solution and size of the cube?

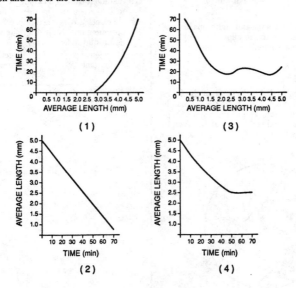

114 Which statement best explains the observed change in size of the potato cubes?
1 Water left the cube by the process of osmosis.
2 Salt left the cube by the process of osmosis.
3 Water entered the cube by the process of diffusion.
4 Salt entered the cube by the process of diffusion.

Group 2

If you choose this group, be sure to answer questions 115–119.

115 The diagram below represents a cell in the field of view of a compound light microscope.

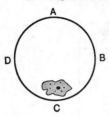

In which direction should the slide be moved on the microscope stage to center the cell in the field of view?

1 toward A   3 toward C
2 toward B   4 toward D

116 Which part of a compound light microscope should a student adjust to allow more light to pass through a specimen?

1 fine adjustment   3 diaphragm
2 ocular   4 stage

117 A student calculated the diameter of the high-power field of a microscope to be approximately 400 micrometers (μm). The diagram below represents the epidermal cells of a leaf observed in this high-power field.

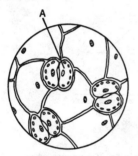

What is the approximate length of cell A?

(1) 0.01 mm   (3) 100 mm
(2) 0.10 mm   (4) 1,000 mm

118 A student placed a slide on the stage of a compound microscope and adjusted the stage clips, as shown in the diagram below.

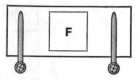

In the space provided on your answer paper, draw with pen or pencil the position of the letter as it would be viewed when observed under low power.

119 A student is using a compound light microscope to observe a wet mount of unstained human cheek cells. Which cell organelle will most likely become more visible after she adds iodine solution to the slide?

1 nucleus   3 cell wall
2 ribosome   4 mitochondrion

400

Group 3
If you choose this group, be sure to answer questions 120–124.

Base your answers to questions 120 through 124 on the information and diagrams below and on your knowledge of biology.

Two laboratory setups were prepared as shown in the diagram below. Each thermos jar contained 250 milliliters of apple juice and each flask contained 100 milliliters of bromthymol blue. Yeast was added to setup A, only. The bromthymol blue turned yellow in setup A, but remained blue in setup B. A daily temperature reading was taken for 5 days and the results are shown in the data table below.

Data Table

Time (Days)	Temperature (°C)	
	Setup A	Setup B
1	22	22
2	25	22
3	24	22
4	23	22
5	22	22

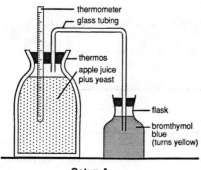

Setup A

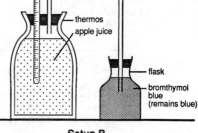

Setup B

Directions (120–122): Using the information in the data table, construct a line graph on the grid provided *on your answer paper*, following the directions below. Pen or pencil may be used for your answer.

The grid below is provided for practice purposes only. Be sure your final answer appears *on your answer paper*.

120 Mark an appropriate scale on the axis labeled "Temperature (°C)."

121 Plot the data for setup A on the grid. Surround each point with a small circle and connect the points.

Example: ⊙–⊙–⊙

122 Plot the data for setup B on the grid. Surround each point with a small triangle and connect the points.

Example: △–△–△

The Effect of Yeast on Temperature

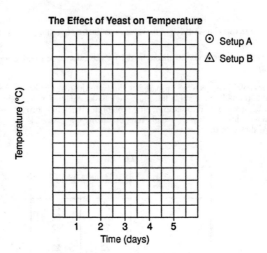

Directions (123–124): Write your answers to questions 123 and 124 in the spaces provided on your answer paper. Your answers must be written in ink.

123 Using one or more complete sentences, state why no yeast was added to setup *B*.

124 Write the name of an end product of the process that took place in the thermos of setup *A*.

Group 4

If you choose this group, be sure to answer questions 125–129.

Base your answers to questions 125 and 126 on the information and diagrams below and on your knowledge of biology.

Four similar bean seedlings were used to study the effects of the growth hormone gibberellic acid on the rate of plant growth. The seedlings were placed in four different flasks, each containing a different hormone concentration in distilled water, as shown below. The change in stem height was recorded every day for 1 week.

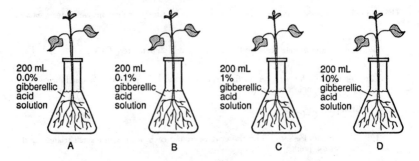

Directions (125-126): Write your answers to questions 125 and 126 in the space provided on the separate answer paper. Your answers must be written in ink.

125 Identify one variable in this investigation.

126 Using one or more complete sentences, state one possible hypothesis that can be tested using this experimental setup.

127 A student divided some insect larvae into four equal groups, each having the same amount of food. Each group was kept at a different temperature, and the average length of the larvae was determined after each shedding of the exoskeleton (molt). The data obtained are shown in the data table below.

Data Table

Molts	Average Length of Larvae (mm)			
	Group 1 (15°C)	Group 2 (20°C)	Group 3 (25°C)	Group 4 (30°C)
1	10.0	10.0	10.0	10.0
2	11.0	11.5	12.0	13.0
3	11.5	12.5	13.5	14.5
4	11.8	14.0	16.3	15.5

According to the data, the most favorable temperature for total growth in these larvae is
(1) 15°C   (3) 25°C
(2) 20°C   (4) 30°C

128 Which laboratory equipment is correctly paired with a unit it measures?
1 metric ruler — centigrams
2 Celsius thermometer — degrees Fahrenheit
3 glass beaker — millimeters
4 graduated cylinder — milliliters

129 Which condition is necessary for an experiment to yield useful data?
1 Similar results should be obtained when the experiment is repeated.
2 Only the expected results should be considered each time the experiment is performed.
3 The hypothesis must be correct.
4 The experimental period must be short.

Group 5
If you choose this group, be sure to answer questions 130–134.

Base your answers to questions 130 through 134 on the reading passage below and on your knowledge of biology.

The Sea Lamprey

Prior to the construction of the Welland Canal over 60 years ago, sea lampreys could travel no farther west than Lake Ontario. When the canal opened, it allowed ships moving up the St. Lawrence River to pass around Niagara Falls, but it also allowed the lampreys to enter Lake Erie, Lake Huron, and eventually all of the Great Lakes.

The lamprey has a thin body similar to that of an eel, a slimy brownish skin, and a tail and two single dorsal fins for swimming. The adult can be 60 centimeters long and weigh about 500 grams. The lamprey's head has a small undeveloped eye on each side and a nasal opening on top leading to a sac containing nerve endings that aid in smell. There are seven oval gill slits on each side of the head that open into pouches containing many feathery gills.

Rather than a jaw, the lamprey has a funnel-like mouth lined with many sharp teeth. The tongue, located in the middle of the mouth, has tooth-like projections. This mouth allows the lamprey to attach to a host fish. Then, using its many teeth, it tears a hole through the body of the fish and sucks out its blood and body fluids. When the host fish dies or weakens, the lamprey moves on. Lampreys feed mainly on lake trout but will attack whitefish, pike, and others. They have greatly reduced the number of desirable fish species in the Great Lakes.

130 The sea lamprey can best be described as
1 a parasite
2 a saprophyte
3 an herbivore
4 an autotroph

131 The movement of the sea lamprey into the Great Lakes as a result of the construction of the Welland Canal is an example of
1 overhunting
2 technological oversight
3 exploitation
4 poor land-use management

132 The lamprey is able to attach to a host fish by means of its
1 gill slits
2 dorsal fins
3 nasal opening
4 funnel-like mouth

133 Since the lamprey uses the blood and body fluids of host fish for nourishment, its digestive tract is probably
1 very complex, in order to hydrolyze the large food molecules the lamprey obtains from the environment
2 nonexistent, since as an adult the lamprey reproduces, but does not eat
3 simple, since much of the lamprey's food is already in a soluble form
4 simple, because the lamprey's many teeth are used to chew food into small pieces

134 Based on the description in the passage, which diagram best represents what a student's laboratory drawing of a sea lamprey should look like?

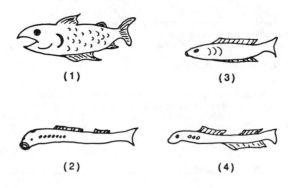

(1)   (3)

(2)   (4)

Part I

Answer all 59 questions in this part. [65]

Directions (1–59): For *each* statement or question, select the word or expression that, of those given, best completes the statement or answers the question. Record your answer on the separate answer paper in accordance with the directions on the front page of this booklet.

1 Short-tailed shrews and ruby-throated hummingbirds have high metabolic rates. As a result, these animals
1 utilize energy rapidly
2 need very little food
3 have very few predators
4 hibernate in hot weather

2 Which activity would *not* be carried out by an organism in order to maintain a stable internal environment?
1 removal of metabolic waste products
2 transport of organic and inorganic compounds
3 production of offspring by the organism
4 regulation of physiological processes

3 Which statement about viruses is true?
1 They carry on aerobic respiration.
2 They reproduce both sexually and asexually.
3 They are photosynthetic organisms.
4 They are an exception to the cell theory.

4 The structural formula below represents urea.

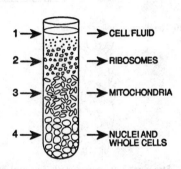

This structural formula indicates that urea is
1 an organic compound
2 an inorganic compound
3 a carbohydrate
4 a nucleic acid

5 Which activity is an example of cyclosis?
1 the movement of water from the soil into a root hair
2 the movement of food vacuoles through the cytoplasm of a paramecium
3 blood cells moving through the capillaries in a goldfish tail
4 the pumping action of a contractile vacuole in an ameba

6 The diagram below represents a sample of crushed onion cells that was centrifuged. Cells and cell components were dispersed in layers as illustrated.

The organelles that act as the sites of protein synthesis are found in the greatest concentration within layer
(1) 1
(2) 2
(3) 3
(4) 4

7 Maltose molecules are formed from glucose by the process of
1 dipeptide synthesis
2 intracellular digestion
3 dehydration synthesis
4 biological oxidation

8 Two species of bacteria produce different respiratory end products. Species A always produces ATP, CO_2, and H_2O; species B always produces ATP, ethyl alcohol, and CO_2. Which conclusion can correctly be drawn from this information?

1 Only species A is aerobic.
2 Only species B is aerobic.
3 Species A and species B are both anaerobic.
4 Species A and species B are both aerobic.

405

9 Two plants were observed to have the characteristics indicated in the chart below. An X indicates that the characteristic was present.

Specimen	Multicellular	Photosynthetic	Vascular Tissue	Roots	Stems	Leaves
Plant A	X	X				
Plant B	X	X	X	X	X	X

According to the chart, which statement about these plants is correct?
1 Plant A is a tracheophyte, and plant B is a bryophyte.
2 Plant A has xylem and phloem, but plant B does not.
3 Plant A could be a pine tree, and plant B could be a moss.
4 Plant A is a bryophyte, and plant B is a tracheophyte.

---

10 The diagram below represents a protist.

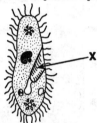

Structure X is most directly involved in the process of
1 extracellular digestion
2 enzymatic hydrolysis
3 ingestion
4 transpiration

11 A fungus is classified as a heterotroph rather than an autotroph because it
1 grows by mitosis
2 absorbs food from the environment
3 manufactures its own food
4 transforms light energy into chemical energy

12 The concentration of nitrates is often higher in plant roots than in the soil around them. Plants maintain this difference in concentration through
1 active transport   3 diffusion
2 osmosis   4 waste egestion

13 A wet-mount slide of photosynthetic protists was prepared and then exposed to light that had been broken up into a spectrum. When viewing this preparation through the microscope, a student would most likely observe that most of the protists had clustered in the regions of
1 yellow and blue light
2 orange and green light
3 green and yellow light
4 red and blue light

14 In an ameba, which process is best represented by the arrows shown in the diagram below?

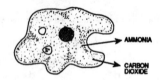

1 absorption by active transport
2 excretion by diffusion
3 respiratory gas exchange
4 egestion of digestive end products

15 Which statement describes a relationship between the human cells illustrated in the diagrams below?

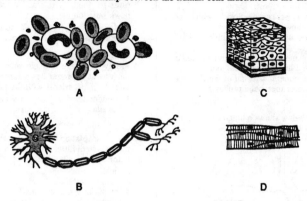

A

C

B

D

(1) B may cause D to contract.
(2) A is produced by D.

(3) C transports oxygen to A.
(4) B is used to repair C.

16 One way in which the intake of oxygen is similar in the hydra and the earthworm is that both organisms
1 absorb oxygen through a system of tubes
2 utilize cilia to absorb oxygen
3 use capillaries to transport oxygen
4 absorb oxygen through their external surfaces

17 The life function of transport in the grasshopper involves
1 an internal gas exchange surface and alveoli
2 an open circulatory system and tracheal tubes
3 moist outer skin and hemoglobin
4 a dry external body surface and hemoglobin

18 Which process is correctly paired with its major waste product?
1 respiration — oxygen
2 protein synthesis — amino acids
3 dehydration synthesis — water
4 hydrolysis — carbon dioxide

19 The diagram below represents a growth response in a plant.

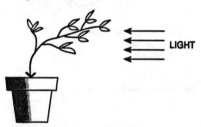

LIGHT

This growth response was most likely due to the effect of light on
1 acetylcholine
2 minerals
3 auxin distribution
4 vascular tissue

20 The diagram below represents three steps of a chemical reaction.

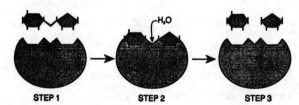

H₂O

STEP 1   STEP 2   STEP 3

This diagram best illustrates the
1 deamination of amino acids
2 emulsification of a fat
3 synthesis of a polysaccharide
4 hydrolysis of a carbohydrate

21 Which statement best describes protein metabolism in the hydra?
  1 It produces excess carbon dioxide, which is recycled for photosynthesis.
  2 It produces urea, which is eliminated by nephridia.
  3 It produces ammonia, which is transported out of the animal into the environment.
  4 It produces mineral salts, all of which are retained for other metabolic processes.

22 The diagram below shows a longitudinal section of the human heart.

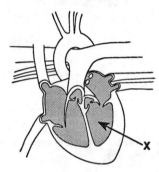

The structure labeled X is known as
  1 a ventricle   3 a valve
  2 an atrium   4 the aorta

23 A hawk sees a field mouse, which it then captures for food. In this activity, the eyes of the hawk function as
  1 effectors   3 stimuli
  2 receptors   4 neurotransmitters

24 Methyl cellulose is a chemical that slows the movement of paramecia on a slide. This chemical most likely interferes with the movement of
  1 pseudopods   3 setae
  2 flagella   4 cilia

25 Which adaptation found within the human respiratory system filters, warms, and moistens the air before it enters the lungs?
  1 clusters of alveoli
  2 rings of cartilage
  3 involuntary smooth muscle
  4 ciliated mucous membranes

26 Food is usually kept from entering the trachea by the
  1 diaphragm   3 villi
  2 epiglottis   4 ribs

27 The nephrons and alveoli of humans are most similar in function to the
  1 nephridia and skin of earthworms
  2 Malpighian tubules and gastric caecae of grasshoppers
  3 nerve nets and gastrovascular cavities of hydras
  4 cilia and pseudopods of protozoa

28 The diagrams below represent stages of a cellular process.

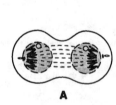         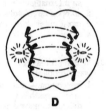

A   B   C   D

Which is the correct sequence of these stages?
  (1) A → B → C → D   (3) C → B → D → A
  (2) B → D → C → A   (4) D → B → A → C

29 Which part of the human central nervous system is correctly paired with its function?
  1 spinal cord — coordinates learning activities
  2 cerebellum — serves as the center for reflex actions
  3 cerebrum — serves as the center for memory and reasoning
  4 medulla — maintains muscular coordination

30 Tendons are best described as
  1 tissue that is found between bones and that protects them from damage
  2 cords that connect bone to bone and that stretch at the point of attachment
  3 striated tissue that provides a wide range of motion
  4 fibrous cords that connect muscles to bones

31 Which statement best describes the division of the cytoplasm and the nucleus in budding?
1 Both the cytoplasm and the nucleus divide equally.
2 The cytoplasm divides unequally, but the nucleus divides equally.
3 The cytoplasm divides equally, but the nucleus divides unequally.
4 Both the cytoplasm and the nucleus divide unequally.

32 *Rhizopus*, a bread mold, usually reproduces asexually by
1 budding   3 regeneration
2 sporulation   4 fission

33 In sexually reproducing species, doubling of the chromosome number from generation to generation is prevented by events that take place during the process of
1 gametogenesis   3 nondisjunction
2 cleavage   4 fertilization

34 Which statement is true about the process of fertilization in both tracheophytes and mammals?
1 It normally results in the production of monoploid offspring.
2 It occurs externally in a watery environment.
3 It is followed by yolk production.
4 It occurs within female reproductive organs.

35 The production of large numbers of eggs is necessary to insure the survival of most
1 mammals   3 fish
2 molds   4 yeasts

36 Mendel developed his basic principles of heredity by
1 microscopic study of chromosomes and genes
2 breeding experiments with drosophila
3 mathematical analysis of the offspring of pea plants
4 ultracentrifugation studies of cell organelles

37 The diagrams below represent the gametes and zygotes associated with two separate fertilizations in a particular species.

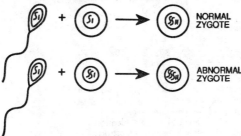

The abnormal zygote is most likely the result of
1 polyploidy   3 chromosome breakage
2 nondisjunction   4 gene linkage

Base your answers to questions 38 and 39 on the diagram below of a flower and on your knowledge of biology.

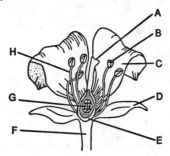

38 Which structures form the stamen?
(1) A and F   (3) C and D
(2) B and H   (4) E and G

39 During pollination, pollen is transferred from
(1) B to A   (3) B to G
(2) C to D   (4) F to H

40 Pea plants heterozygous for both height and color of seed coat (*TtYy*) were crossed with pea plants that were homozygous recessive for both traits (*ttyy*). The offspring from this cross included tall plants with green seeds, tall plants with yellow seeds, short plants with green seeds, and short plants with yellow seeds. This cross best illustrates
1 gene mutation
2 environmental influence on heredity
3 independent assortment of chromosomes
4 intermediate inheritance

41 In raccoons, a dark face mask is dominant over a bleached face mask. Several crosses were made between raccoons that were heterozygous for dark face mask and raccoons that were homozygous for bleached face mask. What percentage of the offspring would be expected to have a dark face mask?
(1) 0%   (3) 75%
(2) 50%   (4) 100%

42 Traits that are controlled by genes found on an X-chromosome are said to be
1 autosomal dominant
2 autosomal recessive
3 codominant
4 sex-linked

43 The diagram below represents possible lines of evolution of primates.

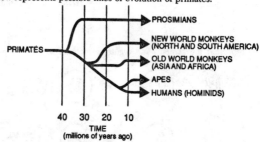

Which inference can best be made based on the diagram?
1 Acquired adaptations for living in trees are inherited.
2 Humans and apes have a common ancestor.
3 The embryos of monkeys and apes are identical.
4 The period of maturation is similar in most primates.

44 Which situation is a result of crossing-over during meiosis?
1 Genes are duplicated exactly, ensuring that offspring will be identical to the parents.
2 Chromatids thicken and align themselves, helping to ensure genetic continuity.
3 Genes are rearranged, increasing the variability of offspring.
4 Chromatids fail to sort independently, creating abnormal chromosome numbers.

45 What is the role of DNA in controlling cellular activity?
(1) DNA provides energy for all cell activities.
(2) DNA determines which enzymes are produced by a cell.
(3) DNA is used by cells for the excretion of nitrogenous wastes.
(4) DNA provides nucleotides for the construction of plasma membranes.

46 The best scientific explanation for differences in structure, function, and behavior found between life forms is provided by the
1 heterotroph hypothesis
2 lock-and-key model
3 theory of use and disuse
4 theory of organic evolution

47 Substances that increase the chance of gene alterations are known as
1 mutagenic agents
2 genetic agents
3 chromosomal agents
4 adaptive agents

48 Fossils of two different organisms, A and B, are found in different undisturbed layers of rock. The layer containing fossil A is located above the layer containing fossil B. Which statement about these fossils is most likely true?
1 Fossil B is older than fossil A.
2 Fossils A and B represent organisms that are closely related and evolved from a common ancestor.
3 Fossil A represents an organism that evolved from fossil B.
4 Fossil B represents an organism that evolved from fossil A.

49 Since the time of Darwin, increased knowledge of heredity has resulted in
1 the addition of use and disuse to Lamarck's theory
2 the elimination of all previous evolutionary theories
3 increased support for the theory of natural selection
4 disagreement with Mendel's discoveries

50 The diagrams below represent homologous structures.

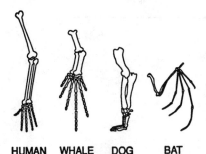

HUMAN   WHALE   DOG   BAT

The study of the evolutionary relationships between these structures is known as comparative
1 cytology   3 anatomy
2 biochemistry   4 embryology

51 Which processes are directly involved in the carbon-hydrogen-oxygen cycle?
1 respiration and photosynthesis
2 transpiration and evaporation
3 nutrition and ecological succession
4 diffusion and alcoholic fermentation

52 Which title would be most appropriate for a textbook on general ecology?
1 *The Interactions Between Organisms and Their Environment*
2 *The Cell and Its Organelles*
3 *The Physical and Chemical Properties of Water*
4 *The Hereditary Mechanisms of Drosophila*

53 Which is an example of an ecosystem?
1 a population of monarch butterflies
2 the interdependent biotic and abiotic components of a pond
3 all the abiotic factors found in a field
4 all the mammals that live in the Atlantic Ocean

54 According to the heterotroph hypothesis, which event immediately preceded the evolution of aerobes?
1 the production of oxygen by autotrophs
2 the production of ammonia by heterotrophs
3 the production of carbon dioxide by autotrophs
4 the production of carbon dioxide by heterotrophs

55 In a self-sustaining ecosystem, which component *cannot* be recycled because it is lost from food chains and becomes unavailable?
1 carbon   3 water
2 nitrogen   4 energy

56 Termites can be found living in dead trees partially buried under soil and stones. Within the tree trunks, the termites feed on the wood fiber, creating passageways having a high humidity. The wood fiber is digested by protozoans living within the digestive tract of the termite.

What are the biotic factors in this habitat?
1 tree trunk, stones, and protozoans
2 soil and humidity
3 termites and protozoans
4 humidity, soil, and stones

57 One theory about the extinction of dinosaurs is that the collision of an asteroid with the Earth caused environmental changes that killed off the dinosaurs in a relatively short time, changing the course of evolution. This theory is an example of which evolutionary concept?
1 gradualism
2 competition
3 the heterotroph hypothesis
4 punctuated equilibrium

58 The cartoon below illustrates a type of nutrition.

"Just think . . . Here we are, the afternoon sun beating down on us, a dead, bloated rhino underfoot, and good friends flying in from all over. . . . I tell you, Frank, this is the best of times."

Frank and the other birds in this cartoon are classified as

1 saprophytes   3 scavengers
2 herbivores   4 producers

59 Which type of organism is *not* represented in the diagram below?

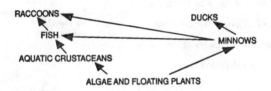

1 secondary consumers   3 carnivores
2 producers   4 decomposers

Part II

This part consists of five groups, each containing ten questions. Choose two of these five groups. Be sure that you answer all ten questions in each group chosen. Record the answers to these questions in accordance with the directions on the front page of this booklet. [20]

Group 1 — Biochemistry

If you choose this group, be sure to answer questions 60–69.

Base your answers to questions 60 through 62 on the structural formulas below and on your knowledge of biology.

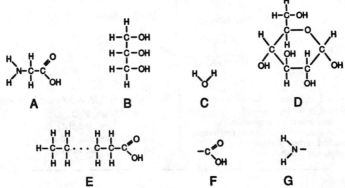

A   **B**   **C**   **D**

E   **F**   **G**

60 By which formula can molecule *D* be represented?

(1) $C_6H_{12}O_6$ (3) $C_3H_5(OH)_3$
(2) $C_5H_{12}O_5$ (4) C_3H_5COOH

61 Which structural formulas represent the building blocks of a lipid?

(1) *A* and *C*   (3) *C* and *E*
(2) *B* and *E*   (4) *F* and *G*

62 A single carboxyl group is represented by
(1) *F*   (3) *C*
(2) *B*   (4) *G*

---

63 According to the summary equations below, what is the net gain of ATP molecules from the complete oxidation of one glucose molecule?

(A) 1 glucose + 2 ATP $\xrightarrow{\text{enzymes}}$
    2 pyruvic acid + 4 ATP

(B) 2 pyruvic acid + oxygen $\xrightarrow{\text{enzymes}}$
    carbon dioxide + water + 34 ATP

(1) 34   (3) 38
(2) 36   (4) 40

64 If an enzyme works best at a neutral pH, in which pH range is that enzyme expected to function?
(1) 1–3   (3) 6–8
(2) 3–5   (4) 10–12

65 Bread dough that contains yeast and sugar expands during alcoholic fermentation as a result of an increase in the
1 production of molecular oxygen
2 absorption of minerals
3 secretion of ATP
4 production of carbon dioxide

Base your answers to questions 66 and 67 on the diagrams below of some stages of an enzyme-controlled reaction and on your knowledge of biology.

A   **B**   **C**   **D**

66 An enzyme-substrate complex is represented by diagram
(1) *A*   (3) *C*
(2) *B*   (4) *D*

67 A nonprotein vitamin required for this reaction would function as a
1 product   3 polypeptide
2 substrate   4 coenzyme

Base your answers to questions 68 and 69 on the diagram below which represents some of the events that take place in a plant cell.

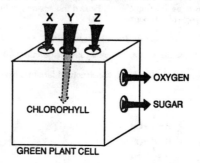

GREEN PLANT CELL

68 The oxygen and sugar leaving the cell were most likely produced by the processes of
1 hydrolysis and anaerobic respiration
2 dehydration synthesis and aerobic respiration
3 photolysis and carbon fixation
4 deamination and fermentation

69 The letters X, Y, and Z most likely represent
(1) N_2, O_2, and H_2O
(2) CO_2, light, and H_2O
(3) light, ammonia, and H_2O
(4) light, O_2, and methane

Group 2 — Human Physiology

If you choose this group, be sure to answer questions 70–79.

Base your answers to questions 70 through 72 on the diagram below and on your knowledge of biology.

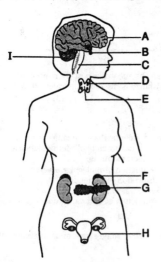

70 Which structure produces secretions that regulate E and H?
(1) A   (3) I
(2) B   (4) D

71 Which structure controls involuntary activities such as breathing and heartbeat?
(1) A   (3) C
(2) B   (4) G

72 Which two structures secrete substances that control the menstrual cycle?
(1) A and F   (3) C and D
(2) B and H   (4) E and I

73 Which letter indicates the location of nephrons in the diagram below?

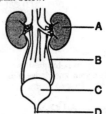

(1) A   (3) C
(2) B   (4) D

74 Which sequence represents the direction of flow of carbon dioxide as it passes out of the respiratory system into the external environment?
1 alveoli → trachea → bronchioles → bronchi → pharynx → nasal cavity
2 alveoli → bronchi → pharynx → bronchioles → trachea → nasal cavity
3 alveoli → pharynx → trachea → bronchioles → bronchi → nasal cavity
4 alveoli → bronchioles → bronchi → trachea → pharynx → nasal cavity

75 An inflammation of the region labeled A in the diagram at the right is known as
1 meningitis
2 arthritis
3 bronchitis
4 tendinitis

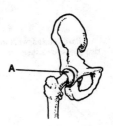

76 Which substances produced in the body are directly responsible for the rejection of a transplanted organ?
1 antigens
2 histamines
3 antibodies
4 excretions

Base your answers to questions 77 through 79 on the diagram below and on your knowledge of biology.

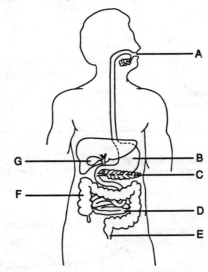

77 Which letter indicates the organ that secretes hydrochloric acid and protease?
(1) A
(2) B
(3) E
(4) D

78 Which letter indicates the organ that produces insulin and glucagon?
(1) E
(2) B
(3) C
(4) F

79 A painful condition resulting from the formation of small stone-like deposits of cholesterol may be treated by surgically removing structure
(1) G
(2) E
(3) F
(4) D

Group 3 — Reproduction and Development
If you choose this group, be sure to answer questions 80–89.

Base your answers to questions 80 through 83 on the diagram below and on your knowledge of biology. The diagram shows stages in the life cycle of a unicellular flagellated green alga.

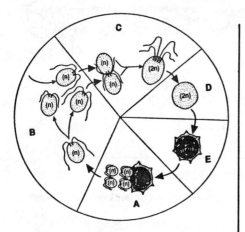

80 The process that takes place at stage *B* normally produces cells with
1 the same chromosome number as the parent cell
2 fewer chromosomes than the parent cell
3 pairs of homologous chromosomes
4 a polyploid number of chromosomes

81 Fertilization involving like gametes takes place at stage
(1) *A*   (3) *C*
(2) *B*   (4) *E*

82 The process that most likely takes place between stages *E* and *A* is
1 mitosis   3 fertilization
2 meiosis   4 cleavage

83 A specialized structure that provides protection from harsh environmental conditions is represented at stage
(1) *E*   (3) *C*
(2) *B*   (4) *D*

Base your answers to questions 84 through 86 on the graph below and on your knowledge of biology. The graph shows the different concentrations of female reproductive hormones during the menstrual cycle of humans.

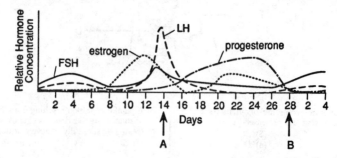

84 Which event normally occurs at *A*?
1 ovulation
2 embryo implantation
3 differentiation
4 follicle formation

85 Which process usually begins at *B*?
1 fertilization
2 embryo development
3 corpus luteum development
4 menstruation

86 Which is a correct inference about an event that occurs prior to day 14?
1 A high level of estrogen may stimulate the production of LH.
2 A high level of LH may stimulate the production of FSH.
3 A low level of FSH inhibits the production of estrogen.
4 A low level of progesterone inhibits the production of estrogen.

416

87 The yolk of a developing bird embryo functions
as a
1 moist respiratory membrane
2 storage site for waste
3 food source
4 fluid environment

88 In humans, the fertilization of two eggs at the
same time usually results in
1 chromosome abnormalities
2 gene mutations
3 identical twins
4 fraternal twins

89 In chicken eggs, the embryonic membrane
known as the allantois functions in the
1 release of oxygen to the atmosphere
2 storage of nitrogenous wastes
3 absorption of nitrogen for use in protein synthesis
4 transport of carbon dioxide directly to the
embryo

Group 4 — Modern Genetics
If you choose this group, be sure to answer questions 90–99.

Base your answers to questions 90 through 92 on the chart below and on your knowledge of biology. The
chart represents the inheritance of Tay-Sachs disease in a family.

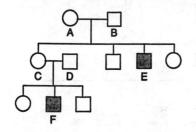

KEY

☐	NORMAL MALE
◯	NORMAL FEMALE
▨	MALE WITH TAY-SACHS DISEASE
◉	FEMALE WITH TAY-SACHS DISEASE

90 What are the genotypes of individuals A and B
with regard to Tay-Sachs disease?
1 One must be homozygous dominant and the
other must be homozygous recessive.
2 One must be homozygous dominant and the
other must be heterozygous.
3 Both must be homozygous.
4 Both must be heterozygous.

91 If individuals C and D have another child, what is
the chance this child will exhibit Tay-Sachs disease?
(1) 0%   (3) 50%
(2) 25%   (4) 100%

92 Which statement is true about individuals E
and F?
1 They are unable to metabolize glucose.
2 They are unable to metabolize phenylalanine
because they lack a specific enzyme.
3 They have an accumulation of excess fatty
material in their nerve tissue.
4 They have an abnormal chromosome number.

93 Which event is not part of the process of DNA
replication?
1 Nitrogenous base pairs are formed.
2 Hydrogen bonds are broken.
3 A double-stranded molecule unwinds.
4 Ribosomes are synthesized.

94 Deoxyribonucleic acid molecules serve as a template for the synthesis of molecules of
1 amino acids   3 messenger RNA
2 carbohydrates   4 lipids

95 Which procedure is usually used to help determine whether a child will be born with Down's
syndrome?
1 amniocentesis
2 cloning
3 microdissection of sperm cells and egg cells
4 analysis of urine samples from the mother

Base your answers to questions 96 through 99 on the information below and on your knowledge of biology.

For many generations, a particular species of snail has lived in an isolated pond. Some members of the species have light-colored shells and some have dark-colored shells. During this time, the species has been producing large numbers of offspring through random mating, and no migration has occurred.

96 Which additional condition must be present if the gene frequencies of these snails are to remain constant?
1 asexual reproduction
2 lack of mutations
3 genetic variation
4 common ancestry

97 A change in the environment of the pond caused the light-colored shells to become an important survival trait, and the number of light-colored snails increased. This situation will most likely cause
1 the addition of a fifth nitrogenous base to the DNA of the snails
2 a change in the frequency of the genes for shell color
3 an increase in the number of ribosomes in the cells of the snail
4 the extinction of this species of snail

98 The total of all the inheritable genes found in these snails is referred to as a
1 pedigree   3 phenotypic ratio
2 karyotype   4 gene pool

99 All of the snails of this species living in the pond may be classified as
1 a population   3 a community
2 an ecosystem   4 a biome

Group 5 — Ecology
If you choose this group, be sure to answer questions 100–109.

Base your answers to questions 100 and 101 on the diagram below and on your knowledge of biology.

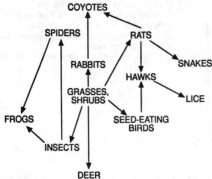

100 Which organisms would contain the greatest amount of available energy?
1 rabbits and deer   3 lice
2 grasses and shrubs   4 hawks

101 The primary consumers include
1 rabbits and snakes
2 insects and seed-eating birds
3 rats and frogs
4 spiders and coyotes

102 The diagram below represents the feeding areas during summer and fall of two populations in the same ecosystem. Both populations feed on oak trees.

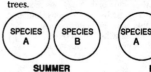

The portion of the diagram labeled X most likely indicates that
1 these populations compete for food in the fall, but not in the summer
2 the species are separated by a geographic barrier in the fall
3 the supply of oxygen is greater in the summer than in the fall
4 random mating occurs between these species in the summer

103 One reason a marine organism may have trouble surviving in a freshwater habitat is that
1 there are more carnivores in freshwater habitats
2 salt water holds more nitrogen than fresh water
3 more photosynthesis occurs in fresh water than in salt water
4 water balance is affected by salt concentration

418

104 The chart below illustrates some methods of pest control.

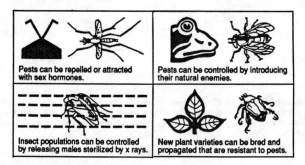

One likely effect of using these methods of pest control will be to
1 prevent the extinction of endangered species
2 increase water pollution
3 reduce pesticide contamination of the environment
4 harm the atmosphere

Base your answers to questions 105 through 107 on the sequence of diagrams below and on your knowledge of biology.

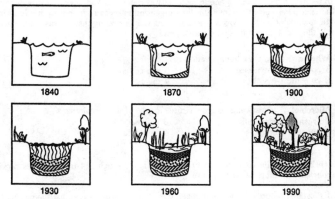

105 This sequence of diagrams best illustrates
   1 ecological succession
   2 organic evolution
   3 the effects of acid rain
   4 a food chain

106 If no human intervention or natural disaster occurs, by the year 2050 this area will most likely be a
   1 lake   3 desert
   2 swamp   4 forest

107 The natural increase in the amount of vegetation from 1840 to 1930 is related to the
   1 decreasing water depth
   2 increasing amount of sunlight
   3 presence of bottom-feeding fish
   4 use of the pond for fishing

108 In the nitrogen cycle, plants use nitrogen compounds to produce

1 glucose   3 lipids
2 starch   4 proteins

109 A flea in the fur of a mouse benefits at the mouse's expense. This type of relationship is known as

1 commensalism   3 saprophytism
2 parasitism   4 mutualism

Part III

This part consists of five groups. Choose three of these five groups. For those questions that are followed by four choices, record the answers on the separate answer paper in accordance with the directions on the front page of this booklet. For all other questions in this part, record your answers in accordance with the directions given in the question. [15]

Group 1

If you choose this group, be sure to answer questions 110–114.

Base your answers to questions 110 through 114 on the reading passage below and on your knowledge of biology.

Viruses

Most viruses are little more than strands of genetic material surrounded by a protein coat. Given the opportunity to enter a living cell, a virus springs into action and is reproduced.

Researchers have long known that viruses reproduce by using some of the cell's enzymes and protein-making structures. However, the precise details of the process remain unclear. Microbiologists have recently enabled viruses to reproduce outside a living cell, in a test-tube medium containing crushed human cells, salts, ATP, amino acids, and nucleotides.

In the test tube, the viral genetic material was replicated and new viral proteins were synthesized. These new proteins were then organized into coats around the newly formed genetic material. Complete viruses were formed, demonstrating that a virus can be active outside the cell if given the right environment.

110 When a virus enters a human cell, it may

1 control photosynthesis   3 reproduce
2 copy the DNA of the cell   4 enlarge

111 Microbiologists were able to grow viruses in a test tube containing

1 crushed human cells   3 glucose
2 nutrient agar   4 ammonia

Directions (112–114): Write your answers to questions 112 through 114 in the spaces provided on your answer paper. Your answers must be written in ink.

112 Using one or more complete sentences, describe a possible reason that the microbiologists added ATP to the test-tube medium.

113 Using one or more complete sentences, explain the function of the new viral proteins.

114 Using one or more complete sentences, state a valid conclusion that can be drawn from this research about viruses.

If you choose this group, be sure to answer questions 115-119.

Base your answers to questions 115 through 119 on the information below and on your knowledge of biology.

To measure glucose use in a human, a blood sample was taken from a vein, and the amount of glucose in the sample was determined. A glucose solution was then ingested by the person being tested. Blood samples were taken periodically for 5 hours and tested to determine the amount of glucose present. Results from the tests were used to construct the data table below.

Data Table

Time (hours)	Glucose (mg/100 dL)
0	80
0.5	170
1	120
2	90
3	80
4	70
5	70

Directions (115-116): Using the information in the data table, construct a line graph on the grid provided *on your answer paper*, following the directions below. The grid on the next page is for practice purposes only. Be sure your final answer appears *on your answer paper*.

115 Mark an appropriate scale on each of the labeled axes.

116 Plot the data from the data table. Surround each point with a small circle and connect the points.

Example: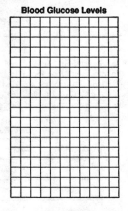

Blood Glucose Levels

Glucose (mg/100 dL)

Time (hours)

Directions (117-119): Write your answers to questions 117 through 119 in the spaces provided on the separate answer paper. Your answers must be written in ink.

117 Using one or more complete sentences, give a possible explanation for the drop in the glucose level between 0.5 and 1 hour after the glucose was ingested.

118 Using one or more complete sentences, state a function of glucose in the human body.

119 Based on the information given, how much glucose would most likely be present in 100 deciliters (dL) of the blood 1.5 hours after the glucose was ingested?

(1) 90 mg   (3) 120 mg
(2) 105 mg   (4) 170 mg

Group 3

If you choose this group, be sure to answer questions 120–124.

Directions (120–121): Write your answers to questions 120 and 121 in the spaces provided on the separate answer paper. Your answers must be written in ink.

120 Forty bean seeds were planted in 40 different pots containing soil of the same composition and moisture level. All seeds were of the same age and plant species. The pots were divided into four groups of 10, and each group was kept at a different temperature: 5°C, 10°C, 15°C, and 20°C, respectively, for a period of 30 days. All other environmental conditions were kept constant.

Using one or more complete sentences, state a problem being investigated in this experimental setup.

121 Choose one of the labeled animal cell parts from the diagram below. In the space provided on the separate answer paper, write the letter of the part you have chosen and, using one or more complete sentences, identify the part and state one of its functions.

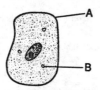

---

122 Twenty-five geranium plants were placed in each of four closed containers and then exposed to the light conditions shown in the data table below. All other environmental conditions were held constant for a period of 2 days. At the beginning of the investigation, the quantity of CO_2 present in each closed container was 250 cubic centimeters. The data table shows the amount of CO_2 remaining in each container at the end of 2 days.

Data Table

Container	Color of Light	CO_2 (cm^3)
1	blue	75
2	red	50
3	green	200
4	orange	150

The variable in this investigation was the
1 type of plant
2 color of light
3 amount of CO_2 in each container at the beginning of the investigation
4 number of days needed to complete the investigation

123 In the diagram at the right, the view of the insect specimen
can best be described as
1 a ventral view, with the posterior end to the right of the page
2 an external view showing the ventral side of the abdomen
3 a dorsal view, with the anterior end to the left of the page
4 an internal view showing the dorsal side of the head region

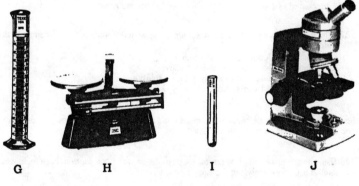

124 In addition to an indicator and proper safety equipment, which pieces of equipment
shown below should be used to test for the presence of glucose in apple juice?

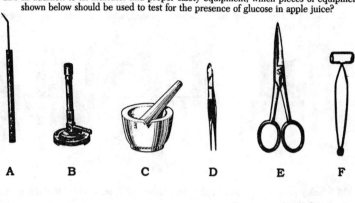

A   B   C   D   E   F

G   H   I   J

(1) A, D, and E   (3) C, G, and H
(2) B, F, and I   (4) A, B, and J

If you choose this group, be sure to answer questions 125–129.

Base your answers to questions 125 through 129 on the information below and on your knowledge of biology.

A human was fed a meal containing measured amounts of proteins, starch, and fats. Eight hours later, a 10-milliliter sample of fluid was removed from the human's small intestine for analysis.

Note that question 125 has only three choices.

125 Based on the relative amounts of nutrients present, which graph best represents the results of the analysis?

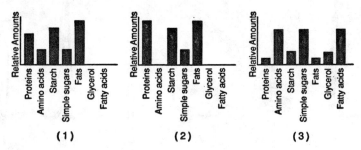

126 Which piece of equipment should be used to accurately measure the 10-milliliter sample for analysis?

1 triple-beam balance   3 large test tube
2 graduated cylinder   4 metric ruler

127 Which indicators could be used to test for the presence of some of the substances in the fluid sample?

(1) Benedict's solution and Lugol's iodine
(2) bromthymol blue solution and pH paper
(3) Fehling's solution and bromthymol blue solution
(4) pH paper and Lugol's iodine

Directions (128–129): Write your answers to questions 128 and 129 in the spaces provided on the separate answer paper. Your answers must be written in ink.

128 Using one or more complete sentences, describe *one* safety precaution that a technician should use while analyzing the sample of intestinal fluid.

129 Using one or more complete sentences, describe *one* way that the results of the analysis would be different if the human was fed a single boiled potato instead of the meal containing measured amounts of proteins, starch, and fats.

If you choose this group, be sure to answer questions 130–134.

Base your answers to questions 130 through 134 on the photograph below and on your knowledge of biology. The photograph shows onion root-tip tissue viewed under the high-power objective of a compound light microscope.

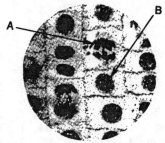

130 The photograph illustrates stages in the process of
1 meiosis in root tips
2 mitotic cell division in plants
3 water conduction in onions
4 chlorophyll production in chloroplasts

Directions (131–133): Write your answers to questions 131 through 133 in the spaces provided on the separate answer paper. Your answers must be written in ink.

131 Identify the structure indicated by arrow *A*.

132 Identify the structure indicated by arrow *B*.

133 Using one or more complete sentences, describe one adjustment that could be made to the microscope to make the field of view brighter.

134 When viewed with a compound light microscope, which letter would best illustrate that the microscope inverts and reverses an image?

(1) A

(2) W

(3) F

(4) D

ANSWERS TO REGENTS EXAMINATIONS

JANUARY 1993

1. 3	18. 2	35. 2	52. 1	69. 2	86. 1	103. 4	120. Chloroplasts
2. 2	19. 2	36. 3	53. 1	70. 4	87. 2	104. 1	
3. 4	20. 4	37. 2	54. 4	71. 2	88. 3	105. 4	121. Nucleus
4. 3	21. 4	38. 3	55. 2	72. 3	89. 4	106. 1	or vacuole
5. 1	22. 1	39. 4	56. 1	73. 4	90. 4	107. 3	122. Cyclosis
6. 3	23. 3	40. 4	57. 2	74. 3	91. 1	108. 3	123. 2
7. 4	24. 1	41. 4	58. 3	75. 1	92. 3	109. 2	124. 1
8. 3	25. 4	42. 3	59. 1	76. 2	93. 2	110. 3	125. 1
9. 1	26. 3	43. 2	60. 3	77. 4	94. 4	111. 2	126. 2
10. 4	27. 2	44. 1	61. 1	78. 3	95. 1	112. 4	127. 2
11. 2	28. 3	45. 1	62. 4	79. 1	96. 2	113. 1	128. essay
12. 3	29. 4	46. 4	63. 2	80. 3	97. 3	114. 4	129. essay
13. 3	30. 2	47. 3	64. 3	81. 2	98. 1	115. 1	130. graph
14. 1	31. 4	48. 4	65. 1	82. 1	99. 4	116. 2	131. graph
15. 2	32. 3	49. 2	66. 4	83. 2	100. 1	117. 1	132. graph
16. 1	33. 1	50. 1	67. 2	84. 4	101. 2	118. 4	133. essay
17. 4	34. 1	51. 4	68. 1	85. 1	102. 2	119. 2	134. 3

JUNE 1993

1. 2	18. 4	35. 2	52. 1	69. 3	86. 3	103. 1	120. 4
2. 1	19. 1	36. 1	53. 4	70. 4	87. 4	104. 3	121. 2
3. 3	20. 4	37. 3	54. 3	71. 3	88. 4	105. 2	122. 3
4. 4	21. 4	38. 4	55. 1	72. 2	89. 3	106. 4	123. 2
5. 3	22. 2	39. 1	56. 2	73. 3	90. 3	107. 2	124. essay
6. 1	23. 3	40. 3	57. 3	74. 1	91. 2	108. 3	125. 1
7. 1	24. 2	41. 1	58. 2	75. 5	92. 1	109. 1	126. 3
8. 4	25. 1	42. 3	59. 2	76. 2	93. 3	110. graph	127. A-pharynx;
9. 2	26. 3	43. 1	60. 4	77. 2	94. 3	111. graph	B-esophagus;
10. 2	27. 4	44. 2	61. 1	78. 4	95. 1	112. 1	C-crop; D-gizzard; E-intestine
11. 1	28. 1	45. 3	62. 2	79. 2	96. 2	113. 2	128. 1
12. 2	29. 1	46. 4	63. 3	80. 2	97. 4	114. essay	129. essay
13. 4	30. 2	47. 1	64. 4	81. 1	98. 4	115. 2	130. 1
14. 3	31. 3	48. 2	65. 4	82. 1	99. 1	116. 3	131. 4
15. 2	32. 4	49. 3	66. 3	83. 2	100. 2	117. 1	132. 2
16. 3	33. 2	50. 4	67. 1	84. 3	101. 4	118. 4	133. 3
17. 3	34. 4	51. 1	68. 2	85. 2	102. 1	119. 2	134. essay

ANSWERS TO REGENTS EXAMINATIONS

JANUARY 1994

1. 2	18. 1	35. 1	52. 3	69. 1	86. 4	103. 4	120. graph
2. 3	19. 3	36. 4	53. 4	70. 1	87. 1	104. 1	121. graph
3. 1	20. 2	37. 3	54. 1	71. 4	88. 3	105. 3	122. graph
4. 4	21. 3	38. 2	55. 4	72. 3	89. 4	106. 2	123. essay
5. 2	22. 1	39. 3	56. 3	73. 4	90. 1	107. 1	124. essay
6. 1	23. 4	40. 1	57. 4	74. 4	91. 1	108. 4	125. essay
7. 1	24. 2	41. 4	58. 4	75. 1	92. 2	109. 2	126. essay
8. 3	25. 3	42. 2	59. 1	76. 2	93. 4	110. 2	127. 3
9. 4	26. 4	43. 1	60. 1	77. 4	94. 2	111. 4	128. 4
10. 1	27. 1	44. 3	61. 4	78. 2	95. 4	112. 3	129. 1
11. 1	28. 2	45. 2	62. 1	79. 3	96. 3	113. 4	130. 1
12. 3	29. 3	46. 4	63. 3	80. 2	97. 4	114. 1	131. 2
13. 2	30. 4	47. 1	64. 4	81. 1	98. 3	115. 3	132. 4
14. 1	31. 3	48. 1	65. 2	82. 5	99. 1	116. 3	133. 3
15. 2	32. 2	49. 2	66. 2	83. 2	100. 4	117. 2	134. 2
16. 4	33. 3	50. 4	67. 3	84. 4	101. 2	118. ⊣	
17. 3	34. 4	51. 3	68. 3	85. 2	102. 1	119. 1	

JUNE 1994

1. 1	18. 3	35. 3	52. 1	69. 2	86. 1	103. 4	120. essay
2. 3	19. 3	36. 3	53. 2	70. 2	87. 3	104. 3	121. essay
3. 4	20. 4	37. 2	54. 1	71. 3	88. 4	105. 1	122. 2
4. 1	21. 3	38. 2	55. 4	72. 2	89. 2	106. 4	123. 3
5. 2	22. 1	39. 1	56. 3	73. 1	90. 4	107. 1	124. 2
6. 2	23. 2	40. 3	57. 4	74. 4	91. 2	108. 4	125. 3
7. 3	24. 4	41. 2	58. 3	75. 2	92. 3	109. 2	126. 2
8. 1	25. 4	42. 4	59. 4	76. 3	93. 4	110. 3	127. 1
9. 4	26. 2	43. 2	60. 1	77. 2	94. 3	111. 1	128. essay
10. 3	27. 1	44. 3	61. 2	78. 3	95. 1	112. essay	129. essay
11. 2	28. 3	45. 2	62. 1	79. 1	96. 1	113. essay	130. 2
12. 1	29. 3	46. 4	63. 2	80. 1	97. 2	114. essay	131. chromo-
13. 4	30. 4	47. 1	64. 3	81. 3	98. 4	115. graph	some
14. 2	31. 2	48. 1	65. 4	82. 2	99. 1	116. graph	132. nucleus or
15. 1	32. 2	49. 3	66. 2	83. 1	100. 2	117. essay	nuclear membrane
16. 4	33. 1	50. 3	67. 4	84. 1	101. 2	118. essay	133. essay
17. 2	34. 4	51. 1	68. 3	85. 4	102. 1	119. 2	134. 3

INDEX

Index

disaccharide, 20
disjunction, 135
DNA, 182, 188-195
dominance, 167, 168
dominant allele, 167
Down's syndrome, 180
Drosophila, 166, 176, 183

E

earthworm
  digestion in, 45-46
  excretion in, 64
  gas exchange in, 60
  locomotion in, 73
  regulation in, 68
  transport in, 54
ecological succession, 257-259
ecology, 240
ecosystem, 240-242, 248
  formation, 257-259
ectoderm, 140-141
effector, 68
egestion, 43
elements, 17
embryo, 140, 215
embryonic development, 140
  external, 141-143
  in humans 149-150
  internal, 143-144
  in plants, 154
emphysema, 100
emulsification, 84
endocrine gland, 70
endocrine system, 105, 109-111
endoderm, 140-141
endoplasmic reticulum, 14
endoskeleton, 113
energy pyramid, 250
enzymatic hydrolysis, 43
enzyme, 24-27, 43, 83, 84
enzyme-substrate complex, 25
epicotyl, 154
epidermis, leaf, 40
epidermis, root, 60, 64
epiglottis, 99
esophagus, 84
essential amino acids, 85
estrogen, 110, 148
evolution, 5, 212-231
excretion, 2, 63-65
  in humans, 102-104
exoskeleton, 73
extensor, 114

F

fallopian tubes, 147
fatty acid, 21
feces, 85
fermentation, 58, 230
fertilization, 139-140
  in humans, 149
  in plants, 154
filament, 153
flagella, 72
flexor, 114
flower structure, 152
fluid-mosaic model, 50
follicle, 147, 148
follicle-stimulating
  hormones (FSH), 109, 147, 148
food chain, 248
food web, 248
fossils, 213-214, 226
Fox, Sidney, 229, 305
freshwater biome, 264-265
fruit, 154, 155
fruit fly, 166
fungi, 5, 6, 43, 60, 129, 243

G

gallbladder, 84
gallstones, 88
gametes, 134, 136
gametogenesis, 134-137
  in humans, 146-147
ganglia, 68
gas exchange
  in earthworm, 60
  in grasshopper, 61
  in humans, 61, 98, 100
  in hydra, 60
gastrula, 140
gastrulation, 140
gene, 167, 172-177
  frequency, 200-201, 223
  linkage, 174
  mutations, 182, 185-186, 195, 223
  pool, 200, 223
gene-chromosome theory, 166-167
genetic code, 190, 191
genetic counseling, 184
genetic engineering, 195-196
genetic research, 195-196
genetic screening, 184
genetic theory, 166-201
genotypes, 168, 169

Index